Structural Geology and Tectonics

Soumyajit Mukherjee
Editor

Structural Geology and Tectonics

Springer

Editor
Soumyajit Mukherjee
Department of Earth Sciences
Indian Institute of Technology Bombay
Mumbai, Maharashtra, India

ISBN 978-981-95-4742-5 ISBN 978-981-95-4743-2 (eBook)
https://doi.org/10.1007/978-981-95-4743-2

This Springer imprint is published by the registered company Springer Nature Singapore Pte Ltd.
The registered company address is: 152 Beach Road, #21-01/04 Gateway East, Singapore 189721, Singapore

Preface: Introduction to Structural Geology and Tectonics

This edited book consists of 19 main chapters dealing with various issues of structural geology and tectonics. In Chapter "Deep Structure and Deformation Structural Style of the Georgian Part of Lesser Caucasus Orogen", Alania et al. (2026) through fieldwork studies document that the main structural styles of the compressional structures in the Lesser Caucasus double wedge orogen are related to several detachments. In Chapter "Rare Field Evidence of Basal Breccia Within the Kruševac Lacustrine Basin: Constraints on the Late Oligocene (Early Miocene) Opening Across Southern Peri-Pannonian Intramontane Basins (Central Serbia)", Spahić et al. (2026) report large crystalline cobbles from the Kruševac lacustrine basin (Serbia) that possibly come from a nearby peri-Pannonian basal sequence. In Chapter "Field Guide to Coorg Block and the Adjoining Shear Zones, Southern Granulite Terrane, South India", Nanda and Rekha (2026) present a field guide for the Coorg Block and nearby shear zones from the Southern Granulite Terrane, India. In Chapter "Geodynamic Evolution of the Indo-Myanmar Ranges, NE India", Ovung and Nayak (2026) from the Indo-Myanmar range describe the geology and structures from the Neotethyan oceanic lithosphere and the flyschoidal deposits. In Chapter "Innovative Technology to Harness Geothermal Energy from Active Ridge Flanks and Oceanic Islands: Advantage for the Oil Industries to Bring Energy Transition", Chandrasekharam (2026) explains how to tap the geothermal energy from the Red Sea area in the most feasible way. In Chapter "Significance of Oriented-Core in Sub-surface Geological Exploration", Goswami (2026) presents how to measure and interpret structural geological data from drill cores (also see Chakraborty and Mukherjee 2020; and Chatterjee and Mukherjee 2023 in this context).

In Chapter "4D Monitoring of Analogue Models in CT Scanners Using Linear Markers of X-ray Highly Absorbent Materials", Pueyo et al. (2026) present the use of highly absorbent X-Ray markers to accurately monitor analogue models in 4D. In Chapter "Structural Form of Stratovolcanoes on Planetary Bodies: A Theoretical Perspectives and Model Validation", Sahoo et al. (2026) extended Dupuit's approximation to model the shape of different volcanoes in the solar system. In Chapter "Deformation Models Induced by Major Regional Scale Structures Beneath the Northeast India and the Contemporary Seismotectonics", Mohanty and Mohanty

(2026) attribute the India-Eurasia collision to be the main controlling factor of the southern Shillong region. On the other hand, the Assam foredeep and the central Shillong region developed without any role of the shallow/crustal-scale structures. In Chapter "Successive Studies of the 2011 Tohoku-Oki Earthquake Signatures from Global Navigation Satellite System Measurements", Ansari (2026a) discusses the 2011 Tohoku-Oki Japan earthquake's manifestations in terms of geodetic data on the Earth surface, the Pacific Ocean and in space. In Chapter "Review of Crustal Deformation Studies in Western India by Using Global Navigation Satellite System Measurements", Ansari (2026b) performs study of Global Navigation Satellite System and proves that the horizontal velocity trend at the southern portion of western India is opposite to that at the northern part. In Chapter "Uranium Geochemistry in Phosphate-Bearing Sedimentary Rocks of Masrana and Kimoi Blocks of the Mussoorie Syncline, Uttarakhand, India", Khan et al. (2026) discuss uranium geochemistry of phosphorites from the Kimoi and the Masrana blocks of the Mussoorie syncline, Lesser Himalaya, India. In Chapter "Structural Style and Deformation of the Gachsaran Formation in the Dezful Embayment of the Zagros Fold Thrust Belt, Chamshir Area, Iran", Tavakolian et al. (2026) present the structural geology of the Gachsaran Formation in the Dezful Embayment. This stratigraphic unit is the cap rock of the Asmari reservoir and therefore is important. In Chapter "Folding of Post-bijawar Dykes of Central India Around Hirapur", Singh et al. (2026a) report structures and mild metamorphism of the post-Bijawar dykes from central India. In Chapter "An Updated Geological Map of Hirapur and Adjoining Areas of Madhya Pradesh, Central India", Singh et al. (2026b) provide an updated map of the Hirapur and surrounding areas where granites and post-Bijawar dykes have been mapped. In Chapter "Field Observations of Dykes and Geometry of Igneous Enclaves in Ladakh Batholith Granitoids Around Taru, Leh, Ladakh, North India", Mir et al. (2026) provide a detailed account of dyke's attitudes from Ladakh area from field. In Chapter "Hydrocarbon Prospectively of the Ganga Basin (India) Based on Plate Tectonic Reconstruction", Mazumder and Kumar (2026) study the Indo-Gangetic plain and bring out a tectonic correlation with a part of Arabian plate and also western Indian basins. The strike-slip Najd Fault system well established in the Arabian plate is traced on the Ganga basin. In Chapter "Preliminary Study of Artificial Slope Facets from Coal Dump Sites", Raha et al. (2026) study few coal dump-sites from eastern India. Aspects of different morphologies of slopes and their stability issues were investigated. In Chapter "Potential Location for Relief Center for Earthquake Hazards Using Bipolar Fuzzy Graphs Example from Kutch Area, Gujarat, India", Kanvinde et al. (2026) apply bipolar fuzzy graph theory to identify the most feasible relief centers (first choice-Bhuj, second choice-Mandvi) in the Kutch region, Gujarat state in western India.

Mumbai, India

Soumyajit Mukherjee
smukherjee@iitb.ac.in

Acknowledgements Springer is thanked for proofreading. Mukherjee (2026) summarized this chapter.

References

Alania, V., Beridze, T., Enukidze, O., Gusmeo, T., Razmadze, A., Tevzadze, N., Vashakidze, G. (2026). Deep structure and deformation structural style of the Georgian part of Lesser Caucasus orogen. In: Mukherjee, S. (Ed.) Structural Geology and Tectonics. Springer. pp. 1–14. ISBN: 978-981-95-4743-2.

Ansari, K. (2026a). Successive Studies of the 2011 Tohoku-Oki Earthquake Signatures from Global Navigation Satellite System Measurements. In: Mukherjee, S. (Ed.) Structural Geology and Tectonics. Springer. pp. 171–185. ISBN: 978-981-95-4743-2.

Ansari, K. (2026b). Review of Crustal deformation measurements by using global Navigation Satellite system Measurements in Western India. In: Mukherjee, S. (Ed.) Structural Geology and Tectonics. Springer. pp. 187–205. ISBN: 978-981-95-4743-2.

Chakraborty, M., Mukherjee, S. (2020). Structural geological interpretations from unrolled images of drill cores. Marine and Petroleum Geology 115, 104241.

Chandrasekharam, D. (2026). Innovative technology to harness geothermal energy from active ridge flanks and oceanic islands: Advantage for the oil industries to bring energy transition. In: Mukherjee, S. (Ed.) Structural Geology and Tectonics. Springer. pp. 65–95. ISBN: 978-981-95-4743-2.

Chatterjee, S., Mukherjee, S. (2023). Review on drilling-induced fractures in drill cores. Marine and Petroleum Geology 151, 106089.

Goswami, S. (2026). Significance of oriented-core in subsurface geological exploration. In: Mukherjee, S. (Ed.) Structural Geology and Tectonics. Springer. pp. 97–112. ISBN: 978-981-95-4743-2.

Kanvinde, M., Manjusha, O. T., Mukherjee, S. (2026). Potential location for relief centre for earthquake hazards using bipolar fuzzy graphs: Example from Kutch area, Gujarat, India. In: Mukherjee, S. (Ed.) Structural Geology and Tectonics. Springer. pp. 335–347. ISBN: 978-981-95-4743-2.

Khan, S. A., Dar, S. A., Patel, S., Mir, A. R. (2026). Uranium geochemistry in phosphate bearing sedimentary rocks of Masrana and Kimoi blocks of Mussoorie syncline, Uttarakhand, India. In: Mukherjee, S. (Ed.) Structural Geology and Tectonics. Springer. pp. 207–226. ISBN: 978-981-95-4743-2.

Mazumder, S., Kumar, S. (2026). Hydrocarbon Prospectively of the Ganga Basin (India) Based on Plate Tectonic Reconstruction. In: Mukherjee, S. (Ed.) Structural Geology and Tectonics. Springer. pp. 291–317. ISBN: 978-981-95-4743-2.

Mir, A. R., Mukherjee, S., Shaida, A., Nazir, S. A., Dar, S. A. (2026). Field observations of dykes and geometry of igneous enclaves in Ladakh batholith granitoids around Taru, Leh, Ladakh, North India. In: Mukherjee, S. (Ed.) Structural Geology and Tectonics. Springer. pp. 271–289. ISBN: 978-981-95-4743-2.

Mohanty, D. D., Mohanty, R. B. (2026). Deformation models induced by major regional scale structures beneath the northeast India and the contemporary seismotectonics. In: Mukherjee, S. (Ed.) Structural Geology and Tectonics. Springer. pp. 155–169. ISBN: 978-981-95-4743-2.

Mukherjee, S. (2026). Introduction to "Structural Geology & Tectonics". In S. Mukherjee (Ed.), *Structural Geology & Tectonics*. Springer. ISBN: 978-981-95-4743-2.

Nanda, S., Rekha, S. (2026). Field guide to Coorg Block and the adjoining shear zones, Southern Granulite Terrane, south India. In: Mukherjee, S. (Ed.) Structural Geology and Tectonics. Springer. pp. 41–51. ISBN: 978-981-95-4743-2.

Ovung, T. N., Nayak, B. R., (2026). Geodynamic evolution of the Indo-Myanmar ranges, NE India. In: Mukherjee, S. (Ed.) Structural Geology and Tectonics. Springer. pp. 53–63. ISBN: 978-981-95-4743-2.

Pueyo, E. L., Román-Berdiel, T., Willingshofer, E., Ros, L. H., Beekman, F., Peiro, A., Soto, R., Rincón, L., Ramón, M. J., Casas, A. M. (2026). 4D monitoring of analogue models in CT scanners using linear markers of X-Ray highly absorbent materials. In: Mukherjee, S. (Ed.) Structural Geology and Tectonics. Springer. pp. 113–130. ISBN: 978-981-95-4743-2.

Raha, A., Das, S., Biswas, M., Mukherjee, S. (2026). Preliminary study of artificial slope facets from coal dump-sites. In: Mukherjee, S. (Ed.) Structural Geology and Tectonics. Springer. pp. 319–333. ISBN: 978-981-95-4743-2.

Sahoo, S., Subhadarshini, S., Jena, A. K., Kundu, B. (2026). Structural form of stratovolcanoes on planetary bodies: A theoretical perspectives and model validation. In: Mukherjee, S. (Ed.) Structural Geology and Tectonics. Springer. pp. 131–154. ISBN: 978-981-95-4743-2.

Singh, G. K., Jain, A., Trivedi, R. K. (2026b). An updated Geological map of Hirapur and adjoining areas of Madhya Pradesh, central India. In: Mukherjee, S. (Ed.) Structural Geology and Tectonics. Springer. pp. 267–270. ISBN: 978-981-95-4743-2.

Singh, G. K., Trivedi, R. K., Jhariya, A. K., Jain, A. (2026a). Folding of Post Bijawar dykes of Central India around Hirapur. In: Mukherjee, S. (Ed.) Structural Geology and Tectonics. Springer. pp. 255–265. ISBN: 978-981-95-4743-2.

Spahić, D., Bojić, Z., Vuković, S. (2026). Rare field evidence of basal breccia within the Kruševac lacustrine basin: Constraints on the Late Oligocene (Early Miocene) opening across southern peri-Pannonian intramontane basins (central Serbia). In: Mukherjee, S. (Ed.) Structural Geology and Tectonics. Springer. pp. 15–40. ISBN: 978-981-95-4743-2.

Tavakolian, E., Pash, R. R., Mukherjee, S., Eghbalpour, E., Ghanavati, M., Khoshnoodkia, M., Mohammadian, M., Sadeghi, G., Bakhtiar, H. A. (2026). Structural Style and Deformation of the Gachsaran Formation in the Dezful Embayment of the Zagros Fold Thrust Belt, Chamshir Region, Iran. In: Mukherjee, S. (Ed.) Structural Geology and Tectonics. Springer. pp. 227–253. ISBN: 978-981-95-4743-2.

Contents

About the Editor

Prof. Soumyajit Mukherjee (IIT Bombay) has published 23 books and special volumes with reputed publishers and special volumes and more than 150 research papers. He presently also works as an Associate Editor in several international journals. He teaches structural geology to M.Sc. applied geology and applied geophysics students in terms of fieldwork, theory studies and practical classes. He also teaches M.Tech. students advanced methods of structural geology. His teaching experience as a faculty member spans a total of 16+ years. He has conducted so far seven Ph.Ds. and more than 35 master's theses. He has reviewed more than 250 manuscripts for more than 40 journals (most of which are international), handled as an Associate Editor more than 200 manuscripts, and published in more than 23 journals, several of which are international and Q1. He has delivered oral presentations in more than 26 international and national conferences and has two NPTEL lecture sets available in YouTube, total lectures 120 in number. He received two internal awards for quality teaching from IIT Bombay. His research interest is mainly ductile shear zones and collisional zone kinematics. He has worked in diverse terrains and tectonic regimes and on rock types in India and Bangladesh.

Deep Structure and Deformation Structural Style of the Georgian Part of Lesser Caucasus Orogen

Victor Alania, Tamar Beridze, Onise Enukidze, Thomas Gusmeo, Alexander Razmadze, Niko Tevzadze, and Giorgi Vashakidze

Abstract The Lesser Caucasus double-wedge orogen accommodates the crustal shortening due to the far-field effects of the collision between the Arabian and Eurasian plates started to collide along the Bitlis thrust zone in the early Miocene. Subsequent convergence of the Arabian and Eurasian plates during the late Alpine time caused extensive intracontinental deformation in the Lesser Caucasus (LC). Here we show a new structural model for the Georgian part of the LC orogen by integrating field observations and re-interpreted published seismic data. These data are interpreted to develop structural cross-sections. We established that the dominant structural styles of the compressional structures are related to multiple detachments. The structural cross-section reveals the presence of the Khrami basement thrust-sheet, crustal-scale duplex, fault-related folds, triangle zones, passive-roof duplex, and passive-roof and passive-back thrusts. Pliocene–Quaternary lava flows are involved in the compressional deformation and are related to an out-of-thrust sequence of the Khrami basement thrust-sheet. According to the structural cross-section, the horizontal displacement of the basement thrust-sheet is ~20 km.

Keywords Lesser Caucasus · Seismic profile · Basement thrust-sheet · Fault-related fold · Triangle zone · Duplex

---

V. Alania (✉) · O. Enukidze
M. Nodia Institute of Geophysics, I. Javakhishvili Tbilisi State University, Tbilisi, Georgia
e-mail: victor.alania@tsu.ge

T. Beridze · A. Razmadze · G. Vashakidze
A. Janelidze Geological Institute, I. Javakhishvili Tbilisi State University, Tbilisi, Georgia

T. Gusmeo
Department of Biological, Geological, and Environmental Sciences, University of Bologna, Bologna, Italy

N. Tevzadze
Georgian Oil and Gas Ltd., Tbilisi, Georgia

S. Mukherjee (ed.), *Structural Geology and Tectonics*,
https://doi.org/10.1007/978-981-95-4743-2_1

1 Introduction

The Caucasus region is located in the northernmost part of the Arabia-Eurasia collision zone (Fig. 1). Inversion of back-arc basins, exhumation, and crustal thickening took place in the far-field zone, forming two orogens, the Lesser Caucasus (LC) and the Greater Caucasus (GC) (Fig. 1a, b) (e.g. Adamia et al., 2011a, 2011b; Alania et al., 2021a, 2021b; Cavazza et al., 2024; Mosar et al., 2022; Nemcok et al., 2013; Sosson et al., 2016). Continuous convergence between the LC and GC caused incremental deformation of the Rioni and Kura foreland basins (Alania et al., 2017, 2022; Banks et al., 1997; Corrado et al., 2021; Gusmeo et al., 2021, 2022; Tari et al., 2018; Tibaldi et al., 2017).

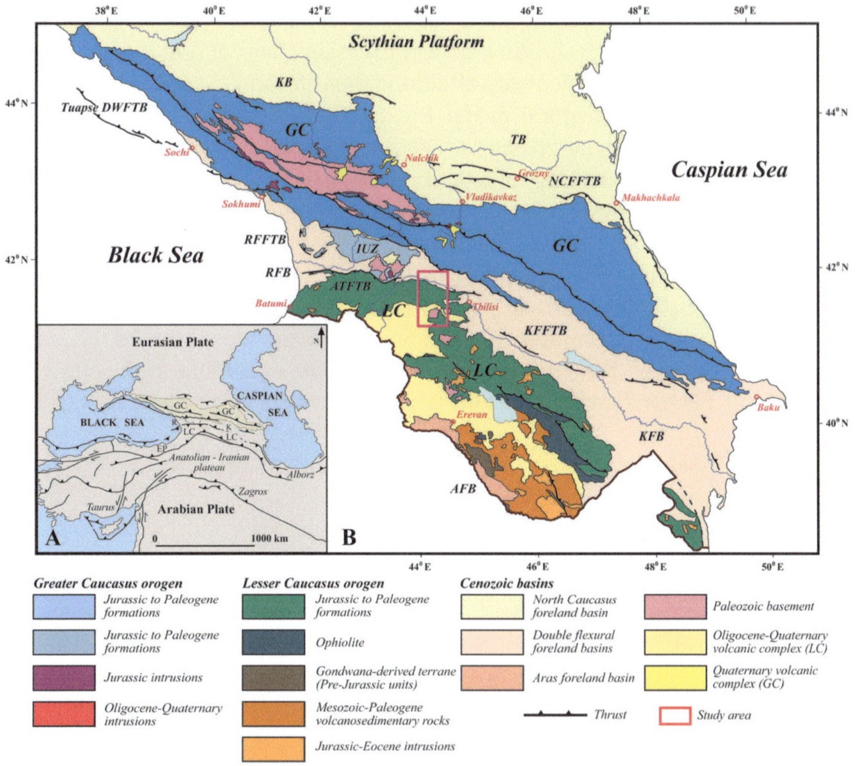

Fig. 1 **a** Tectonic map of the Arabia-Eurasia collision zone (modified from Sosson et al., 2016). **b** Tectonic map of the Caucasus (modified from Adamia et al., 2011a, 2011b; Alania et al., 2021a, 2021b, 2022; Mosar et al., 2022). Abbreviations: AFB-Araks foreland basin; ATFTB-Achara-Trialeti fold-and-thrust belt; DWFTB-Deepwater fold-and-thrust belt; GC-Greater Caucasus; IUZ-Imereti uplift zone; KB-Kuban Basin; KFB-Kura foreland basin; KFFTB-Kura foreland fold-and-thrust belt; LC-Lesser Caucasus; NCFFTB-North Caucasus foreland fold-and-thrust belt; RFB-Rioni foreland basin; RFFTB-Rioni foreland fold-and-thrust belt, TB-Terek Basin

The structure of the central part of the LC has been extensively studied through field surveys (e.g. Adamia et al., 2011a; Banks et al., 1997; Papava, 1967; Sosson et al., 2016). Unlike the central part, most structural studies in the frontal part of the LC are based on the interpretation of seismic reflection profiles (e.g. Alania et al., 2017; Tari et al., 2021). This is largely due to oil industrial interest. Nevertheless, some aspects of the deformation structural style, deep structure, and their kinematic evolution of LC remain contentious. The structure of the LC is very complex and controversial as well. Yilmaz et al. (2014) suggest that the uplift of pre-Jurassic crystalline rocks within the LC is related to south-vergent thrust systems. According to the model presented by Sosson et al. (2016), the Achara-Trialeti is a thick-skinned fold-and-thrust belt and basin inversion was caused by the Oligocene-early Miocene reactivation of the Paleocene-Eocene normal faults. As indicated by Banks et al. (1997), the deformation in the Georgian part of the LC is interpreted as the result of a blind basement thrust emplacement.

The study area includes the Khrami massif and the southern part of the Achara-Trialeti fold-and-thrust belt (ATFTB) of the LC orogen (Figs. 1b and 2). The present study aims to address two main questions about the structural architecture of the Georgian part of the LC orogen: (1) What is the deformation structural style? And (2) What is the structure lying beneath? We answer these by constructing a new structural model based on the surface geological data and the interpretation of seismic profile.

2 Geological Setting

2.1 Tectonics

Recent works on the LC demonstrate that it is a double-wedge orogen (e.g. Alania et al., 2023; Cavazza et al., 2024) formed by the ongoing collision between the Arabian and Eurasian plates that took place 20 Ma back (Okay et al., 2010) (Fig. 1a). The LC is represented by pro- and retro-wedges actively propagating into the Kura foreland sedimentary basin to the North and the LC foreland basin to the South (Alania et al., 2017, 2021a, 2021b; Cavazza et al., 2024; Corrado et al., 2021; Gusmeo et al., 2021, 2022).

In the LC orogen (Fig. 1b), towards north three main domains are distinguished: (1) the Gondwana-derived terrane (South Armenian Block-SAB), (2) the ophiolitic Sevan-Akera suture zone, (3) the Eurasian plate, which was affected by three main tectonic events: (1) Upper Cretaceous ophiolite obduction onto the SAB, (2) Paleocene-Eocene collision of the SAB with Eurasia, and (3) Oligocene–Miocene Arabia-Eurasia collision. Obduction occurred during the late Coniacian to Santonian and is responsible for the widespread ophiolitic nappe outcrop in front of the suture zone (Sosson et al., 2016, 2019).

Within the study area (Fig. 2), the rocks involved in the deformation range from Paleozoic basement rocks to Mesozoic-Tertiary rocks (e.g. Banks et al., 1997; Sosson

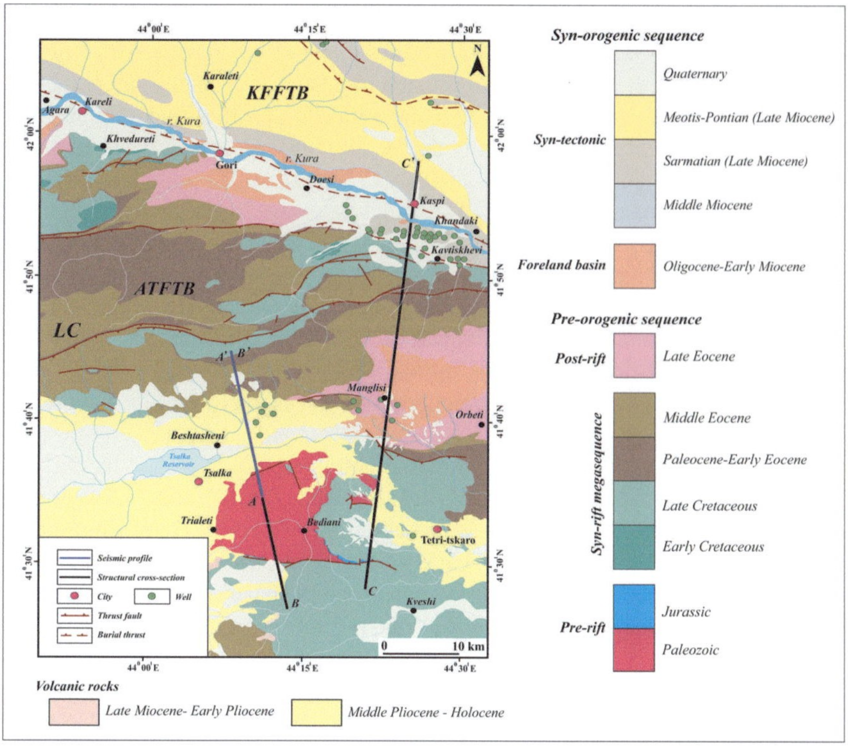

Fig. 2 Geological map of study area and the surroundings (modified from Papava, 1967). Seismic line along AA^I and structural cross-section along BB^I are available. Abbreviations: ATFTB-Achara-Trialeti fold-and-thrust belt; KFFTB-Kura foreland fold-and-thrust belt; LC-Lesser Caucasus

et al., 2016). According to Gusmeo et al. (2021), north-vergent basement (Khrami massif) thrusting is possibly related to the Paleocene continental accretion along the Sevan-Akera suture to the south and started being exhumed in the Early Eocene. The northward thrusting of the LC basement wedge during the late Alpine (late Miocene-Pliocene) time, the major structural trends are formed by ramp-anticline that transfers shortening from the mountain belt towards the north into the Kura foreland basin (KFB) (Banks et al., 1997) (Fig. 3).

Three subzones were recognized in the study area; towards north, these are: (1) the basement culmination (Khrami massif), (2) the central synformal zone, and (3) the frontal culmination (antiformal stack duplex) (Banks et al., 1997). The low-temperature thermochronologic evolution of the Achara-Trialeti back-arc basin during the Neogene refers to Miocene cooling (Gusmeo et al., 2021, 2022).

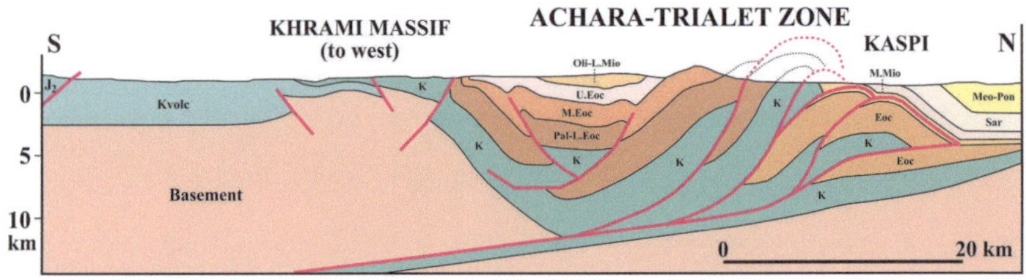

Fig. 3 Geological cross-section (C-C^{I}) through the Khrami massif and Achara-Trialeti zone (Banks et al., 1997). Location is shown in Fig. 2

2.2 *Stratigraphy*

The sedimentary cover of the LC retro-wedge is commonly >7 km thick. It is predominantly composed of shallow and deep marine rocks (e.g. Adamia et al., 2002, 2011a) and is divided by pre- and syn-orogenic sequences (Alania et al., 2021a, 2021b, 2023). The latter is summarized in the tectonostratigraphic chart below (Fig. 4).

The pre-orogenic sequences consist of Paleozoic and Jurassic-late Eocene shallow and deep marine rocks strata. The Paleozoic rocks of the Khrami massif are made up mainly of late Paleozoic (Variscan) granites and early Paleozoic gneisses, diorites, and tonalites (Adamia et al., 2011a; Zakariadze et al., 2007). The early Jurassic (about 80–200 m thick) sandstones and mudstones, with conglomerates at the base overlain the basement rocks and represent the oldest formation exposed south of the study area (Khrami massif). The middle Jurassic is made up of about 100–500 m Bajocian volcanogenic rocks. Early Cretaceous limestones, marls, and mudstones, with conglomerates at the base, transgressively rest on the late Jurassic limestones. Their total thickness is 30–100 m (Adamia et al., 2011a; Papava, 1967). About 900–1500 m thick late Cretaceous strata are represented by limestones, marls, volcanogenic deposits, and sandstones. During the Cretaceous and Paleogene, the extensional basin was filled with ∼∼2500–3000 m thick sediments (Adamia et al., 2002). Paleogene is represented by Paleocene (alternation of mudstones, siltstones, and sandstones—50–300 m), early Eocene (shales and turbidites—100–400 m), middle Eocene (debrites and volcanogenic turbidites; thickness is 200–600 m), and late Eocene (turbidites and shales-total thickness of the unit is 300–800 m) rocks strata (Adamia et al., 2002; Beridze, 2019; Tari et al., 2021).

The syn-orogenic sequences are composed of the foreland basin (Oligocene-early Miocene) and syn-tectonic strata. The Oligocene-Neogene stratigraphy of the ATFTB is dominated by siliciclastic sequences. Syn-tectonic strata of middle-late Miocene, Pliocene, and Pleistocene are represented by shallow marine and continental, predominantly terrigenous clastic rocks.

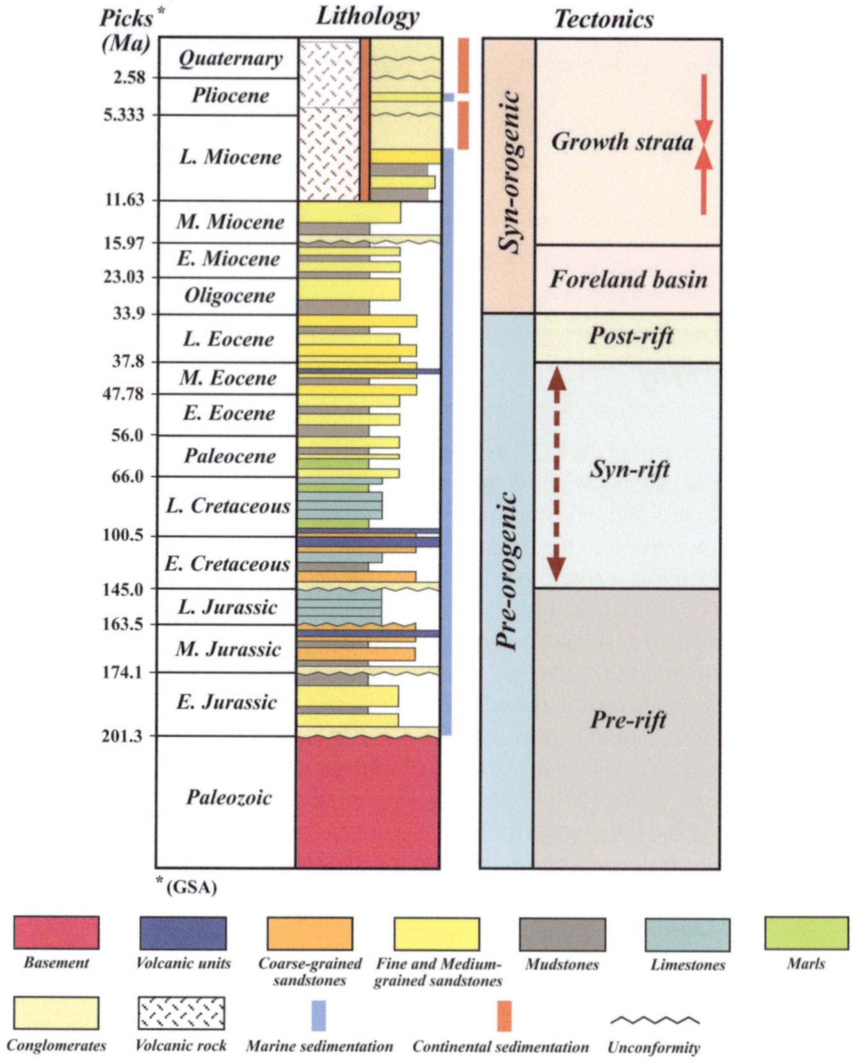

Fig. 4 Tectonostratigraphic chart of the Khrami massif and surrounding area (modified from Papava, 1967; Adamia et al., 2002, 2010, 2011a; Lebedev et al., 2008; Alania et al., 2017, 2023)

3 Data and Method

The surface geological information was obtained from the 1:100,000 geological map (Papava, 1967), relevant papers (Adamia et al., 2011a; Banks et al., 1997; Lebedev et al., 2008) and our own field data from the study area (Fig. 2). To constrain the subsurface structure of the structural cross-section of the Khrami massif and surrounding area, we used post-stack time migrated seismic profile AA^{I} (Fig. 5).

Fault-related folding and wedge thrust folding theories (Shaw et al., 2005) were used in the interpretation of seismic reflection profile (AA^I) (Fig. 5) and construction of the structural cross-section ($B\text{-}B^I$) (Fig. 6). For the interpretation of the seismic profile and construction of the structural cross-section, the Move 2022.1 software (Petroleum Experts Limited) was used.

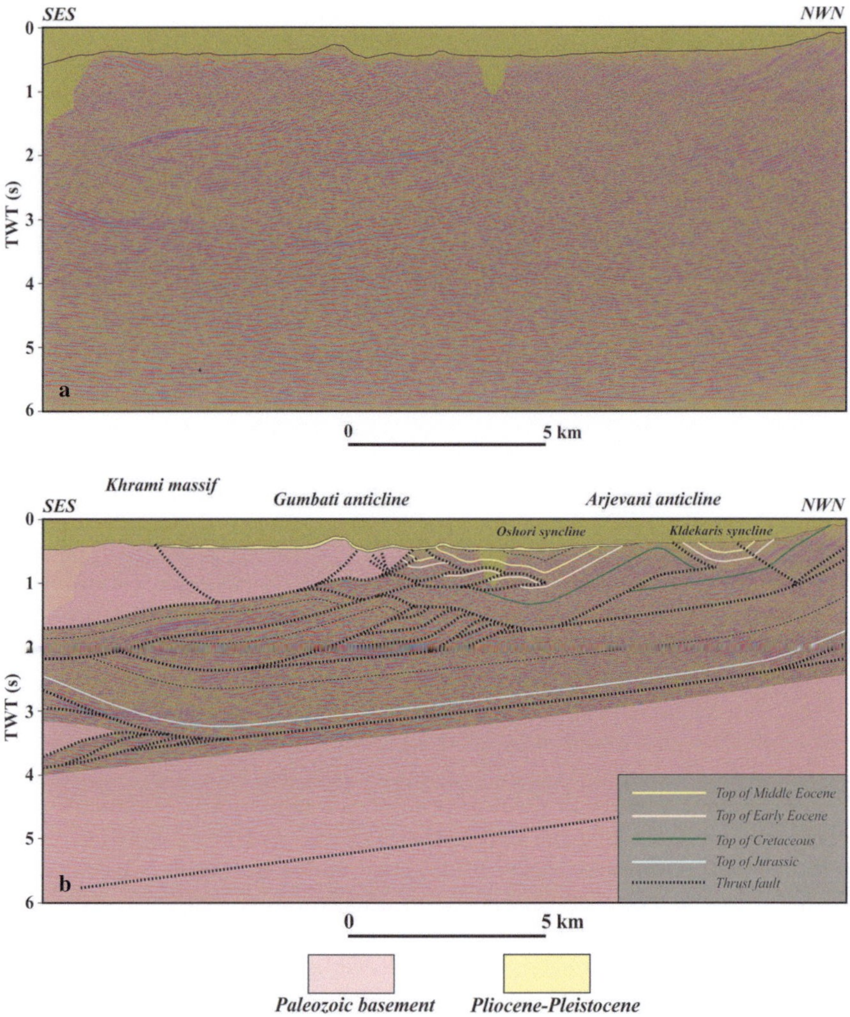

Fig. 5 **a** Uninterpreted and **b** interpreted seismic reflection profiles AA^I. Location is shown in Fig. 2

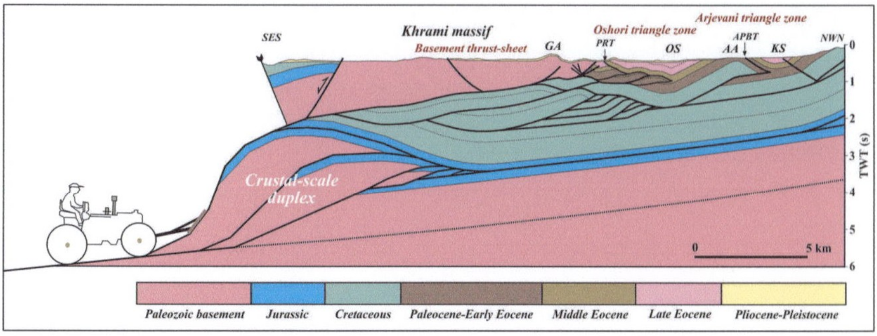

Fig. 6 Structural cross-section BB^{I}. Location is shown in Fig. 2. Abbreviations: AA-Arjevani anticline; APBT- Arjevani passive-back thrust; GA-Gumbati anticline; KS-Kldekari syncline; OS-Oshori syncline; PRT-Passive-roof thrust

4 Results

In this section, we will detail the surface and subsurface structures of the study area. A deformation structural style and detachment levels are found in the seismic profile AA^{I} (Fig. 5). Seismic reflection data (Fig. 5) and structural cross-section (Fig. 6) have revealed the presence of the basement thrust-sheet, crustal-scale duplexes, fault-related folds, and triangle zones. The structural interpretation comprises four horizons (i.e. top of Jurassic, top of Cretaceous, top of early Eocene, and top of middle Eocene). Imbricate thrusts, duplexes and fault-related folds form the main structure of the study area and exhibit characteristics associated with multiple detachments (Fig. 6).

The most important observation from seismic data and structural cross-section is that the Khrami massif is a basement thrust-sheet (Figs. 5 and 6). Various types of tectonic structures are observed, from crustal-scale duplexes to fault-related folds.

The structural inversion is complicated by the interaction between several detachments from the mid-crustal detachment to the uppermost detachment along the late Eocene shales. According to the seismic profile and structural cross-section, the crustal-scale duplex was recognized under the basement thrust-sheet, which propagates northward along the early Jurassic shale layers (Figs. 5 and 6).

Within the LC two main stages of volcanic activity are distinguished: (1) late Miocene-early Pliocene and (2) Pliocene–Quaternary (e.g. Adamia et al., 2011a; Lebedev et al., 2008). The main part of the study area is covered with Pliocene to Quaternary volcanic formations, which lie over the Paleozoic, Cretaceous, and Paleogene rocks (Fig. 2).

Most of the Gumbati anticline is covered by the Quaternary lava flow (Fig. 7a–d). Dip of Quaternary lava interflow horizons and basaltic lavas of the forelimb of Gumbati anticline varies along strike between 30 and 80^{0} (Fig. 7b). K/Ar dating of these basaltic lava flows yields ages from 3.25 $\pm$ 0.25 Ma to 2.45 $\pm$ 0.25 Ma (Lebedev et al., 2008). The formation of basement-cored Gumbati anticline in the frontal part of the basement thrust-sheet is related to out-of-sequences thrusting.

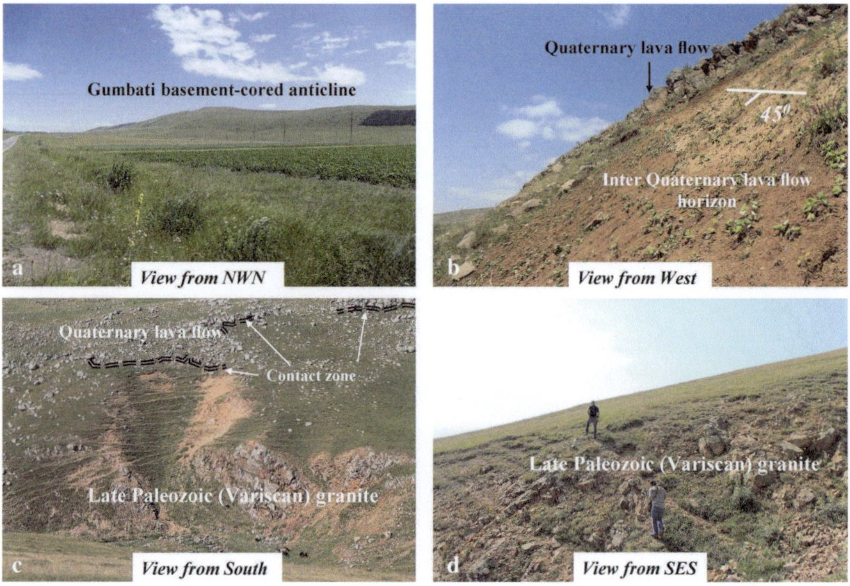

Fig. 7 **a** Panorama picture of the basement-cored Gumbati anticline, **b** Forelimb of the basement-cored Gumbati anticline, **c** Contact zone between Quaternary lava flow and Paleozoic granite, **d** Back limb of the basement-cored Gumbati anticline

According to these observations, the deformation structural style is discussed, and it is proposed that the Quaternary lava flow is involved in deformation and is related to the out-of-sequence activity of the basement thrust-sheet. Based on the structural cross-section, the horizontal displacement of the basement thrust-sheet is ~20 km (Fig. 6).

Two triangle zones are distinguished in the study area. The first Oshori triangle zone was recognized under the Quaternary lava flow. A roof thrust is located within the Paleogene section, defining passive-roof duplex and passive-roof thrust (Figs. 5 and 6). This structure in Fig. 8a is interpreted as a multiple fault-bend fold wedges model (Shaw et al., 2005) and is a clear example of a ramp-dominated triangle zone (Hagke & Malz, 2018). The second triangle zone, the Arjevani structure is defined by detachment and a roof thrust, constituting a fault-bend fold wedge with a passive-back thrust (Figs. 5 and 6). This structure in Fig. 8b is interpreted as a simple wedge model (Shaw et al., 2005) and represents also a ramp-dominated triangle zone (Hagke & Malz, 2018).

The middle Miocene-Pliocene structural inversion of the Achara-Trialeti back-arc basin based on low-temperature thermochronologic data (Gusmeo et al., 2021) and the coeval syn-sedimentary deformation within the Kura and Rioni foreland basins (Alania et al., 2017, 2022; Tibaldi et al., 2017) is interpreted as an effect of far-field stress transmission from the Arabia-Eurasia collision zone into the hinterland (Cavazza et al., 2024). Published information about historical and recent earthquake

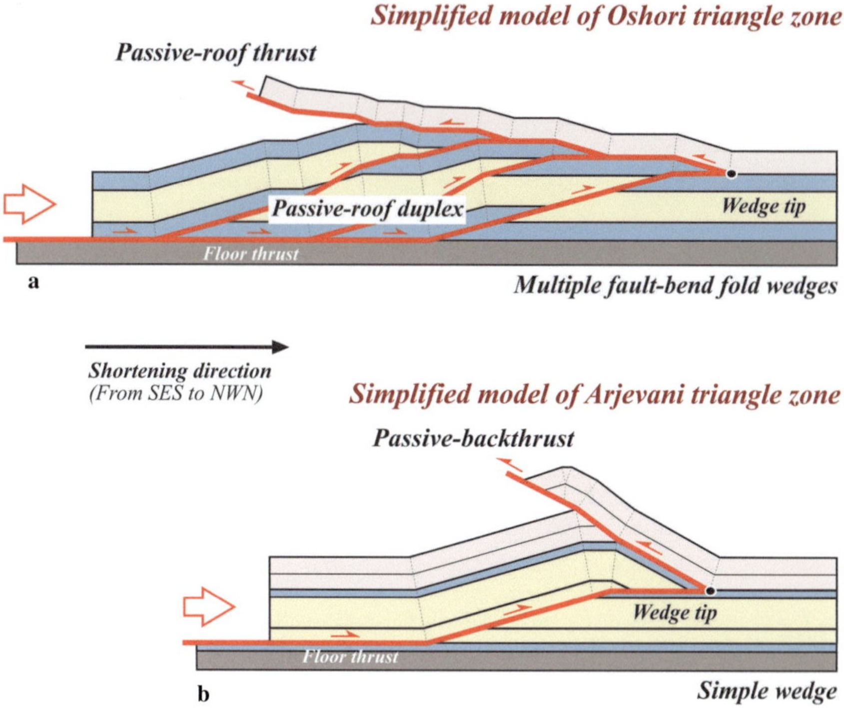

Fig. 8 Simplified sketch of the types of triangle zones of the study area and their main structural geometries associated with the **a** multiple fault-bend fold wedges (Shaw et al., 2005) and **b** simple fault-bend fold wedge (Shaw et al., 2005) and are a clear example of ramp-dominated triangle zones (Hagke & Malz, 2018)

data (Adamia et al., 2004, 2017; Tsereteli et al., 2016), and absolute ages of deformed volcanic rocks (Pliocene–Quaternary) from the study area (Lebedev et al., 2008) indicate that LC retro-wedge is tectonically active. Seismic reflection data (Fig. 5) and structural cross-section (Fig. 6) suggest that our model is similar to a basement-involved thin-skinned fold-and-thrust belt style (Pfiffner, 2017).

The structural style proposed in this chapter offers a promising geological setting for potential structural traps in the Georgian part of the LC. The major structural trends are formed by fault-bend, fault-propagation folds, and duplexes that transfer shortening from the mountain belt towards the north into the KFB. Duplexes and imbricate thrust systems commonly form important structural hydrocarbon traps (e.g. Shaw et al., 2005). The potential productive reservoirs of the study area are of the Cretaceous and Paleogene ages and have ramp-flat duplex structures of triangle zones. In future, we hope that a petroleum industry will drill a well in this zone taking into account the results of our research, and that it will turn out to be productive.

5 Conclusions

We have presented a new structural model across the Georgian part of the LC orogen retro-wedge. The results of our surface and subsurface interpretations have important implications for how this fold-and-thrust belt formed, in addition to the effect of structural style on active tectonics in the retro-wedge of the LC orogen. Our primary results have shown the following.

- Dominant structural styles of the compressional structures are related to multiple detachments and are represented by Khrami basement thrust-sheet, crustal-scale duplex, fault-related folds, triangle zones, passive-roof duplex, passive-roof thrust, and passive-back thrust.
- Triangle zones are interpreted as simple and multiple fault-bend fold wedges models and are a good example of ramp-dominated triangle zone.
- The formation of basement-cored Gumbati anticline in the frontal part of the basement thrust-sheet is related to out-of-sequences thrusting activity during the Quaternary. The horizontal displacement of the basement thrust-sheet is about 20 km.

Acknowledgements The work was supported by the Shota Rustaveli National Science Foundation of Georgia (SRNSFG) [Grant number FR-22-19681-Volcanic facies architecture of lava flows and associated interflow horizons from Javakheti Volcanic Province (South Georgia)]. We are grateful to Georgian Oil and Gas Ltd for providing the seismic profiles for our interpretation. We are thankful to Petroleum Expert (www.petex.com) for providing the academic licence of MOVE software. We are fairly grateful to Prof. Soumyajit Mukherjee (IIT Bombay) for his helpful review comments and suggestions, which have improved this chapter. Mukherjee (2026) summarized this work.

Abbreviations

AA	Arjevani anticline
AFB	Araks foreland basin
APBT	Arjevani passive-back thrust
ATFTB	Achara-Trialeti fold-and-thrust belt
DWFTB	Deepwater fold-and-thrust belt
GA	Gumbati anticline
GC	Greater Caucasus
IUZ	Imereti uplift zone
KB	Kuban Basin
KFB	Kura foreland basin
KS	Kldekari syncline
KFFTB	Kura foreland fold-and-thrust belt
LC	Lesser Caucasus
NCFFTB	North Caucasus foreland fold-and-thrust belt

OS	Oshori syncline
RFB	Rioni foreland basin
RFFTB	Rioni foreland fold-and-thrust belt
PRT	Passive-roof thrust
SAB	South Armenian Block
TB	Terek Basin

References

Adamia, S., Alania, V., Tsereteli, N., et al. (2017). Postcollisional tectonics and seismicity of Georgia. *Geol Soc Amer Spec Pap, 525*, 535–573. https://doi.org/10.1130/2017.2525(17)

Adamia, S., Zakariadze, G., Chkhotua, T., et al. (2011a). Geology of the Caucasus: A Review. *Turkish J Earth Sci, 20*, 489–544. https://doi.org/10.3906/yer-1005-11

Adamia, S., Alania, V., Chabukiani, A., et al. (2011b). Great Caucasus (Cavcasioni): Longliving Northtethyan Back-Arc Basin. *Turkish J Earth Sci, 20*, 611–628. https://doi.org/10.3906/yer-1005-12

Adamia, S., Alania, V., & Chabukiani, A. et al. (2010). Evolution of the Late Cenozoic basins of Georgia (SW Caucasus): A review. *Geological Society, London, Special Publications, 340*, 239–259. https://doi.org/10.1144/SP340.11.

Adamia, S., Alania, V., & Chabukiani, A. et al. (2004). Seismic source zone models of recent earthquakes of Georgia. In A. A. Chatzipetros, S. B. Pavlides (Eds.), *Proceedings of the 5th international symposium on eastern mediterranean geology, Greece, Thessaloniki*, pp 545–551.

Adamia, S., Alania, V., Ananiashvili, G., et al. (2002). Late Mesozoic-Cenozoic geodynamic evolution of the eastern Georgian oil-gas bearing basins. *Geol Carp, 53*, 155–159.

Alania, V., Enukidze, O., & Razmadze, A. et al. (2023). Interpretation and analysis of seismic and analog modeling data of triangle zone: a case study from the frontal part of western Kura foreland fold-and-thrust belt. *Georgia. Frontiers in Earth Sciences, 11*, 1195767. https://doi.org/10.3389/feart.2023.1195767.

Alania, V., Melikadze, G., Pace, P., et al. (2022). Deformation structural style of the Rioni foreland fold-and-thrust belt, western Greater Caucasus: Insight from the balanced cross-section. *Frontiers in Earth Science, 10*, 10968386. https://doi.org/10.3389/feart.2022.968386

Alania, V., Beridze, T., & Enukidze, O. et al. (2021a). The geometry of the two orogens convergence and collision zones in central Georgia: new data from seismic reflection profiles. In F. L. Bonali et al (Eds.), *Building knowledge for geohazard assessment and management in the Caucasus and other Orogenic Regions, NATO science for peace and security series C: environmental security*, pp. 73–88. Springer. https://doi.org/10.1007/978-94-024-2046-3_6.

Alania, V., Tibaldi, A., & Bonali, F. et al. (2021b). Structural architecture of the western Greater Caucasus orogen: New data from a crustal-scale structural cross-section. In F.L. Bonali et al. (Eds.), *Building knowledge for Geohazard assessment and management in the Caucasus and other Orogenic regions, NATO Science for Peace and Security Series C: Environmental Security*, pp. 59–71. Springer. https://doi.org/10.1007/978-94-024-2046-3_5.

Alania, V., Chabukiani, A., & Chagelishvili, R. et al. (2017). Growth structures, piggyback basins and growth strata of Georgian part of Kura foreland fold and thrust belt: implication for Late Alpine kinematic evolution. *Geological Society, London, Special Publications, 428*, 171–185. https://doi.org/10.1144/SP428.5.

Banks, C., Robinson, A., & Williams, M. (1997). Structure and regional tectonics of the Achara-Trialeti fold belt and the adjacent Rioni and Kartli foreland basins. Republic of Georgia. In A. G. Robinson (Ed.), *Regional and petroleum geology of the black sea and surrounding region*, no. 68, pp. 331–36. American Association of Petroleum Geologists Memoir.

Beridze, T. (2019). Middle Eocene submarine mass-transport deposits of Tbilisi environs (eastern Achara-Trialeti fold and thrust belt, Georgia). In *Abstracts of the IAS 34-th meeting on Sedimentology, Rome, Italy*, 10–13 Sep 2019

Cavazza, W., Gusmeo, T., Zattin, M., et al. (2024). Two-step exhumation of Caucasian intraplate rifts: A proxy of sequential plate-margin collisional orogenies. *Geoscience Frontiers, 15*, Article 101737. https://doi.org/10.1016/j.gsf.2023.101737

Corrado, S., Gusmeo, T., Schito, A., et al. (2021). Validating far-field deformation styles from the Adjara-Trialeti fold-and-thrust belt to the Greater Caucasus (Georgia) through multi-proxy thermal maturity datasets. *Marine and Petroleum Geology, 130*, Article 105141. https://doi.org/10.1016/j.marpetgeo.2021.105141

Gusmeo, T., Schito, A., Corrado, S., et al. (2022). Tectono-thermal evolution of central Transcaucasia: Thermal modelling, seismic interpretation, and low-temperature thermochronology of the eastern Adjara-Trialeti and western Kura sedimentary basins (Georgia). *Journal of Asian Earth Sciences, 237*, Article 105355. https://doi.org/10.1016/j.jseaes.2022.105355

Gusmeo, T., Cavazza, C., Alania, V., et al. (2021). Structural inversion of back-arc basins-The Neogene Adjara-Trialeti fold-and-thrust belt (SW Georgia) as a far-field effect of the Arabia-Eurasia collision. *Tectonophysics, 803*, Article 228702. https://doi.org/10.1016/j.tecto.2020.228702

Hagke, C., & Malz, A. (2018). Triangle zones—Geometry, kinematics, mechanics, and the need for appreciation of uncertainties. *Earth-Science Reviews, 177*, 24–42. https://doi.org/10.1016/j.earscirev.2017.11.003

Lebedev, V. A., Bubnov, S. N., Dudauri, O. Z., & Vashakidze, G. T. (2008). Geochronology of Pliocene Volcanism in the Dzhavakheti Highland (the Lesser Caucasus). Part 2: Eastern Part of the Dzhavakheti Highland. Regional Geological Correlation. *Stratigraphy and Geological Correlation, 16*, 101–123.

Mosar, J., Mauvilly, J., & Koiava, K. et al. (2022). Tectonics in the Greater Caucasus (Georgia–Russia): From an intracontinental rifted basin to a doubly verging fold-and-thrust belt. *Marine and Petroleum Geology, 140*, 105630. https://doi.org/10.1016/j.marpetgeo.2022.105630.

Mukherjee, S. (2026). Introduction to “Structural Geology & Tectonics”. In S. Mukherjee (Ed.), *Structural Geology & Tectonics*. Springer. ISBN: 978-981-95-4743-2

Nemcok, M., Glonti, B., Yukler, A., & Marton, B. (2013). Development history of the foreland plate trapped between two converging orogens: Kura Valley, Georgia, case study. *Geological Society, London, Special Publications, 377*, 159–188. https://doi.org/10.1144/SP377.9

Okay, A. I., Zattin, M., & Cavazza, W. (2010). Apatite fission-track data for the Miocene Arabia-Eurasia collision. *Geology, 38*, 35–38. https://doi.org/10.1130/G30234.1

Papava, P. (1967). Some question of eastern part of the Trialeti Ridge and perspective of oil-bearing of Cretaceous and Paleogene deposits. In P. Gamkrelidze (Ed.), *Geology of the Georgian SSR*, pp. 188–204. NEDRA (in Russian)

Pfiffner, O. A. (2017). Thin-skinned and thick-skinned tectonics: A global perspective. *Geosciences, 7*(71). https://doi.org/10.3390/geosciences7030071.

Shaw, J. H., Connors, C. D., & Suppe, J. (Eds.) (2005). *Seismic interpretation of contractional fault-related folds*. AAPG.

Sosson, M., Stephenson, R., & Adamia, S. et al. (2019). The Eastern Black Sea and Caucasus Domain Origin and Its Tectonic Evolution: New Insights from Results of a Decade of Field Works and of Geophysical Research. In F Rossetti et al. (Eds.), *The Structural Geology Contribution to the Africa-Eurasia Geology: Basement and Reservoir Structure, Ore Mineralization and Tectonic Modelling. Proceedings of the 1st Springer Conference of the Arabian Journal of Geosciences (CAJG-1), Tunisia*, pp. 303–305.

Sosson, M., Stephenson, R., Sheremet, Y., et al. (2016). The Eastern Black Sea-Caucasus region during Cretaceous: New evidence to constrain its tectonic evolution. *Compte-Rendus Geosci, 348*, 23–32. https://doi.org/10.1016/j.crte.2015.11.002

Tari, G., Vrsic, A., Gumpenberger, T., Mekonnen, E., et al. (2021). Eocene volcaniclastics in the Kartli basin, Georgia: A fractured reservoir sequence. *Journal of Petroleum Geology, 44*(3), 413–433. https://doi.org/10.1111/jpg.12795

Tari, G., Vakhania, D., Tatishvili, G., et al. (2018). Stratigraphy, structure and petroleum exploration play types of the Rioni Basin, Georgia. *Geological Society, London, Special Publications, 464*, 403–438. https://doi.org/10.1144/SP464.14

Tibaldi, A., Alania, V., & Bonali, F. L., et al. (2017). Active inversion tectonics, simple shear folding and back-thrusting at Rioni Basin, Georgia. *Journal of Structural Geology, 96*, 35–53. https://doi.org/10.1016/j.jsg.2017.01.005.

Tsereteli, N., Tibaldi, A., Alania, V., et al. (2016). Active tectonics of central-western Caucasus, Georgia. *Tectonophysics, 691*, 328–344. https://doi.org/10.1016/j.tecto.2016.10.025S0040-1951(16)30478-4

Yilmaz, A., Adamia, S., & Yilmaz, H. (2014). Comparisons of the suture zones along a geotraverse from the Scythian platform to the Arabian Platform. *Geoscience Frontiers, 5*(6), 855–875. https://doi.org/10.1016/j.gsf.2013.10.004

Zakariadze, G., Dilek, Y., Adamia, S., et al. (2007). Geochemistry and geochronology of the Neoproterozoic Pan-African Transcaucasian Massif (Republic of Georgia) and implications for island-arc evolution of the late Precambrian Arabian Nubian Shield. *Gondwana Research, 11*, 97–108. https://doi.org/10.1016/j.gr.2006.05.012

Rare Field Evidence of Basal Breccia Within the Kruševac Lacustrine Basin: Constraints on the Late Oligocene (Early Miocene) Opening Across Southern Peri-Pannonian Intramontane Basins (Central Serbia)

Darko Spahić, Zoran Bojić, and Srđan Vuković

Abstract We study the oldest evolutionary stage of the Miocene Kruševac lacustrine basin at a site exposing rarely observed latest Oligocene to early Neogene sediments. The recent historically lowest water levels in the Morava River (central Serbia) have played a crucial role in unveiling remarkable outcrops, providing a unique perspective on the earliest stages of basin evolution. The rarely exposed latest Oligocene to early Neogene sediments, particularly those of the Kruševac intramontane paleolake, have made a substantial contribution to our understanding of the peri-Pannonian syn-rift opening of a number of intramontane basins. This study delves into the early paleolake development stage in connection with the adjoining Aleksinac basin, both of which are accommodated on top of the former Neotethyan suture, its overriding crystalline continental plate (Mesozoic European foreland; central and eastern Serbia). A number of intramontane basins is, nevertheless, also developed on top of Neotethyan subducting counterpart, its formerly descending oceanic plate (Dinarides and Vardar Zone; Croatia, Bosnia & Herzegovina, Serbia). Our research methodology, which involved mapping an exposed area of the Morava riverbed uncovered by the record low water level during spring 2021 (Kruševac basin, central Serbia), has led to the discovery of a unique sequence with breccia composed of large angular crystalline clasts. We found large angular basement-derived gneiss-bearing Serbo-Macedonian cobbles embedded within the rocks of the presumed Oligocene–Miocene interval. The study highlights the importance of the presence of these large crystalline cobbles, which suggests the vicinity of the sediment sources responsible for supplying the initial syn-rift peri-Pannonian basal sequence. Drawing this analogy, such breccias appear to be deposited during

D. Spahić (✉)
University of Belgrade, Institute for Multidisciplinary Research, Kneza Višeslava 1, Belgrade, Serbia
e-mail: darko.spahic@imsi.bg.ac.rs; darkogeo2002@hotmail.com

Z. Bojić · S. Vuković
Geological Survey of Serbia, Belgrade, Serbia

S. Mukherjee (ed.), *Structural Geology and Tectonics*,
https://doi.org/10.1007/978-981-95-4743-2_2

the opening of numerous intramontane basins across Serbia and the Balkans. The initial transport from the crystalline source must be short, presumably gravitational. Shortly after, a shallow-water environment succeeded the early alluvial stage. It is worth noting that the brecciated sequence is devoid of fossils, which poses a challenge in determining the Late Oligocene age of the cobble-bearing sequence.

Keywords Peri-Pannonian extension · Kruševac basin · Extraordinary riverbed outcrop · Late Oligocene breccia · Crustal stretching

1 Introduction

An abrupt switch from compressional Alpine orogen-related tectonic processes to widespread extension across Central Europe and the western Balkans occurred during the Late Oligocene (e.g., Ziegler & Cloetingh, 2004; Scettino & Turco, 2011). The change in the regional stress field resulted in a transfer from the collisional crustal thickening process into the rapid orogenic relaxation (and extension-driven subsidence). The Late Oligocene–Neogene regional extension and formation of several crustal-scale depressions across the Alpine Carpathian-Balkan-Dinaride orogen (western Balkan Peninsula) provided suitable play-to-prospect conditions for the development of oil and gas-bearing basins (Obradović et al., 2000; Čokorilo et al., 2009; Boote et al., 2018; Sachsenhofer et al., 2018a, 2018b; Burazer et al., 2021a, 2021b; Bartha et al., 2022; and many others; Fig. 1a, b). The two most prominent basins developed during this time are well-explored central European basins belonging to the Paratethys: Pannonian and Dacian basins (e.g., Popov et al. 2002; Marović et al., 2007c; Krézsek et al., 2013; Rundić et al., 2019; Krézsek & Olariu, 2021; Fig. 1b). Processes of syn-rift stretching, transtension, and early subsidence of the southern peri-Pannonian basin area (e.g., Tari et al., 1992; Marović et al., 2002; Spahić et al., 2023b; Tari et al., 2023; Fig. 2) yielded sufficient accommodation space for the Oligocene to lowermost Miocene terrestrial lacustrine sequences (e.g., Lazarević et al., 2019; Radisavljević et al., 2024; Burazer et al., 2025). One of the central yet underexplored Neogene graben-shaped basinal features distributed along the southern Peri-Pannonian realm is the Kruševac depression of central Serbia (Fig. 1b). The Neogene deposits of the Kruševac basin overlie the complex Paleotethyan-Neotethyan Mesozoic amalgmation (e.g., Dimitrijević, 2001; Kräutner & Krstić, 2002; Karamata, 2006; Spahić & Gaudenyi, 2020a) and its much older Serbo-Macedonian basement (e.g., Antić et al., 2016; Spahić et al., 2021; Spahić & Gaudenyi, 2020b) in a transgressive manner (Figs. 1c, 2 and 3).

Together with the onset and opening of Pannonian and Aegean large basins during the late Oligocene (e.g., Anđelković, 1986; Dumurdžanov et al., 2005), continuing into early and middle Miocene (e.g., Prelogović et al. 1998; Mandic et al., 2012; Rundić et al., 2013), there was an intervening central large elevated morphostructure referred to as the *"Dinarian land"* (Sachsenhofer et al., 2018a). This intramontane area represents a barrier between two large basins: *(i)* the developing Paratethys

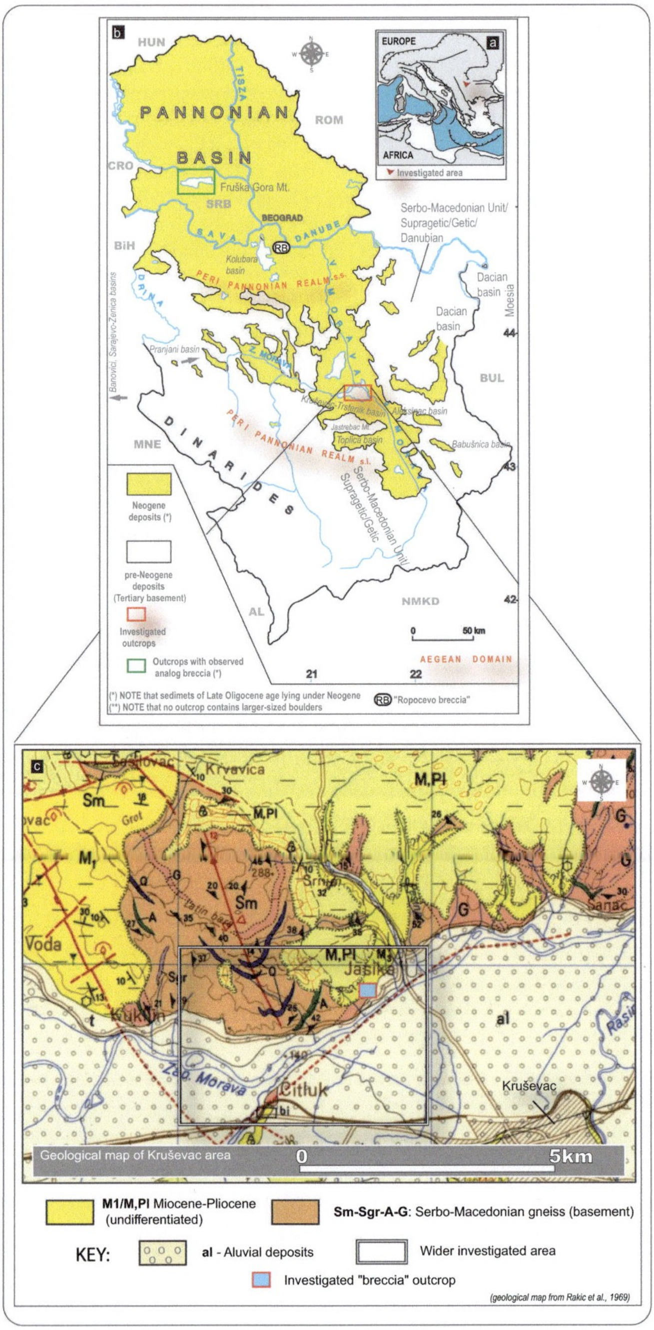
b
HUN
PANNONIAN
BASIN
ROM
CRO
Fruška Gora Mt.
SRB
BEOGRAD
SAVA
DANUBE
TISZA
BiH
Kolubara basin
PERI PANNONIAN REALM s.s.
DRINA
Banovici, Sarajevo-Zenica basins
Pranjani basin
Z. MORAVA
MORAVA
Kruševac-Trstenik basin
Aleksinac basin
Jastrebac Mt.
Toplica basin
Babušnica basin
MNE
DINARIDES
PERI PANNONIAN REALM s.l.
Serbo-Macedonian Unit/ Supragetic/Getic
Serbo-Macedonian Unit/ Supragetic/Getic/ Danubian
Dacian basin
Dacian basin
Moesia
BUL
44
43
42
NMKD
AL
AEGEAN DOMAIN
0 50 km
21
22
EUROPE
a
AFRICA
Investigated area
Neogene deposits (*)
pre-Neogene deposits (Tertiary basement)
Investigated outcrops
Outcrops with observed analog breccia (*)
(*) NOTE that sedimets of Late Oligocene age lying under Neogene
(**) NOTE that no outcrop contains larger-sized boulders
RB "Ropocevo breccia"
c
Krvavica
M,Pl
Sm
M1
Sm
Voda
Jasika
Sanac
al
Kruševac
Zap. Morava
Čitluk
Geological map of Kruševac area
0
5km
M1/M,Pl Miocene-Pliocene (undifferentiated)
Sm-Sgr-A-G: Serbo-Macedonian gneiss (basement)
KEY:
al - Aluvial deposits
Wider investigated area
Investigated "breccia" outcrop
(geological map from Rakic et al., 1969)

◂**Fig. 1** **a** Investigated location within Alpine Europe. The inset provides a broader European context and detailed views of the specific Kruševac basin region. **b** Geographical map showing the Pannonian Basin and surrounding regions, including parts of Hungary, Serbia, and Romania. The position of the Kruševac basin within the southern Peri-Pannonian realm, including the position of the Dacian basin. The map highlights Neogene deposits above the pre-Neogene undifferentiated basement (including the approximate location of Createceous Ropočevo breccia, RP), with specific areas marked for investigated outcrops. Note that combined Paleozoic–Mesozoic basement units occupy white areas, with the N-S striking Alpine pile of nappes, stacked from the west to the east (former Paleotethys, Neotethys, and Alpine Tethys): Drina-Ivanjica block, Jadar block as a segment of Vardar Zone, and across eastern Serbia, Serbo-Macedonian Unit/Supragetic/Getic/Danubian/Moesia (for details see e.g., Dimitrijević, 2001; Kräutner & Krstić, 2002; Antić et al., 2016; Spahić, 2022a, 2022b, 2022c; Spahić & Gaudenyi, 2020b, 2020c). **c** Geological map of broader investigated area (after Rakić et al., 1969, modified). A key indicates symbols for Miocene-Pliocene deposits, Serbo-Macedonian gneiss, alluvial deposits, and position of the investigated area

and *(ii)* the proto-Mediterranean Sea (de Leeuw et al., 2011; Mandić et al., 2011; Sachsenhofer et al., 2018a). Despite being elevated above the paleo-sea level, this large paleo-island experienced subsidence in its inner realm, followed by development of alluvial, lacustrine freshwater basins, and a widespread late Paleogene-early Neogene volcanism (Popov et al. 2002; Obradović et al., 1997, 2000; Burazer et al., 2020; Spahić, 2022c; Andrić-Tomašević et al., 2021; Spahić et al., 2023b). The timing of subsidence onset remains uncertain (Late Oligocene or Early Miocene), whereby the exact mechanism of crustal-driven downshifting is still connected with numerous controversies (rotation, trench retreat, or slab rollback; Horváth, 1993; Andrić et al., 2018; Andrić-Tomašević et al., 2021; Balling et al., 2021; de Leeuw et al., 2012; Spahić & Gaudenyi, 2020a). A similar situation is expected in the more southern peri-Pannonian areas, which were additionally under the interference of slighly earlier south Balkan-Aegean crustal stretching lasting from Eocene to Oligocene–Miocene (Burchfiel et al., 2000; Dumurdžanov et al. 2005; Burchfiel et al., 2008; Jolivet et al., 2021).

Interestingly, in most instances, these initial basin opening graben systems remained deeply buried, with their infill rarely exposed to surface locations (e.g., Obradović et al., 1997; Burazer et al., 2020, 2021a, 2021b; Radisavljević et al., 2024). However, as study will show, a surface mapping fieldwork depicted an outcrop, exposing the well-concealed Oligocene/Miocene (alluvial) equivalents. This stage is presumably in connection to the early syn-rift and basin opening stage (Fig. 1c). Such surface-exposed early basin opening equivalents are a rarity across the southern peri-Pannonian area (e.g., Anđelković, 1986, 1988a; Andrić-Tomašević et al., 2021; Lazarević & Milivojević, 2010; Lazarević et al., 2019; Fig. 1). The assumption is that stretching of the thickened crust (gneissic basement in this case) (Fig. 3) created an accommodation space affecting the formation of high-angle grabens (normal faults), followed by erosion and denudation of the exposed basement crust. In this manner, the early extension affected previosly exposed basement units providing new space for the subsidence-affected earliest shallow-depth depocenters. The confined early accommodation space allowed a short-distance source-to-sink process. As observed in most intermontane basins (e.g.,

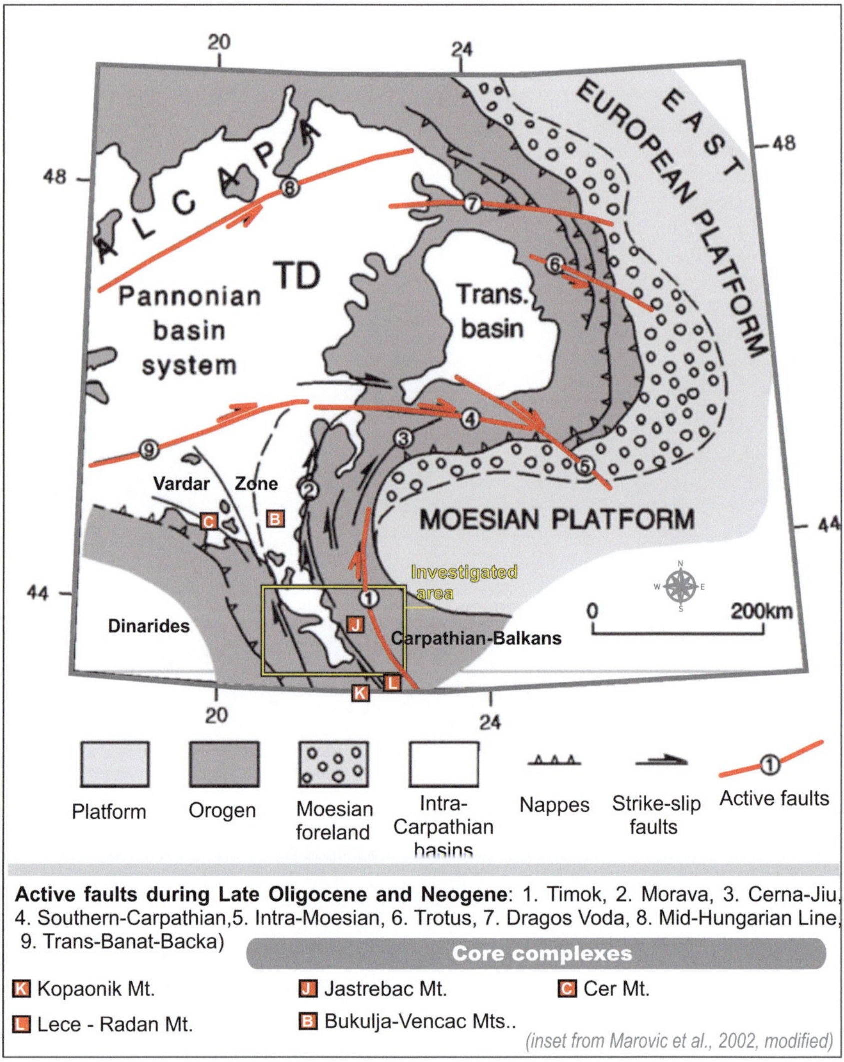

Fig. 2 Simplified sketch of the Late Paleogene and Neogene kinematics of the peri-Pannonian system (modified after Marović et al., 2002). Coordinates and a compass rose are present for orientation. Key regions include the Pannonian basin system, the Transylvanian basin, and the Moesian platform. A yellow box highlights the investigated area. Red lines with numbers indicate active strike-slip faults, with labels for specific faults such as Timok and Morava. The map shows that the investigated basinal features were between two major strike-slip faults. The map includes a legend for platforms, Carpathian-Balkan orogen, Moesian foreland, intra-Carpathian basins, core complexes (red rectangles), nappes overprinted by younger strike-slip system, and active faults

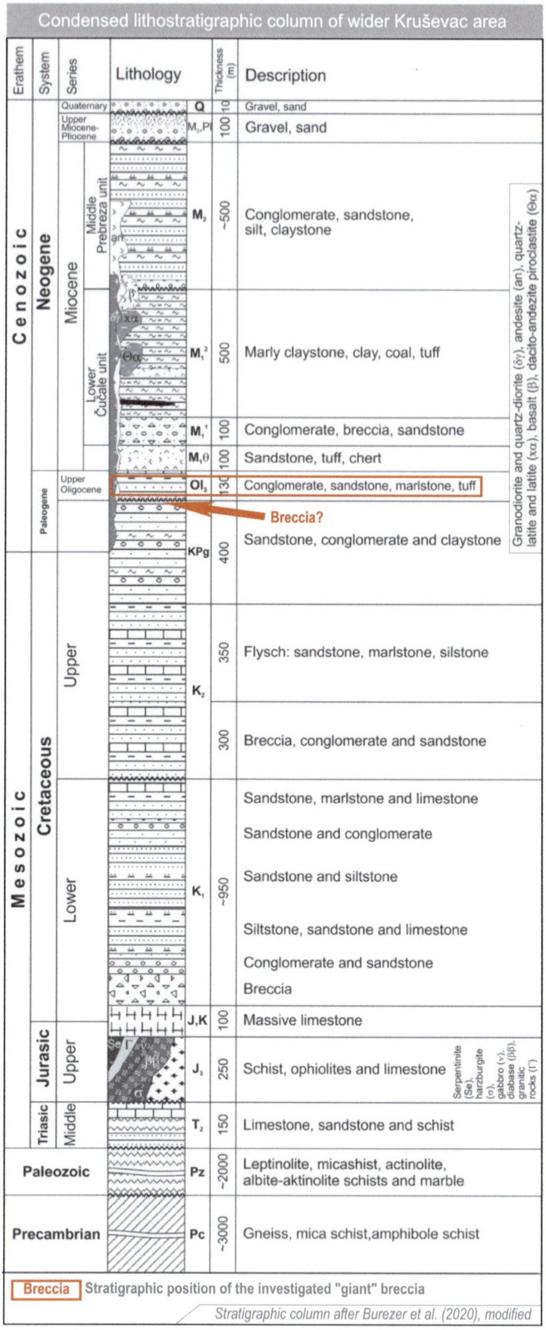

Fig. 3 Condensed lithostratigraphic column of the wider Kruševac area, detailing regionally distributed geological layers from Precambrian to Cenozoic eras (after Burazer et al., 2020, slightly modified). The column includes a highlighted section of the investigated sequence labeled "Breccia?" in the Upper Oligocene layer, indicating the stratigraphic position of the investigated "giant" breccia

Pranjani basin; Andrić-Tomašević et al., 2021; Fig. 1b), the erosion and alluvial flow transported basement grains into the nearby graben-related features, mirroring the Kruševac basin, nearby Toplica depression and other basinal features (e.g., Anđelković, 1986; Radisavljević et al., 2024).

The investigated outcrop area near Kruševac city, typically covered with river water or dense floral veneer, presents a unique opportunity during historically low water levels in the spring of 2021. Three geologists mapped the bedrock on behalf of the Geological Survey of Serbia (Fig. 4a–d). The exposed bedrock revealed remarkable breccia with large angular crystalline cobbles embedded into the matrix. This breccia-bearing agglomerated sequence of the hitherto uncertain age presumably represents the post-Rupelian (Scettino & Turco, 2011) basal series. Once present within the basin, such breccia's are capable of outlining the opening of peri-Pannonian basins (Fig. 4e).

2 Geology

After shifting lithospheric stress from compression to extension across the Dinaride and Carpathian-Balkan regions, such a major change in the tectonic-depositional environment caused widespread syn-rift tectonics throughout the (Peri) Pannonian area. While some authors explain a sea-level drop at the Eocene–Oligocene boundary responsible for Paratethys isolation (e.g., Sachsenhofer et al., 2018a, and references therein), the formation of the Pannonian Basin and the entry of Paratethys waters began in the early Miocene and continued through the middle Miocene (in Serbia predating the Langhian/Serravallian boundary; Mandic et al., 2019; Sremac et al., 2022). This event, regionally interpreted as the Neoalpine stage, extends for approximately 3 million years, lasting until the late middle Miocene (for example, in the Morava depression, eastern Serbia, Mandić et al., (2019); for the "Neoalpine" stage, see Hsü, 1989; Marović et al., 2007c; Fig. 1b). The main stage of Neogene crustal-scale extension (e.g., Marović et al., 2007c; Andrić-Tomašević et al., 2021, and references cited therin) is, in turn, characterized by significant horizontal mobility and rotation of these structures (Marović et al., 2002), block-wise basin differentiation with some evidence of tectonic uplifting (Radivojević et al., 2010). At the same time, a large number of the intramontane minibasins was initiated along the southern realm of the Pannonian basin (Fig. 1a). The regional extension additionally contributed to the formation of a number of core complexes formed across the area of Dinarides, Vardar Zone (as a composite Mesozoic suture), with particularly well-exposed exhumed systems spread across the frontal fragments of former European foreland or East Vardar Zone/Serbo-Macedonian Unit (e.g., Bukulja Mt., Jastrebac Mt., Cer Mt., Kopaonik Mt., Lece-Radan Mts., Knežević et al., 1994; Marović et al., 2007a, 2007b; Stojadinovic et al., 2013, 2017; Brun et al., 2018; Erak et al., 2017; Löwe et al., 2021; van Unen et al., 2019; Tančić et al., 2021; Marinković et al., 2023, 2024; Poznanović Spahić et al., 2023; Figs. 1b and 2). However, with a few exceptions suggesting strike-slip character of the leading

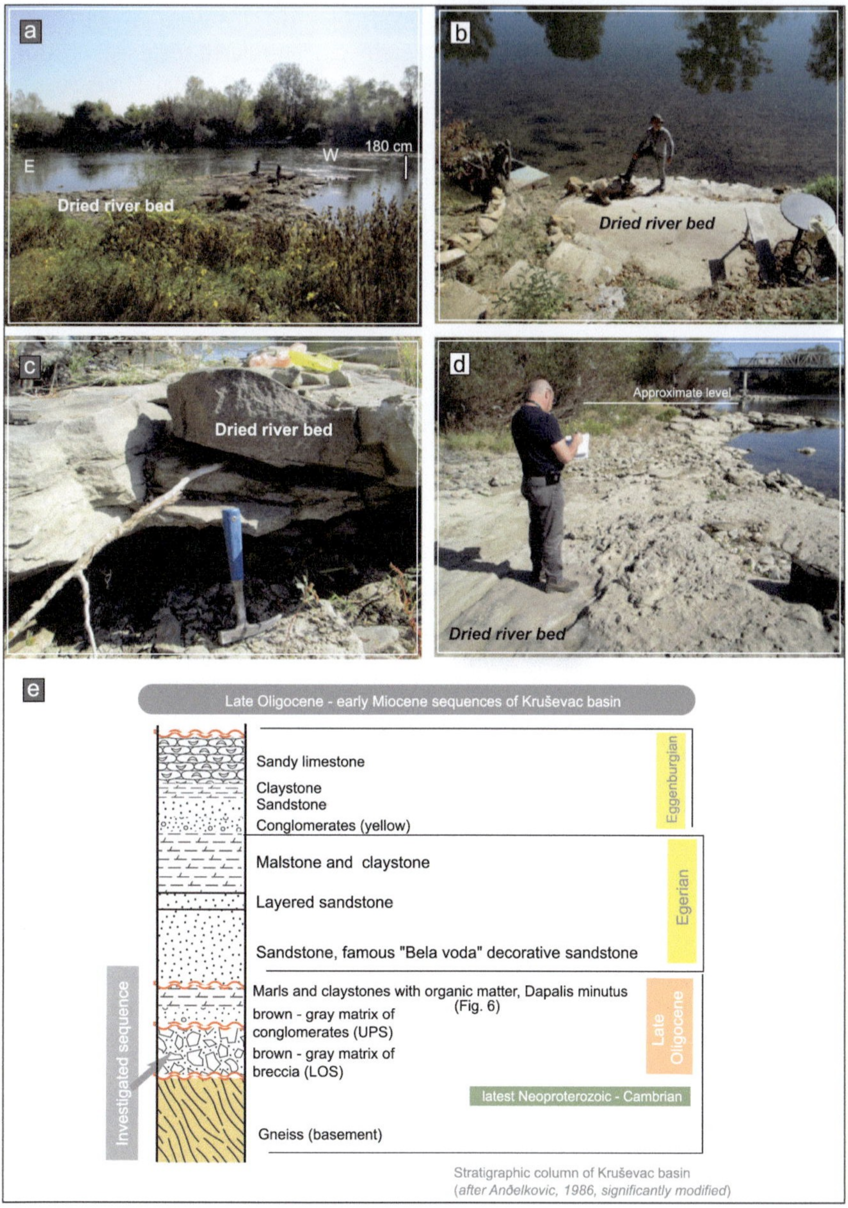
a
E
W
180 cm
Dried river bed
b
Dried river bed
c
Dried river bed
d
Approximate level
Dried river bed
e
Late Oligocene - early Miocene sequences of Kruševac basin
Sandy limestone
Claystone
Sandstone
Conglomerates (yellow)
Malstone and claystone
Layered sandstone
Sandstone, famous "Bela voda" decorative sandstone
Marls and claystones with organic matter, Dapalis minutus (Fig. 6)
brown - gray matrix of conglomerates (UPS)
brown - gray matrix of breccia (LOS)
latest Neoproterozoic - Cambrian
Gneiss (basement)
Eggenburgian
Egerian
Late Oligocene
Investigated sequence
Stratigraphic column of Kruševac basin
(after Anđelkovic, 1986, significantly modified)

◀**Fig. 4** The image is a panel figure with five subfigures labeled **a** to **e** showing exposed outcrop in the Morava River: Subfigure (**a**) shows a dried river bed with a scale indicating 180 cm, and directions marked as E and W. Vegetation and trees are visible in the background; Subfigure (**b**) shows exposed outcrop or segment of river bed; Subfigure (**c**) features a close-up of rock layers in a dried river bed, with a tool placed for scale; Subfigure (**d**) outlines the main exposed riverbed section including visible mega-brecciated cobbles embedded into the rock matrix (riverbed floor). Subfigure (**e**) new stratigraphic column illustrating the Late Oligocene to early Miocene sequences of the Kruševac basin, detailing various rock types such as sandy limestone, claystone, and sandstone, with labels indicating different geological periods, including basement. The new lithostratigraphic column was constructed on the basis of this study and the data from Anđelković (1986)

regional faults (e.g., Marović et al., 2002; Vangelov et al., 2016; Fig. 2), information about the nature of the Late Oligocene extension and the mechanism of opening of numerous lacustrine basins is missing (see e.g., Dumurdžanov et al., 2005; Rundić et al., 2013; Spahić, 2022c, for a discussion).

The early stage of Late Oligocene extension and graben-type basin formation influenced the onset of regional subsidence affecting regional landmasses. The subsidence onset is visible across numerous intramontane basins in Dinarides (e.g., van Unen et al., 2019, and reference cited therein; Spahić, 2022c), including the basins positioned on top of the complex N-S striking Vardar Zone/Serbo-Macedonian interface area (e.g., investigated Kruševac- and the nearby Topica basins (Fig. 2a, b). As suggested, the intensity of the early stage of post-orogenic crustal extension can be inferred by using the local example of the central Veliki Jastrebac core complex (Figs. 1b, 2; Jastrebac Mt.; Marović et al., 2007a; Erak et al., 2017). In addition to syn-extensional uplift, abrupt change in the regional post-orogenic stress field mirrors the extensional reactivation of older pre-Late Oligocene nappes, or Alpine thrust faults, often concealed beneath the Neogene cover (Andrić et al., 2018; Tari et al., 2021). As a result, the subsidence and lifespan of most of these earliest central and peripheral intramontane basins was controlled by a limited, rather narrow, accommodation space. Furthermore, young minibasins were mobile, exposed to changing extensional-, extensional uplift-dominated, and regional compression episodes. The abrupt interchanges of regional stress stages, vertical and horizontal mobility, often led to incomplete sedimentary record of minibasins evolution (Anđelković, 1986; Györfi et al., 1999; Marović et al., 2002; Burazer et al., 2020; Krstekanić et al., 2020; Bartha et al., 2022; Radisavljević et al., 2024; Fig. 4e).

The suitable humid paleogeographic and paleoclimatic conditions across the former Paratethyan "Dinaride-Anatolian" island resulted in the formation of widespread, early-stage polycyclic terrestrial mainly isolated intramontane lacustrine-type basins (Obradović et al., 1997, 2000; Šajnović et al., 2009; Burazer et al., 2020; Sachsenhofer et al., 2018a, 2018b; Andrić-Tomašević et al., 2021; Radisavljević et al., 2024; Burazer et al., 2025). The Neogene intramontane freshwater lakes experienced subsidence trends of similar speed and paleoclimatic conditions, controlled by the cyclicity of the alluvial, swamp, and lacustrine facies (Obradović et al., 1997; Šajnović et al. 2009; Andrić-Tomašević et al., 2021; Burazer et al., 2021a, 2021b, 2025; Spahić et al., 2023b). The following large number of intramontane basins are accommodated on top of Dinarides and Vardar Zone, and on top of the amalgamated south European microplates (Figs. 1b and 2): the central Kruševac-Trstenik or Kruševac basin, Valjevo-Mionica, Jadar, Sokobanja, Pranjani, Aleksinac, Ugljevik, Banovići, Sarajevo-Zenica-Jablanica, Kolubara, Babušnica-Koritnik, Žagubica, Toplica, Kosovo and Metohija basins, and a few others. (e.g., Anđelković, 1986; Obradović et al., 1997; Marović et al., 2002, 2007c; de Leeuw et al., 2010, 2011, 2012; Mandić et al. 2011; Andrić et al., 2015, 2017; Burazer et al., 2020, 2021a, 2021b; Andrić-Tomašević et al., 2021; Spahić, 2022c; Radisavljević et al. 2024; Ahnelt et al., 2024; Burazer et al., 2025).

2.1 Main Features of the Kruševac Basin

At the investigated WNE-ESE striking Kruševac-Trstenik-Aleksinac basin, the "Neoalpine" subsidence reached over 2000 m (Marović et al., 2007c; Fig. 1c). The investigated Late Oligocene–Miocene 2000 m thick assembly lies on top of the crystalline basement or Serbo-Macedonian Unit. The underlying Serbo-Macedonian basement configuration, together with the crustal segment belonging to the former Neotethyan Mesozoic amalgamation, is exhumed being visible in neighboring Jastrebac Mt. core-complex; Figs. 2 and 3). The (half graben) area was developed on top of the underlying Vardar / Serbo-Macedonian N-S striking interface containing three main depocenters – Trstenik, Aleksinac, and Kruševac. Deposition with mini-depocenters reachead max. ~2000 m in depth. Central depression is characterized by the western half of the depression, which is steeper than the eastern. The faults that control the depression's geometry have strike directions NNW–SSE (main downthrow normal faults) and NE–SW (diagonal faults; Marović et al., 2002, 2007c; Fig. 2). However, just a few older reports indicate a Late Oligocene age of the earliest deposits in the Kruševac basin (Anđelković, 1986, 1988b). As per most authors, the initial, typical Neogene subsidence, was generated during the Ottnangian/Karpatian to be later tectonically inverted. The regional structural inversion of subsided depocenters can be corroborated by a finding of a few terrestrial Neogene mammal fossils (Radović & Bradić-Milinović, 2018).

In the Kruševac depression, the early Miocene is characterized by the famous decorative Bela reka sandstone or "*Belorečki peščar*" (Egerian; Fig. 4e).[1] This early Miocene sequence is overlain by clastic, claystone, and limestone deposits (Eggenburgian) having an unconformity at the top (Anđelković, 1986; Fig. 4c). The central depression of the (Kruševac) Aleksinac basin is sub-rounded and was filled with freshwater sediments during the late stage of Lower Miocene subsidence. The thickness of Neogene deposits (ca. over 1250 m) indicates, however, a limited subsidence caused by the early extensional process, allowing for the development of such relatively shallow depths (in comparison with the central Pannonian basin).

[1] Belorečki peščar proizvodnja kvarcnog peska—Jugokaolin assessed on 08/07/2024.

The upper Miocene sediments, known as the "Red Clastic Series" of the Aleksinac basin (Fig. 1b), rest unconformably over the Lower Miocene (clastic) series (Oberauer, 2017). In the nearby Toplica basin, a few exploratory wells penetrate the oldest Eggenburgian–Karpatian Čučale sediments at ~500 m, reaching total depths at ~670 m. The oldest drilled Čučale sediments are represented by a clastic sequence (Burazer et al., 2021a; Fig. 3). By using an analogy, early Miocene volcanics and Čučale clastic sediments should be at the top of the brecciated horizon observed in the Kruševac area, i.e., its Toplica equivalent (Figs. 3 and 4a–e).

3 Results: Geology of the Outstanding Field Outcrops of Oligocene/Lower Miocene Basin Opening Sequence

Due to a higher early subsidence rate controlling the early "Neoalpine" extension (Spahić et al., 2023b), the evidence of crustal-scale opening of the intramontane basins is often obscured by thick overburden. By analogy, the oldest succession should have alluvial character because alluvial deposits are already observed in a few intramontane basins, e.g. Pranjani and Toplica basins (being mainly drilled; Andrić-Tomašević et al., 2021; Burazer et al., 2020; Radisavljević et al., 2024; Fig. 1b). This earliest sequence in Pranjani basin comprises the Kamenica Series (Late Oligocene–Miocene), which starts with red alluvial clastics, breccia, and conglomerate (Anđelković et al., 1991; Andrić-Tomašević et al., 2021). The Kamenica sequence grades into lacustrine sandstones, dolostones, dolomitic limestones, claystones, and bituminous marlstone-bearing fossil plant and debuted fish remains. The succession includes the late Oligocene paleofloral assemblage (Anđelković, 1986; Andrić-Tomašević et al., 2021; Lazarević et al., 2019). Despite being a prominent regional tectonic marker, such an agglomerated cobble-bearing sequence has not yet been analyzed from the perspective of the crustal-scale basin-opening mechanisms and their intensity.

3.1 The Kruševac Breccia

The investigated basal sequence comprises several sets of coarse-grained clastic rocks, with large subangular fragments of crystalline rocks (Fig. 5a–d). The extrapolated total thickness of the investigated brecciated sequence reaches < 120 m (Fig. 4a–d). The limited time exposure in the riverbed allowed us to separate the two following sub-facies (Fig. 4e): (i) the prominent basal breccia unit that exposes limited thickness (ca. 1 m across the outcrop) and (ii) the topping rather conglomeratic section (Fig. 5a–d). This stratigraphically lower-positioned breccia sequence starts with the extraordinary exposure of Serbo-Macedonian angular cobbles embedded into the

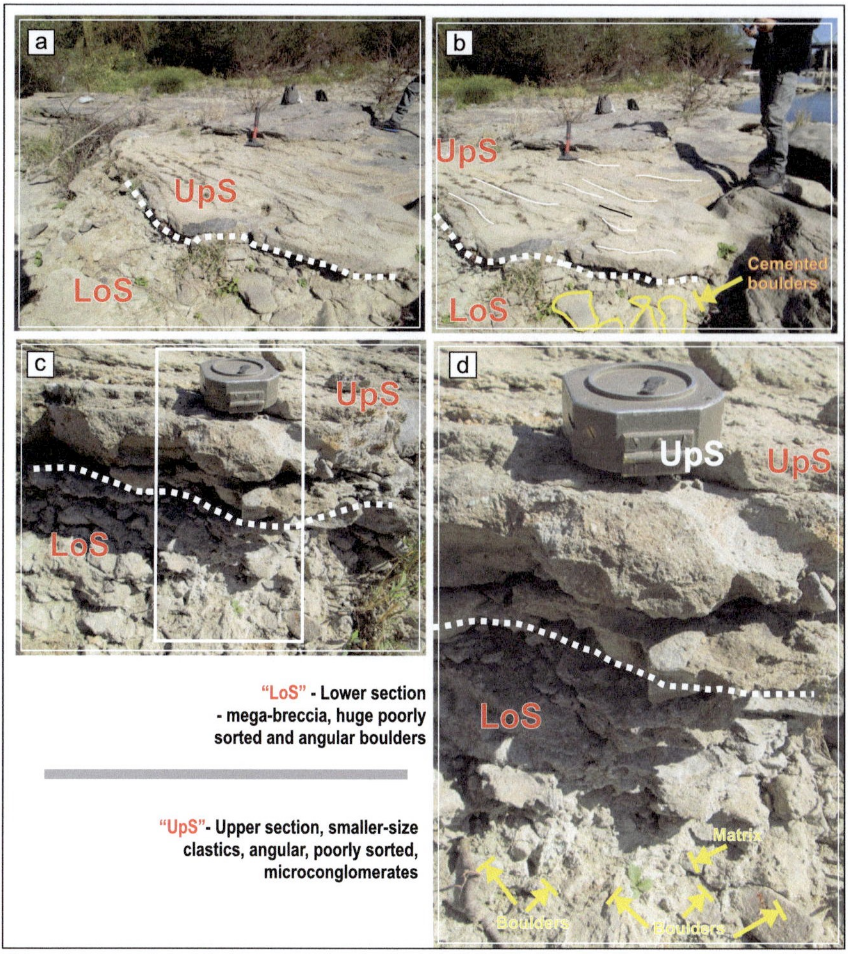

Fig. 5 Photographic figure showing geological formations with four panels labeled **a**, **b**, **c**, and **d**. Each panel highlights rock layers with dotted lines separating the "LoS" (Lower section) and "UpS" (Upper section). The "LoS" is described as a mega-breccia with large, poorly sorted, angular boulders, while the "UpS" consists of smaller clastics, angular, poorly sorted microconglomerates. Yellow arrows in panels **b** and **d** point to cemented boulders and matrix, respectively. The boundary between two subsequences (**c**, **d**) separates coarse breccia on top and more fine but still coarse-grained clastic sediments

matrix (designated as "LoS" in Fig. 5a–d). The lower brecciated part comprises a fine-grained matrix holding tight crystalline cobbles with > 20 cm diameter. This cobble-bearing sub-sequence gradually grades into the (micro)conglomerate, coarse but sorted clastic sequence, with the smaller size of grains relative to the breccia (designated as "UpS"; Fig. 4e). The matrix contains poorly sorted grains, exposing the variation in gradual vertical sorting. The lower or clastic part ("LoS") comprises mainly breccia or "agglomerate" fitting into the Oligocene description of Anđelković (1986;

Fig. 4e). The upper part ("UpS") is represented by mixed, better-sorted conglomerate breccias, with smaller grains and sandstone having gray-reddish color. The observed sequence entirely fits with the earlier description of the conglomeratic section by Anđelković (1986). The clastic cobble-rich and conglomerate-sandstone sequences resemble strongly with the Pranjani basin and the Kamenica Series (Lazarević & Milivoljević, 2010; Andrić-Tomašević et al., 2021). The clastic sequence containing cobbles gradually transitions into thin-layered, tabular claystone that contains organic matter with *Dapalis minutes* Blainville (similar to the Kamenica sequence; Fig. 6a–d). Another similarity with the Pranjani basin reflects this overlying beige-colored claystone sequence rich in organic matter (Fig. 4e; Andrić-Tomašević et al., 2021) additionally.

As mentioned, one of the most exciting features of the exposed basal breccia sequence is the dominance of remarkable crystalline cobbles (Serbo-Macedonian gneiss; Fig. 5) relative to other nearby (basement) rocks like the Mesozoic ophiolites, limestones, and turbidites (for non-crystalline basement details see, e.g., Karamata, 2006; Schmid et al., 2008; Spahić & Gaudenyi, 2022; Spahić et al., 2023a). These cobbles are large angular gneiss-bearing grains that allow immediate onsite characterization (Fig. 7a–e). This mainly whitish-colored matrix comprises intraclasts of large brownish gneiss fragments of the underlying Serbo-Macedonian Unit. Despite not being uncovered, the breccia unit likely overlies the erosional surface toward the older basement rocks near or slightly beneath the riverbed outcrops (Serbo-Macedonian basement is exhumed across the area; Fig. 1c). This unit of the latest Neoproterozoic-Cambrian age together with the East Vardar Zone represents a segment of the sutured margin of the Neotethyan Vardar Ocean (Schmid et al., 2008; Spahić & Gaudenyi, 2022; Spahić et al., 2023b). The presence of large angular brecciated grains shows that the peripheral Serbo-Macedonian basement underwent a significant amount of extension. During the syn-rift stage and the production of significant accommodating, these conditions facilitated the formation of numerous core complexes, the uplift of the extensional grabens, their erosion, and the onset of graben-sourced terrestrial sedimentation (e.g., Marović et al., 2007a, 2007b, 2007c; Burazer et al., 2020; Spahić et al., 2023b; Radisavljević et al., 2024; Fig. 8).

4 Discussions

4.1 *Regional Breccia Analog Seqeunces*

Using an analogy with a higher sedimentation rate in the early Toplica basins, large cobbles may indeed be indicative of short material transport, mainly from eroded parts of uplifted shoulders (Fig. 8). The development of adjoining extensional Jastrebac core complex and tectonic exhumation of this part of former margin is in line with the proposed syn-extensional (footwall) uplift, basement elevation, erosion, and basin

Fig. 6 Three-panel image **a**, **b**, and **c** showing platy thin-bedded lacustrine-type claystone outcrop positioned at ca. 100 m from the main outcrop. Panel (**a**): Shows a low water level that uncovered tabular claystone, which is rich with organic matter. Panel (**b**): A closer view of the sedimentary layers, showing fractured and weathered tabular claystone surfaces. Panel (**c**): A detailed view of the rock layers, highlighting the texture and stratification. A hammer is placed on the rocks for scale

Fig. 7 **a–e** A series of photographs showing geological formations with a hammer for scale. The images highlight a large, angular, brecciated grain composed of crystalline schists (gneiss). Labels identify features such as "Los," "matrix," "gravel," "other rocks," and "crystalline schists"

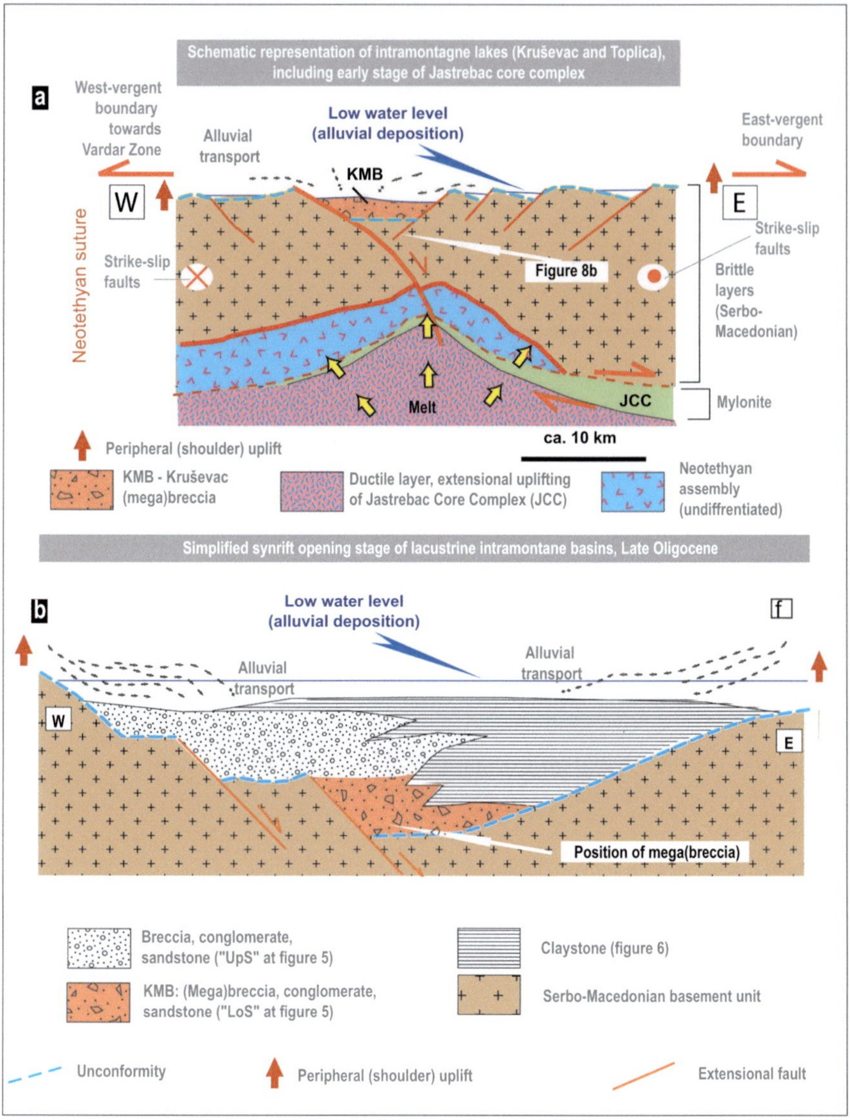
Schematic representation of intramontagne lakes (Kruševac and Toplica), including early stage of Jastrebac core complex
a
West-vergent boundary towards Vardar Zone
Alluvial transport
Low water level (alluvial deposition)
KMB
East-vergent boundary
W
E
Neotethyan suture
Strike-slip faults
Figure 8b
Strike-slip faults
Brittle layers (Serbo-Macedonian)
Melt
JCC
Mylonite
Peripheral (shoulder) uplift
ca. 10 km
KMB - Kruševac (mega)breccia
Ductile layer, extensional uplifting of Jastrebac Core Complex (JCC)
Neotethyan assembly (undiffrentiated)
Simplified synrift opening stage of lacustrine intramontane basins, Late Oligocene
b
Low water level (alluvial deposition)
f
Alluvial transport
Alluvial transport
W
E
Position of mega(breccia)
Breccia, conglomerate, sandstone ("UpS" at figure 5)
KMB: (Mega)breccia, conglomerate, sandstone ("LoS" at figure 5)
Claystone (figure 6)
Serbo-Macedonian basement unit
Unconformity
Peripheral (shoulder) uplift
Extensional fault

◀**Fig. 8** Schematic representation of the formation of intramontane lakes, illustrating the early stage of the Jastrebac core complex, and development of Neogene graben-type basins of central Serbia, Kruševac, and Toplica basins. Panel (**a**) shows a cross-section with alluvial transport, low water levels, and peripheral uplift. Key features include the Kruševac mega-breccia (KMB), ductile layers, and mylonite. The sketch illustrates the main detachment fault, extension, graben structures, and uplifted Serbo-Macedonian basement shoulders (source of alluvial deposits, including mega-breccia). The initial shallow depressions contain early alluvial material, including the breccia, which indicates high-energy shallow-water environments. *—Mylonite and shear zones are well-documented across the Jastrebac Mt. (Erak et al., 2017; Marović et al., 2007a. Panel (**b**) depicts a similar cross-section with breccia, conglomerate, and sandstone layers, highlighting the position of mega-breccia and extensional faults. The breccia is deposited directly on top of the Serbo-Macedonian basement, and the narrow Oligo-Miocene depression is filled with material from graben shoulders. Both panels emphasize geological processes like alluvial deposition and extensional uplifting

material supply. The extensional core complex can expose the (metamorphic) basement rocks formerly buried at 2–4 km or 7–8 km depth (e.g., Marović et al., 2007a, 2007b; Kounov et al. 2018; Löwe et al., 2021; Fig. 8a, b). In that manner, a voluminous extension causing the rise of the core complexes could also have affected the contemporaneous subsidence of adjoining basins, like Kruševac and Toplica depressions.

The extension and the early basin opening stage occurred under humid climatic conditions controlling the Late Oligocene and Miocene "Dinaric land" (Andrić-Tomašević et al., 2021; Burazer et al., 2025). Abundant precipitation coupled with the extensional footwall uplift and erosion of the crystalline basement rocks (gneiss), provide suitable paleoenvironmental conditions for developing early terrestrial depositional (alluvial) systems (Fig. 8). In regional scale, analog basal breccia-bearing units can be find as (i) the Late Oligocene sequence belonging to the "Promina series" in Dinarides (the youngest part; Vlahović et al., 2012) and (ii) the "Kamenica series" in the Pranjani intramontane basin of western Serbia (Andrić-Tomašević et al., 2021; Lazarević et al., 2019; Obradović et al., 1997; Fig. 1b). The Kamenica series comprises various clastics, conglomerates, breccia, and gravel. These basal breccia sequences lie either over different pre-Oligocene basement segments: Mesozoic turbidites and Jurassic ophiolites (like in the case of the Pranjani basin, Kamenica series; Andrić-Tomašević et al., 2021; Lazarević et al., 2019). The overlying much younger Neogene rocks can rest on top of crystalline basements, like in our Kruševac case, providing a veneer masking the crystalline rocks of Serbo-Macedonian Unit. Thus, we believe that the investigated crystalline breccia, once discovered, can serve as marker of the underlying late Paleogene-Neogene unconformity developed during the investigated terrestrial stage (also in Obradović et al., 1997). The comparison with other similar age sequences shows that the investigated breccia has the best fit with the Kamenica series of Pranjani intramontane basin: (i) the lower part is composed of coarse and medium-grained alluvial clastics (conglomerates, breccias, and sandstones). Moreover, the upper part of the Kamenica sequence is consisting of bituminous marlstones (same as in Kruševac basin; Fig. 6) involving claystones with fish remains as important stratigraphic dating material (Obradović et al., 1997; Andrić-Tomašević et al., 2021).

4.2 Age Constraints on Breccia Sequence

Along the strike of the East Vardar Zone–Serbo-Macedonian interface, to the north, there is another nearby location exposing a peculiar breccia deposit referred to as the "Ropočevo breccia" (Kurešević et al., 2022; Fig. 1a). However, the investigated Kruševac breccia significantly differs from this older, underlying, recrystallized, limestone-bearing "Ropočevo breccia" of the tentative (Tithonian) Berriasian age (East Vardar Zone; Spahić et al., 2023a; Fig. 1a). As discussed, the stratigraphic position of the presumably much younger Kruševac breccia is consistent with the topping claystone and tabular marlstone with organic matter, almost identical as identified within the Pranjani basin (Figs. 4c, 6 and 7; Anđelković, 1988a, 1988b; Andrić-Tomašević et al., 2021; Obradović et al., 1997). The sampling of the Kruševac breccia matrix yielded no stratigraphic age, probably because of the terrestrial character and intense syndepositional reworking of the sedimentary matrix and fossils (Fig. 8). Nevertheless, previous fossil fish findings, e.g., *Dapalis microcantus* Agassis, *Dapalis macrorurus* Agassis, *Dapalis minutus* (Blainville), and *Leuciscus papiraceus* Agassis, provided some preliminary insight into the age of the investigated mega(breccia). The fish genus was earlier called the "*Smerdis*" and became the genus *Dapalis* (Ahnelt et al., 2024). The new data on the fish genus *Dapalis* indicate the European brackish and freshwater stage spanning the entire Oligocene until the late Early Miocene (Ottnangian). In the investigated Kruševac basin case, the presence of *Dapalis minutus* should indicate fresh-brackish waters with the maximum Late Oligocene–Early Miocene age. *Dapalis minutus* belongs to the Late Oligocene timeframe (Reichenbacher, 1988; Schultz, 2000; Ahnelt et al., 2024). Some authors, however, argue insisting on the absence of the *Dapalis minutes* from all intramontane basins (Andrić-Tomašević et al., 2021, and references cited therein). Recent otolith data from the nearby Babušnica basin (Serbia; Fig. 1b) indicate, in turn, the presence of significantly older the Lower Oligocene *Dapalis* genus in intramontane paleolakes (Ahnelt et al., 2024). Others separate the oldest sequence as of the Early Miocene age, referred to as the "Trstenik sequence" or "*Trstenička serija*" (Pavlović, 1977).

To make matters even more intricate, a review of existing literature focused on similar breccia agglomerates with embedded large cobbles reveals a few cases of younger, post-Late Oligocene agglomerates. In the Dinarides and Danube basin, the Lower Miocene breccia, including Badenian-Sarmatian carbonate platform breccia, occurs as debris flow breccias with much smaller non-crystalline grains (Saftić et al., 2003; Harzhauser & Piller, 2004; Vrsaljko et al., 2006; Cvetković et al., 2019; Nováková et al. 2020; Sebe et al., 2021). Another yet rare Middle Miocene post-unconformity breccia-type deposit in the Pannonian realm is recorded. However, this agglomerate has neither such large nor typical crystalline basement grains (Sebe et al., 2021). Similar breccias of Middle Miocene were neither drilled nor interpreted within drilled well or on available seismic data depicting subsurface geology of Pannonian realm (Radivojević et al., 2010; Matenco & Radivojević, 2012).

The data synthesis further indicates that the presumed terrestrial Late Oligocene alluvial deposition predates the principal marine flooding event recorded across central Serbian paleolakes (occurred at ~14 Ma, slightly predating the Langhian/Serravallian boundary; 13.82 Ma; Mandić et al., 2019; Andrić-Tomašević et al., 2021; and many others). Consistency with the suggested Late Oligocene timeframe includes almost the same sedimentological properties and lithostratigraphic members as in the Kamenica series of the Pranjani basin (see Andrić-Tomašević et al., 2021). The here suggested slight pre-Miocene age of the investigated brecciated basal sequence is further consistent with the Paleogene flora and fauna discovered within the terrestrial clastic rocks of the Pranjani basin (Andrić-Tomašević et al., 2021; Lazarević et al., 2019). In addition, the Fruška Gora inselberg exposes the similar-age Late Oligocene deposition, which comprises a similar-age breccia (Rundić et al., 2013; Fig. 1b, green rectangle).

5 Summary and Conclusions

By mapping the rarely exposed extraordinary riverbed outcrop that unraveled basement-topping cobble-bearing agglomerated sediments, we provide new insight into the earliest stages of the southern peri-Pannonian (mini) basin(s). A half-graben-type opening across the peri-Pannonian realm is fitting into the syn-rift Pannonian basin framework (Balázs et al., 2021; Bartha et al., 2018). After the initial deposition of the investigated terrestrial sequence filled with crystalline boulders, ongoing yet slower crustal-driven subsidence (Spahić et al., 2023b) of the enclosed Kruševac intramontane basin culminated with the production of lacustrine thin-layered claystone (terminal slower subsidence stage). A similar early syn-rift sequence with large grains is observed within the Pranjani basin, which is more external than the Inner Dinarides, towards Vardar Zone (Andrić-Tomašević et al., 2021). The results show that the onset of Balkan-wide regional extension, erosion, transport, and lacustrine-type deposition of large cobbles embedded within breccia, could represent a marker of the widespread earliest Late Oligocene extension. The early extension was followed by the opening of a number of Neogene southern peri-Pannonian (mini) basins, and activation of a number core complexes. This brecciated basal sequence and its Late Oligocene age apply to Kruševac but probably to Pranjani, Aleksinac, Babušnica-Koritnik, and other intramontane basins as well. The slight pre-Miocene basin opening to the south of the Pannonian basin is consistent with the recently proposed westward- or foreland-directed younging of extensional deformation (e.g., Andrić et al., 2018; Matenco & Radivojević, 2012; Spahić & Gaudenyi, 2020a). Interestingly, in a similar fashion, the late Oligocene framework fits well with the case of the Aegean-Hellenic Oligocene–Miocene developments, or the southward-younging of extensional deformation (e.g., Dumurdžanov et al., 2005).

Acknowledgements This study was part of the Geological Survey of Serbia seasonal field mapping work in 2021. We thank Soumyajit Mukherjee for inviting, handling, and reviewing this work. Mukherjee (2026) summarized this chapter.

Funding Declaration This research received no specific grant from funding agencies in the public, commercial, or not-for-profit sectors.

Data Availability Declaration All used data are in the paper itself.

Competing Interest Declaration The authors declare that they have no known competing financial interests or personal relationships that could have appeared to influence the work reported in this paper.

References

Ahnelt, H., Bradić-Milinović, K., & Schwarzhans, W. (2024). *Dapalis pauciserratus*, a new species of freshwater glassfishes (Teleostei, Ambassidae) from the Lower Oligocene of the Central Paratethys (early view). *Cybium*. https://doi.org/10.26028/CYBIUM/2024-006

Anđelković, M. (1986). Geodinamički procesi i stratigrafsko-paleogeografski razvoj oligocenskih i miocenskih slatkovodnih basena središnje i zapadne Srbije [Geodynamic processes and stratigraphic-palaeogeographic development of Oligocene and Miocene freshwater basins in central and western Serbia]. *Geološki Anali Balkanskoga Polusostrva, 50*, 1–74. (in Serbian with English abstract).

Anđelković, J. (1988a). Oligocen u Srbji na osnovu fosilnih riba [The Oligocene in Serbia determined using fossil fishes]. *Geološki Anali Balkanskoga Polusostrva, 51*, 245–251.

Anđelković, M., Eremija, M., Pavlović, M., Anđelković, J., & Mitrović-Petrović, J. (1991). In M. Anđelković (Ed.), *Paleogeography of Serbia—Tertiary. University of Belgrade, Faculty of Mining and Geology, Institute for Regional Geology and Paleontology*, p. 237 (in Serbian with English summary).

Anđelković, M. (1988b). *Geologija Jugoslavije* [Geology of Yugoslavia], p. 692. Beograd: IRO Građevinska kniga.

Andrić, N., Fügenschuh, B., Životic, D., & Cvetković, V. (2015). The thermal history of the Miocene Ibar Basin (Southern Serbia): New constraints from apatite and zircon fission track and vitrinite reflectance data. *Geologica Carpathica, 66*(1), 37.

Andrić, N., Sant, K., Matenco, L., Mandić, O., Tomljenović, B., Pavelić, D., & Ooms, J. (2017). The link between tectonics and sedimentation in asymmetric extensional basins: Inferences from the study of the Sarajevo-Zenica Basin. *Marine and Petroleum Geology, 83*, 305–332.

Andrić, N., Vogt, K., Matenco, L., Cvetković, V., Cloetingh, S., & Gerya, T. (2018). Variability of orogenic magmatism during Mediterranean-style continental collisions: A numerical modeling approach. *Gondwana Research, 56*, 119–134.

Andrić-Tomašević, N., Simić, M. O., Životić, D., Suárez, M., & García-Romero, E. (2021). An arid phase in the Internal Dinarides during the early to middle Miocene: Inferences from Mg-clays in the Pranjani Basin (Serbia). *Palaeogeography, Palaeoclimatology, Palaeoecology, 562*, Article 110145.

Antić, M., Peytcheva, I., von Quadt, A., Kounov, A., Trivić, B., Serafimovski, T., Tasev, G., Gerdjikov, I., & Wetzel, A. (2016). Pre-Alpine evolution of a segment of the North-Gondwanan margin: Geochronological and geochemical evidence from the central Serbo-Macedonian Massif. *Gondwana Research, 36*, 523–544. http://dx.doi.org/10.1016/j.gr.2015.07.020

Balázs, A., Maţenco, L., Granjeon, D., Alms, K., François, T., & Sztanó, O. (2021). Towards stratigraphic-thermo-mechanical numerical modelling: Integrated analysis of asymmetric extensional basins. *Global and Planetary Change, 196*, Article 103386.

Balling, P., Grützner, C., Tomljenović, B., Spakman, W., & Ustaszewski, K. (2021). Post-collisional mantle delamination in the Dinarides implied from staircases of Oligo-Miocene uplifted marine terraces. *Scientific Reports, 11*(1), 2685.

Bartha, A., Balázs, A., & Szalay, Á. (2018). On the tectono-stratigraphic evolution and hydrocarbon systems of extensional back-arc basins: Inferences from 2D basin modelling from the Pannonian basin. *Acta Geodaetica Et Geophysica, 53*(3), 369–394.

Bartha, I. R., Botka, D., Csoma, V., Katona, L. T., Tóth, E., Magyar, I., & Sztanó, O. (2022). From marginal outcrops to basin interior: A new perspective on the sedimentary evolution of the eastern Pannonian Basin. *International Journal of Earth Sciences, 111*(1), 335–357. https://doi.org/10.1007/s00531-021-02117-6

Boote, D. R. D., Sachsenhofer, R. F., Tari, G., & Arbouille, D. (2018). Petroleum provinces of the Paratethyan region. *Journal of Petroleum Geology, 41*(3), 247–297.

Brun, J. P., Sokoutis, D., Tirel, C., Gueydan, F., Van Den Driessche, J., & Beslier, M. O. (2018). Crustal versus mantle core complexes. *Tectonophysics, 746*, 22–45.

Burazer, N., Šajnović, A., Vasić, N., Kašanin-Grubin, M., Životić, D., Mendonça Filho, J. G., & Jovančićević, B. (2020). Influence of paleoenvironmental conditions on distribution and relative abundance of saturated and aromatic hydrocarbons in sediments from the NW part of the Toplica basin, Serbia. *Marine and Petroleum Geology, 115*, Article 104252.

Burazer, N., Šajnović, A., Kašanin-Grubin, M., Gajica, G., Orlić, J., Radisavljević, M., & Jovančićević, B. (2021a). Early-Middle Miocene paleoenvironmental and paleoclimate changes in the Toplica Basin (Serbia) inferred from plant biomarkers, biochemical and elemental geochemical proxies. *Geologica Carpathica, 72*(5), 406–424.

Burazer, N., Šajnović, A., Kašanin-Grubin, M., Radisavljević, M., & Jovančićević, B. (2021b). Polycyclic aromatic hydrocarbons and their relationship to maturity and paleoenvironmental settings in lacustrine sediments of the Neogene Toplica Basin, Serbia. *Journal of Paleolimnology, 66*(3), 187–205.

Burazer, N., Šajnović, A., Spahić, D., Tančić, P., Grba, N., & Jovančićević, B. (2025). Unveiling the paleosalinity constraints on southern peri-Pannonian Lower Miocene lacustrine systems in Serbia and Bosnia and Herzegovina: Lopare (Dinaride Lake System) versus Toplica basin (Serbian Lake System). *Chemical Geology, 671*, 122475. https://doi.org/10.1016/j.chemgeo.2024.122475

Burchfiel, C. B., Nakov, R., Tzankov, T., & Royden, L. H. (2000). Cenozoic extension in Bulgaria and northern Greece: The northern part of the Aegean extensional regime. *Geological Society, London, Special Publications, 173*(1), 325–352.

Burchfiel, B. C., Nakov, R., Dumurdzanov, N., Papanikolaou, D., Tzankov, T., Serafimovski, T., & Nurce, B. (2008). Evolution and dynamics of the Cenozoic tectonics of the South Balkan extensional system. *Geosphere, 4*(6), 919–938.

Čokorilo, V., Lilić, N., Purga, J., & Milisavljević, V. (2009). Oil shale potential IN Serbia. *Oil Shale, 26*(4), 451–462.

Cvetković, M., Troskot-Čorbić, T., Sachsenhofer, R., Ćorić, S., Rukavina, D., Kolenković Močilac, I., & Saftić, B. (2019). Middle and Upper Miocene source rock facies of Dilj Mt., Sava Depression, Pannonian Basin. In *Paratethys petroleum systems Between Central Europe and the Caspian Region*, pp. 31–31.

De Leeuw, A., Mandic, O., Vranjković, A., Pavelić, D., Harzhauser, M., Krijgsman, W., & Kuiper, K. F. (2010). Chronology and integrated stratigraphy of the Miocene Sinj Basin (Dinaride Lake System, Croatia). *Palaeogeography, Palaeoclimatology, Palaeoecology, 292*(1–2), 155–167. https://doi.org/10.1016/j.palaeo.2010.03.040

De Leeuw, A., Mandic, O., Krijgsman, W., Kuiper, K., & Hrvatović, H. (2011). A chronostratigraphy for the Dinaride Lake system deposits of the Livno-Tomislavgrad Basin: The rise and fall of a long-lived lacustrine environment in an intra-montane setting. *Stratigraphy, 530–531*(2), 286–298. https://doi.org/10.1016/j.tecto.2012.01.004

De Leeuw, A., Mandic, O., Krijgsman, W., Kuiper, K., & Hrvatović, H. (2012). Paleomagnetic and geochronologic constraints on the geodynamic evolution of the central Dinarides. *Tectonophysics, 530–531*(2), 286–298. https://doi.org/10.1016/j.tecto.2012.01.004

Dimitrijević, M. D. (2001). Dinarides and the Vardar Zone: A short review of the geology. *Acta Volcanologica, 13*(1/2), 1–8.

Dumurdzanov, N., Serafimovski, T., & Burchfiel, B. C. (2005). Cenozoic tectonics of Macedonia and its relation to the South Balkan extensional regime. *Geosphere, 1*(1), 1–22.

Erak, D., Matenco, L., Toljić, M., Stojadinović, U., Andriessen, P. A., Willingshofer, E., & Ducea, M. N. (2017). From nappe stacking to extensional detachments at the contact between the Carpathians and Dinarides-The Jastrebac Mountains of Central Serbia. *Tectonophysics, 710–711*, 162–183. https://doi.org/10.1016/j.tecto.2016.12.022

Györfi, I., Csontos, L., & Nagymarosy, A. (1999). Early Tertiary structural evolution of the border zone between the Pannonian and Transylvanian Basins. *Geological Society, London, Special Publications, 156*(1), 251–267.

Harzhauser, M., & Piller, W. E. (2004). *The Early Sarmatian-Hidden Seesaw Changes* (pp. 89–112). Courier-Forschungsinstitut Senckenberg.

Horváth, F. (1993). Towards a mechanical model for the formation of the Pannonian basin. *Tectonophysics, 226*(1–4), 333–357.

Hsü, K. J. (1989). Time and place in Alpine orogenesis—the Fermor Lecture. *Geological Society, London, Special Publications, 45*(1), 421–443.

Jolivet, L., Arbaret, L., Le Pourhiet, L., Cheval-Garabédian, F., Roche, V., Robillard, A., & Labrousse, L. (2021). Interactions of plutons and detachments: A comparison of Aegean and Tyrrhenian granitoids. *Solid Earth, 12*, 1357–1388.

Karamata, S. (2006). The geological development of the Balkan Peninsula related to the approach collision and compression of Gondwanan and Eurasian units. *Geological Society, London, Special Publications, 260*(1), 155–178. https://doi.org/10.1144/GSL.SP.2006.260.01.0

Knežević, V., Karamata, S., & Cvetković, V. (1994). Tertiary granitic rocks along the southern margin of the Pannonian Basin. *Acta Mineralogica-Petrolographica, 35*, 71–80.

Kounov, A., Gerdjikov, I., Vangelov, D., Balkanska, E., Lazarova, A., Georgiev, S., & Stockli, D. (2018). First thermochronological constraints on the Cenozoic extension along the Balkan fold-thrust belt (Central Stara Planina Mountains, Bulgaria). *International Journal of Earth Sciences, 107*, 1515–1538.

Kräutner, H. G., & Krstić, B. (2002). Alpine and Pre-Alpine structural units within the Southern Carpathians and Eastern Balkanides. In *Proceedings of XVII congress of Carpathian-Balkan geological association, Bratislava* (Vol. 53, pp. 1–5). Geologica Carpathica.

Krézsek, C., & Olariu, C. (2021). Filling of sedimentary basins and the birth of large rivers: The lower Danube network in the Dacian Basin. *Romania. Global and Planetary Change, 197*, Article 103391.

Krézsek, C., Lăpădat, A., Maţenco, L., Arnberger, K., Barbu, V., & Olaru, R. (2013). Strain partitioning at orogenic contacts during rotation, strike-slip and oblique convergence: Paleogene-Early Miocene evolution of the contact between the South Carpathians and Moesia. *Global and Planetary Change, 103*, 63–81.

Krstekanić, N., Matenco, L., Toljić, M., Mandic, O., Stojadinovic, U., & Willingshofer, E. (2020). Understanding partitioning of deformation in highly arcuate orogenic systems: Inferences from the evolution of the Serbian Carpathians. *Global and Planetary Change, 195*, Article 103361.

Kurešević, L., Septfontaine, M., Vušović, O., & Delić-Nikolić, I. (2022). Contribution to geology and genetic pathway of the Ropočevo breccia–an "orphan" olistolithic body within the upper Cretaceous flysch near Sopot (central Serbia). *Geologica Macedonica, 36*(1), 5–18. https://doi.org/10.46763/GEOL22361005k

Lazarević, Z., Milovanović, L., Milivojević, J., & Vasiljević, I. (2019). Palaeoflora of Kamenica (Pranjani Basin, western Serbia). *Geoloski Anali Balkanskoga Poluostrva, 80*(2), 53–63.

Lazarević, Z., & Milivojević, J. (2010). Early Miocene flora of the intramontane Žagubica Basin (Serbian Carpatho-Balkanides). *Neues Jahrbuch für Geologie und Paläontologie, Abhandlungen, 256*, 141–150.

Löwe, G., Schneider, S., Sperner,B., Balling, P., Pfänder, J. A., & Ustaszewski, K. (2021). Torn between two plates: Exhumation of the Cer massif (internal Dinarides) as a farfield effect of Carpathian slab rollback inferred from $^{40}Ar/^{39}Ar$ dating and cross section balancing. *Tectonics, 40*, e2021TC006699. https://doi.org/10.1029/2021TC006699

Mandic, O., de Leeuw, A., Vuković, B., Krijgsman, W., Harzhauser, M., & Kuiper, K. F. (2011). Palaeoenvironmental evolution of lake Gacko (southern Bosnia and Herzegovina): Impact of the middle Miocene climatic optimum on the dinaride lake system. *Palaeogeography Palaeoclimatology Palaeoecology, 299*(3–4), 475–492.

Mandic, O., De Leeuw, A., Bulić, J., Kuiper, K., Krijgsman, W., & Jurišić-Polšak, Z. (2012). Paleogeographic evolution of the Southern Pannonian Basin: $^{40}Ar/^{39}Ar$ age constraints on the Miocene continental series of northern Croatia. *International Journal of Earth Sciences, 101*, 1033–1046.

Mandic, O., Rundić, L., Ćorić, S., Pezelj, Đ., Theobalt, D., Sant, K., & Krijgsman, W. (2019). Age and mode of the middle Miocene marine flooding of the Pannonian Basin—constraints from Central Serbia. *Palaios, 34*(2), 71–95.

Marinković, G., Papić, P., Spahić, D., Andrijašević, J., & Spahić, M. P. (2023). Case study of mountainous geothermal reservoirs (Kopaonik Mt., southwestern Serbia): Fault-controlled fluid compartmentalization within a late Paleogene-Neogene core-complex. *Geothermics, 114*, 102799. https://doi.org/10.1016/j.geothermics.2023.102799

Marinković, G., Spahić, D., Poznanović Spahić, M., Papić, P., Magazinović, M., & Obradović, N. (2024). Structural control on the mountainous geothermal reservoirs distributed across the Oligocene andesite-type volcanic complex (Lece massif southern Serbia). *Geoenergy Science and Engineering, 235*, 212701. https://doi.org/10.1016/j.geoen.2024.212701

Marović, M., Đoković, I. M., Milićević, V., Toljić, M., & Gerzina, N. S. (2002). Paleomagnetism of the late Paleogene and Neogene rocks of the Serbian Carpatho-Balkanides: Tectonic implications. *Geološki Anali Balkanskoga Poluostrva, 64*(1), 1–12.

Marović, M., Djoković, I., Toljić, M., Spahić, D., & Milivojević, J. (2007a). Extensional unroofing of the Veliki Jastrebac dome (Serbia). *Geološki anali Balkanskoga poluostrva, 68*, 21–27. https://doi.org/10.2298GABP/0701021M

Marović, M., Đoković, I., Toljić, M., Milivojević, J., & Spahić, D. (2007b). Paleogene-early Miocene deformations of Bukulja-Venčac crystalline (Vardar zone, Serbia). *Geološki Anali Balkanskoga Poluostrva, 68*(1), 9–20. https://doi.org/10.2298/GABP0701009M

Marović, M., Toljić, M., Rundić, L., & Milivojević, J. (2007c). Neoalpine tectonics of Serbia. *Belgrade Sr. Monographie Socite Serbe de Geologie, 1*, 1126.

Matenco, L., & Radivojević, D. (2012). On the formation and evolution of the Pannonian Basin: Constraints derived from the structure of the junction area between the Carpathians and Dinarides. *Tectonics, 31*, TC6007.

Mukherjee, S. (2026). Introduction to "Structural Geology & Tectonics". In S. Mukherjee (Ed.), *Structural Geology & Tectonics*. Springer. ISBN: 978-981-95-4743-2.

Novakova, P., Rybar, S., Šarinová, K., Nagy, A., Hudáčková, N., Jamrich, M., & Kováč, M. (2020). The late Badenian-Sarmatian (Serravallian) environmental transition calibrated by sequence stratigraphy (eastern Danube Basin, Central Paratethys). *Geologica Carpathica, 71*(4), 291–313.

Oberauer, K. (2017). *Depositional Environment and Hydrocarbon Potential of Coal and Oil Shale in the Miocene Aleksinac Basin (Serbia)*. PhD Thesis. Universoty of LEoben, Austria. Retrieved: Richtlinien für die Erstellung einer wissenschaftlichen Arbeit am Lehrstuhl für Geomechanik, Tunnelbau und Konstruktiven Tiefb (unileoben.ac.at)

Obradović, J., Djurdjević-Colson, J., & Vasić, N. (1997). Phytogenic lacustrine sedimentation-oil shales in Neogene from Serbia, Yugoslavia. *Journal of Paleolimnology, 18*, 351–364.

Obradović, J., Vasić, N., Kasanin-Grubin, M., & Grubin, N. (2000). Neogene lacustrine sediments and authigenic minerals geochemical characteristics. *Ann Geol Penins Balk, 63*, 135–154.

Pavlović, P., Eremija, M., & Stevanović, P. (1977). *Kruševačko-trstenički basen.* Geologija Srbija: II-3. Beograd. (in Serbian with English abstract)

Popov, S. V., Akhmetiev, M. A., Bugrova, E. M., Lopatin, A. V., Amitrov, O. V., Andreyeva-Grigorovich, A. S., & Shcherba, I. G. (2002). Biogeography of the Northern Peri-Tethys from the Late Eocene to the Early Miocene: Part 2. *Early Oligocene. Paleontological Journal, 36*(3), S185–S259.

Poznanović Spahić, M., Marinković, G., Spahić, D., Sakan, S., Jovanić, I., Magazinović, M., & Obradović, N. (2023). Water–rock interactions across volcanic aquifers of the Lece Andesite Complex (Southern Serbia): Geochemistry and environmental impact. *Water, 15*(20), 3653. https://doi.org/10.3390/w15203653

Prelogović, E., Saftić, B., Kuk, V., Velić, J., Dragaš, M., & Lučić, D. (1998). Tectonic activity in the Croatian part of the Pannonian basin. *Tectonophysics, 297*(1–4), 283–293.

Radisavljević, M., Burazer, N., Šajnović, A., Spahić, D., Gajica, G., Kovač, S., Gajić, V., & Jovančićević, B. (2024). Towards an understanding of southern peri-Pannonian lacustrine depositional cycles: Interplay of sediment delivery and shifting intrabasinal height a case study of drilled Neogene sediments from northwest Toplica Basin (Central Serbia). *International Journal of Sediment Research, 39*(3), 401–420. https://doi.org/10.1016/j.ijsrc.2024.03.006

Radivojević, D., Rundić, L., & Knežević, S. (2010). Geology of the Čoka structure in northern Banat (central Paratethys, Serbia). *Geologica Carpathica, 61*(4), 341–352. https://doi.org/10.2478/v10096-010-0020-5

Radović, P., & Bradić-Milinović, K. (2018). A new elephantoid dental specimen from the Miocene of Kruševac Basin in Central Serbia. *Geoloski Anali Balkanskoga Poluostrva, 79*(2), 1–10.

Rakić, M., Hadži-Vuković, M., Kalenić, M., Marković, V. & Milovanović, L. J. (1969). *Osnovna geološka karta SFRJ 1:100 000—list kruševac* [Basic Geologic Map of former Yugoslavia1:100 000, Sheet kruševac—in Serbian]. Beograd: Savezni geološki zavod.

Reichenbacher, B. (1988). Die Fischfauna der Kirchberger Schichten (Unter-Miozan) an der Typuslokalitat Illerkirchberg bei Ulm. Stuttgerter Beitrage zur Naturkunde. *Serie B (Geologie und Palaontologie), 139*, 1–53.

Rundić, L., Vasić, N., Banješević, M., Prelević, D., Gajić, V., Kostić, B., & Stefanović, J. (2019). Facies analyses, biostratigraphy and radiometric dating of the Lower-Middle Miocene succession near Zaječar (Dacian basin, eastern Serbia). *Geološki Anali Balkanskoga Poluostrva, 80*(2), 13–37.

Rundić, L., Knežević, S., & Rakijaš, M. (2013). Middle Miocene badenian transgression: New evidences from the Vrdnik coal basin (Fruška Gora Mt., Northern Serbia). *Geološki anali Balkanskoga poluostrva, 74*(1), 9–23.

Sachsenhofer, R. F., Popov S. V., Bechtel, A., Coric, S., Francu, J., Gratzer, R., Grunert, P., Kotarba, M., Mayer, J., Pupp, M., Rupprecht, B. J., & Vincent, S. J. (2018a). Oligocene and Lower Miocene source rocks in the Paratethys: Palaeogeographical and stratigraphic controls. *Geological Society, London, Special Publications, 464*(1), 267–306. https://doi.org/10.1144/SP464.1

Sachsenhofer, R. F., Popov, S. V., Coric, S., Mayer, J., Misch, D., Morton, M. T., & Tari, G. (2018b). Paratethyan petroleum source rocks: An overview. *Journal of Petroleum Geology, 41*(3), 219–245.

Saftić, B., Velić, J., Sztano, O., Juhasz, G., & Ivković, Ž. (2003). Tertiary subsurface facies, source rocks and hydrocarbon reservoirs in the SW part of the Pannonian Basin (northern Croatia and south-western Hungary). *Geologia Croatica, 56*(1), 101–122.

Šajnović, A., Stojanović, K., Jovančićević, B., & Golovko, A. (2009). Geochemical investigation and characterisation of Neogene sediments from Valjevo-Mionica Basin (Serbia). *Environmental Geology, 56*(8), 1629–1641. https://doi.org/10.1007/s00254-008-1260-x

Schettino, A., & Turco, E. (2011). Tectonic history of the western Tethys since the Late Triassic. *Bulletin, 123*(1–2), 89–105.

Schmid, S. M., Bernoulli, D., Fügenschuh, B., Matenco, L., Schefer, S., Schuster, R., & Ustaszewski, K. (2008). The Alpine-Carpathian-Dinaridic orogenic system: Correlation and evolution of tectonic units. *Swiss Journal of Geosciences, 101*, 139–183.
Schultz, O. (2000). Ein Zackenbarsch (Epinephelus, Serranidae, Pisces) aus dem Mittel-Miozän von Retznei, Steiermark. *Joannea Geol Paläont, 2*, 5–56.
Sebe, K., Konrád, G., & Sztanó, O. (2021). An exceptional surface occurrence: the middle to upper Miocene succession of Pécs-Danitzpuszta (SW Hungary). *Földtani Közlöny 151*(3, 4): 235–235
Spahić, D. (2022a). Missing link on the western Paleotethys configuration: Stratigraphic constraints on the truncated Triassic "Gornjak" sequence (eastern Serbia Balkan/Carpathian hinterland). *Italian Journal of Geosciences, 141*(2), 278–292. https://doi.org/10.3301/IJG.2022.14
Spahić, D. (2022b). Towards the Triassic configuration of western Paleotethys. *Journal of Earth Science, 33*(6), 1494–1512. https://doi.org/10.1007/s12583-021-1578-9
Spahić, D. (2022c). Remarks on the Oligo-Miocene extensional episode (s) in Inner Dinarides: Towards the tectonic constraints of the origin of intramontane basins. *Geološki Anali Balkanskoga Poluostrva, 83*(1), 39–56. https://doi.org/10.2298/GABP220416002S
Spahić, D., Bojić; Z., Popović, D., & Gaudenyi, T. (2021). Vestiges of Cambro-Ordovician continental accretion in the Carpathian-Balkan orogen: First evidence of the 'Cenerian event' in the central Serbo-Macedonian Unit. *Acta Geologica Polonica, 71*(2), 219–247. https://doi.org/10.24425/agp.2020.134558
Spahić D, & Gaudenyi, T. (2020a). The role of the Zvornik suture for assessing the number of Neotethyan oceans: Surface-subsurface constraints on the fossil plate margin (Vardar Zone vs. Inner Dinarides). *Geoloski anali Balkanskoga poluostrva, 81*(2), 63–86. https://doi.org/10.2298/GABP200918008S
Spahic, D., & Gaudenyi, T. (2020b). 60 years of the Serbo-Macedonian Unit concept: From Cadomian towards alpine tectonic frameworks. *Geoloski anali Balkanskoga poluostrva Annales geologiques de la Peninsule balkanique, 81*(1), 41–66. https://doi.org/10.2298/GABP191018004S
Spahić, D., & Gaudenyi, T. (2020c). Reconsidering paleozoic differences between the Jadar block and the Drina-Ivanjica unit. *Geoloski anali Balkanskoga poluostrva Annales geologiques de la Peninsule balkanique, 81*(1), 1–9. https://doi.org/10.2298/GABP191014002S
Spahić, D., & Gaudenyi, T. (2022). On the Sava Suture Zone: Post-Neotethyan oblique subduction and the origin of the Late Cretaceous mini-magma pools. *Cretaceous Research, 131*, Article 105062. https://doi.org/10.1016/j.cretres.2021.105062
Spahić, D., Kurešević, L., & Cvetković, Ž. (2023a). The paleokarst origin of the carbonate "Ropočevo breccia" and a closing Neotethys: Regional geological constraints on the Vardar zone s.s. (Belgrade area Central Serbia). *Carbonates and Evaporites, 38*, 51. https://doi.org/10.1007/s13146-023-00863-z
Spahić, D., Šajnović, A., Burazer, N., Radisavljević, M., & Jovančićević, B. (2023b). Neogene subsidence rates of the southern Peri Pannonian realm (1D basin modeling): Constraints on the extensional geodynamic drivers of the asymmetric Toplica basin (central-southern Serbia). *Geoenergy Science and Engineering, 226*, Article 211714. https://doi.org/10.1016/j.geoen.2023.211714
Sremac, J., Bošnjak, M., Velić, J., Malvić, T., & Bakrač, K. (2022). Nearshore pelagic influence at the SW margin of the Paratethys Sea—examples from the Miocene of Croatia. *Geosciences, 12*(3), 120. https://doi.org/10.3390/geosciences12030120
Stojadinovic, U., Matenco, L., Andriessen, P. A. M., Toljić, M., & Foeken, J. P. T. (2013). The balance between orogenic building and subsequent extension during the Tertiary evolution of the NE Dinarides: Constraints from low temperature. *Glob Planet Chang, 103*, 19–38. https://doi.org/10.1016/j.gloplacha.2012.08.004
Stojadinovic, U., Matenco, L., Andriessen, P., Toljić, M., Rundić, L., & Ducea, M. N. (2017). Structure and provenance of Late Cretaceous-Miocene sediments located near the NE Dinarides margin: Inferences from kinematics of orogenic building and subsequent extensional collapse. *Tectonophysics, 710*, 184–204.

Tančić, P. I., Spahić, D. N., Jovanović, D., Ćirić, A., Poznanović-Spahić, M., & Vasić, N. (2021). Occurrences and characterization of alunite group minerals from the Lece-Radan Oligo-Miocene volcanic complex (Serbia). *Geological Quarterly, 65*. https://doi.org/10.7306/gq.1587

Tari, G., Bada, G., Beidinger, A., Csizmeg, J., Danišik, M., Gjerazi, I., & Szafián, P. (2021). The connection between the Alps and the Carpathians beneath the Pannonian Basin: Selective reactivation of Alpine nappe contacts during Miocene extension. *Global and Planetary Change, 197*, Article 103401.

Tari, G., Connors, C., Flinch, J., Granath, J., Pace, P., Sobornov, K., & Soto, J. I. (2023). Negative structural inversion: An overview. *Marine and Petroleum Geology, 152*, 106223. https://doi.org/10.1016/j.marpetgeo.2023.106223

Tari, G., Horvath, F., & Rumpler, J. (1992). Styles of extension in the Pannonian Basin. In P. A. Ziegler (Ed.), *Geodynamics of Rifting, Volume I. Case History Studies on Rifts: Europe and Asia* (Vol. 208, pp. 203–219). Tectonophysics.

van Unen, M., Matenco, L., Nader, F. H., Darnault, R., Mandic, O., & Demir, V. (2019). Kinematics of foreland-vergent crustal accretion: Inferences from the Dinarides evolution. *Tectonics, 38*, 49–76. https://doi.org/10.1029/2018TC005066

Vangelov, D., Pavlova, M., Gerdjikov, I., & Kounov, A. (2016). Timok fault and Tertiary strike-slip tectonics in part of western Bulgaria. *Annual of the University of Mining and Geology "St Ivan Rilski", 59*, 112–117.

Vlahović, I., Mandic, O., Mrinjek, E., Bergant, S., Cosovic, V., de Leeuw, A., & Vranjkovic, A. (2012). Marine to continental depositional systems of Outer Dinarides foreland and intramontane basins (Eocene-Miocene, Croatia and Bosnia and Herzegovina). *Journal of Alpine Geology, 54*, 405–470.

Vrsaljko, D., Pavelić, D., Miknić, M., Brkić, M., Kovačić, M., Hećimović, I.,& Kurtanjek, N. (2006). Middle Miocene (Upper Badenian/Sarmatian) palaeoecology and evolution of the environments in the area of Medvednica Mt.(North Croatia). *Geologia Croatica, 59*(1), 51–63.

Ziegler, P. A., & Cloetingh, S. (2004). Dynamic processes controlling evolution of rifted basins. *Earth-Science Reviews, 64*(1–2), 1–50.

Field Guide to Coorg Block and the Adjoining Shear Zones, Southern Granulite Terrane, South India

S. Nanda and S. Rekha

Abstract This chapter investigates the tectonics of the Coorg Block and adjoining shear zones in Southern Granulite Terrane (SGT) located along the southwestern part of India. The Coorg Block exposed along the Kasaragod district of Kerala and parts of Kodagu district in Karnataka states covers ~ 3000 km^2. It is the northernmost block of the SGT and is separated from the Western Dharwar Craton to the north by a major lineament called the Kasargod-Mercara Shear Zone (KMSZ) and from the Nilgiri Block to the south by the Moyar Shear Zone (MSZ). Charnockites and tonalite-trondhjemite-granodiorite (TTG) gneisses are the dominant rock types along with granites, syenites, anorthosites, metapelitic schists/gneisses, metagabbros, quartzo-feldspathic gneisses, etc. Structurally, the Coorg Block is a high-grade terrane that has undergone at least three phases of deformations. The KMSZ to the north is composed of charnockites, mylonites, mafic granulites, metagabbros, TTG gneisses, kyanite/sillimanite-bearing metapelites, anatectic granitoids, syenites, quartz-mica schists, etc. The dextral shear zone trends E-W in the western part of the stretch, and it curves to the south in the middle segment and terminates at the MSZ in the south near Iritty. The major lithologies in the MSZ to the south of Coorg Block are pyroxene granulites, metapyroxenites, quartzites, hornblende-biotite gneisses, deformed/foliated granites, gabbros, and mylonites. MSZ is a steep-dipping shear zone with dextral kinematics where three sets of deformations were identified.

1 Introduction

The Southern Granulite Terrane (SGT), located along the southern part of India covering the northern part of Kerala and southern part of Karnataka, consists of Neoarchean to Early Paleoproterozoic granulite facies gneisses and foliated orthopyroxene-bearing felsic intrusives traversed by network of tens of km wide E-W trending Neoproterozoic to Early Palaeozoic high-strain zones (Drury &

S. Nanda · S. Rekha (✉)
Department of Earth and Atmospheric Sciences, National Institute of Technology, Rourkela, Odisha, India
e-mail: srekha@nitrkl.ac.in

S. Mukherjee (ed.), *Structural Geology and Tectonics*,
https://doi.org/10.1007/978-981-95-4743-2_3

Holt, 1980; Chetty et al., 2012, 1980) (Fig. 1). The northernmost block of the SGT next to the Indian west coast, named as the Coorg Block is separated from the Western Dharwar Craton (WDC) by a major lineament—the Kasargod-Mercara Shear Zone, KMSZ (Chetty et al., 2012; Rekha et al., 2013) (Fig. 1). The Coorg Block is separated from the Nilgiri Block to the south by the Moyar Shear Zone, MSZ (Drury & Holt, 1980; Santosh et al., 2015, 2016) (Fig. 1). The NW–SE trending KMSZ merges with ~ E-W trending MSZ, which represents the coalescence of crustal blocks in Neoarchean to Paleoproterozoic and continues eastwards as the MSZ (Fig. 1).

The study presented here will aid in the fieldwork in and around the Coorg Block and the adjoining shear zones of the SGT.

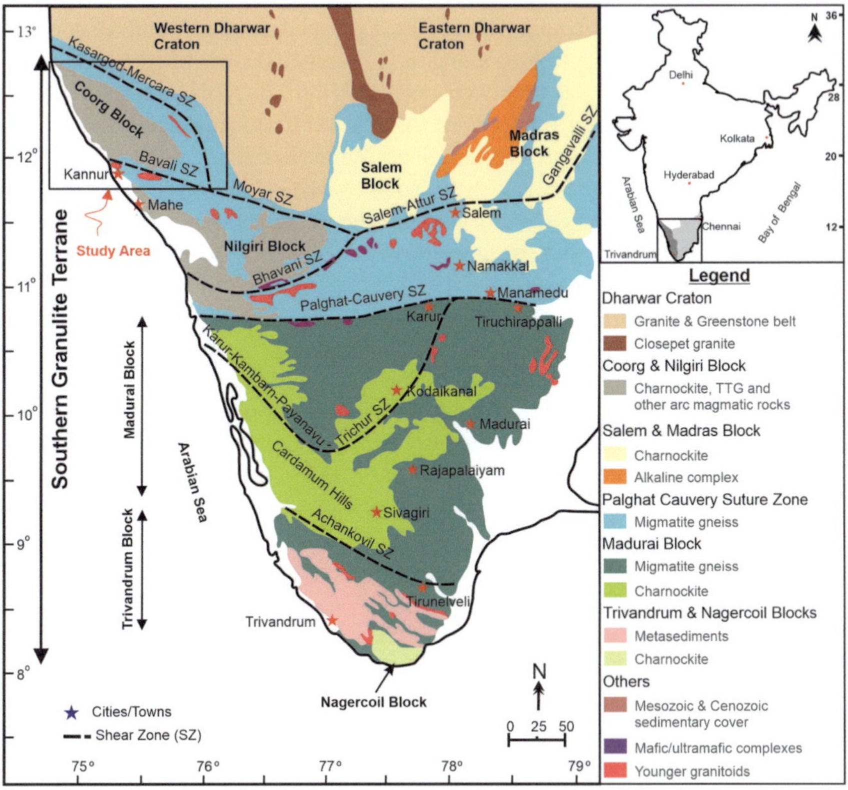

Fig. 1 Tectonic framework of the southern India showing the major crustal blocks and intervening shear/suture zones (after Plavsa et al., 2012). Box: study area. Top right inset map: location of SGT within India

2 Geology

The Coorg Block was considered as the high-grade southern extension of the WDC by previous workers but recent studies show that the Coorg Block is a distinct crustal block and separated from the WDC by the KMSZ (Chetty et. al., 2012; Rekha et al., 2014; Santosh et al., 2015, 2016). It is a granulite facies terrane dominated by arc-magmatic rocks including charnockites and mafic granulites occur as large massifs (Asha Manjari and Malur, 1997; Santosh et al., 2015). The Coorg Block along with other crustal blocks present in the northern part of the SGT has undergone a common late Archean (ca. 2.5 Ga) granulite facies metamorphism (Rao et al., 2003; Sato et al., 2011). Dominantly Mesoarchean ages (3.5–3.0 Ga) were reported from the Coorg Block (Yang et al., 2023). Charnockites, garnetiferous granulites, and migmatitic gneisses from the CB provided Sm–Nd whole-rock isochron dates of 2.9–3.7 Ga (Peucat et al., 2013). U–Pb zircon ages from charnockites, granitoids, mafic enclaves, diorite, and felsic tuff yielded concordant ages of 3.28–3.15 Ga (Santosh et al., 2015). Yang et al. (2023) suggested magmatic emplacement ages of $\sim$ 3.15 Ga closely followed by metamorphism at $\sim$ 3.0 Ga for the charnockites and gabbroic suites from the Coorg Block.

The Coorg Block is bounded by two shear zones—at north by the WNW-ESE trending KMSZ and at south by the E-W trending MSZ. The KMSZ is a curvilinear transpressional shear zone that converges with the MSZ at east (Chetty et al., 2012; Santosh et al., 2015). Earlier Krishnaraj et al. (1994) and Rekha et al. (2014) defined the northern segment of the KMSZ as the Jalsoor-Mercara shear zone and Manjeshwar-Sulya shear zone, respectively. The strongly sheared and deformed gneisses in the ~ 20 km wide, curved KMSZ vary in strike from E-W to NW–SE to N-S towards south, and they dip to the S to SW to W with dextral slip (Chetty et al., 2012). The shear zone wraps around the Coorg Block along its northern edge, extending for > 100 km in strike and 20–30 km in breadth with a dextral slip (Chetty et al., 2012; Santosh et al., 2015). The KMSZ is primarily composed of charnockites, mafic granulites, metagabbros, tonalite-trondhjemite-granodiorite (TTG) gneisses, sillimanite-bearing metapelites (khondalites) and quartz-mica schists (Santosh et al., (2015). As per earlier research, the Rb–Sr ages of the KMSZ gneisses are ~ 2.5 Ga (Spooner & Fairbairn, 1970) whereas the Rb–Sr whole rock ages of the granite and syenite in the KMSZ vary within 0.64–0.71 Ga (Ghosh et al., 2004). Younger intrusive rocks such as Thalur granite (0.71 Ga), Sulya syenite, and Angadimogar syenite (0.64 Ga) occur along the KMSZ (Ravindra & Janardhan, 1981; Santosh et al., 2015).

The MSZ is a high-grade gneissic terrane that is considered as the collision zone between the Coorg Block and Nilgiri Block in the west and between the WDC and the Nilgiri Block towards east (Fig. 1). It is a 200 km long and 20 km wide terrane boundary (Drury & Holt, 1980; Pratheesh et al., 2013) of Proterozoic age (Srikantappa et al., 1986; Yellappa et al., 2012) with a dextral strike-slip and WNW-ESE trending foliation (Drury et al., 1984). Previous studies show that the MSZ has experienced a dextral strike-slip of ~ 70 km (Drury & Holt, 1980; Raith et al., 1990). It was appraised as a suture zone (Chetty et al., 2011; Srikantappa, 1993), a ductile shear zone (Meißner et al., 2002), a transcrustal shear zone (Chetty and Bhaskar Rao, 2006); Chetty et al., 2012) as well as a transpressional shear zone (Swain et al., 2023). The general tectonic trend of the steeply dipping MSZ varies from N70° to N110° (Naha & Srinivasan, 1996; Bhadra 2000). Naha and Srinivasan (1996) and Bhadra and Nasipuri (2017) suggested thrusting of the southern block along the MSZ. According to Raith et al. (1999), the MSZ formed during amalgamation of the SGT with the southern margin of the Dharwar Craton at 2.5 Ga. Zircons from the MSZ enderbites with ~ 2.48 Ga indicate late Archean metamorphic crystallization (Raith et al., 1999). According to Meißner et al. (2002), the 2.4 Ga Sm–Nd garnet age obtained from basic granulite of the MSZ shows the time of metamorphic garnet growth, whereas the 0.54 to 0.62 Ga Rb–Sr mica ages indicates a cooling stage following the last thermal event.

3 Field Works

3.1 Coorg Block

Fieldwork was carried out in two stages taking both N-S and E-W traverses across the CB covering the area within latitude N12°00' to N 12°25' and longitude E75°10' to E75°50' (Fig. 2). In this investigation, the first phase of work was confined within the southern part of the block. In the traverses through Payyannur-Perinthatta-Mathamangalam-Vellora regions, vertical, sub-vertical, horizontal, sub-horizontal, and inclined exposures are studied. These outcrops along the SW region of the Coorg Block are confined to the abandoned quarries, road cuts or alongside canals. Most of the lithologies from here are anorthosites (Perinthatta anorthosite) and associated rocks such as ferrodiorites, syenites, and charnockites. Hornblende granites, quartz-rich mylonites, and mafic xenoliths are also present. Field investigations along the SW included the traverse in and around Iritty-Naduvil-Sreekandapuram regions. The lithologies encountered here vary from medium- to high-grade metamorphic rocks such as quartz-kyanite schists, garnet-sillimanite-quartz-mica schists, quartz-mica schists, mafic granulites, leucogabbros, charnockites, and anorthosites. The eastern part of the Coorg Block was inaccessible as it is a densely vegetated reserved forest area located to the west of Virarajendrapet (Virajpet).

In the second phase of the field investigation, northern part of the Coorg Block was covered taking various traverses, including regions in and around Kasargod-Muttathody-Nekraje-Badiyadka-Belluru-Bandadka-Chittarikal-Perumballa-Chengala-Mulliyar-Bendichal-Munnad, etc. The lithologies in the areas are mostly high-grade metamorphic rocks such as mafic granulites, charnockites and anatectic quartzo-feldspathic gneisses (Fig. 2). In some of the areas, younger intrusives, e.g. anorthosites, foliated granites, and mafic intrusive, are also encountered. Structurally the Coorg Block seems to be a high-grade deformed terrane undergone at least three phases of deformations (D_1, D_2, and D_3) (Figs. 3a–d). The E-W trending S_2

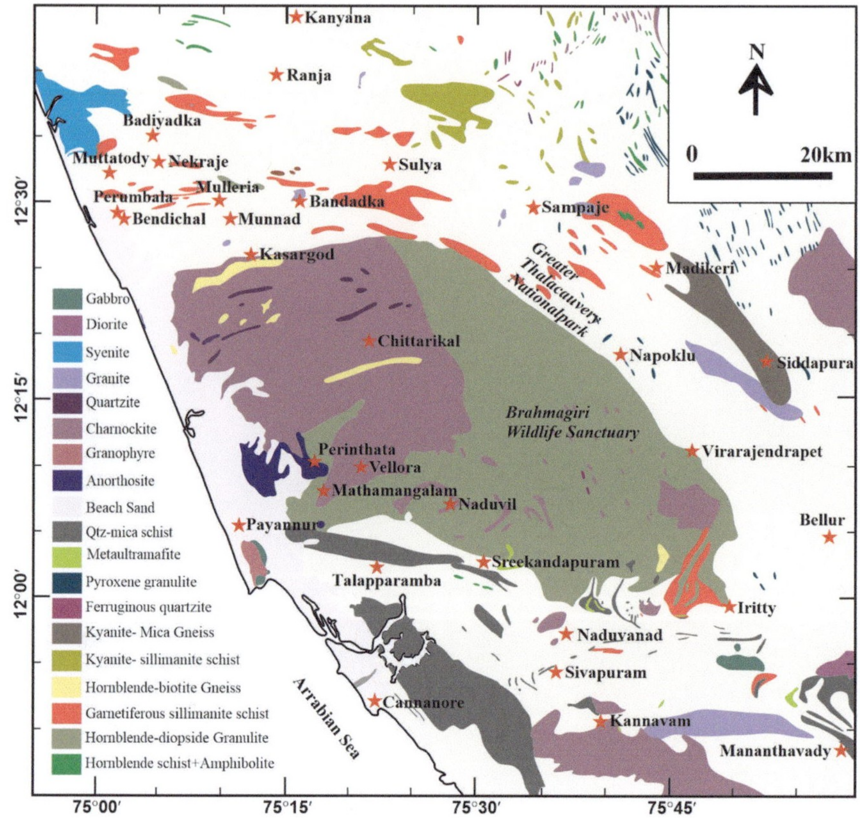

Fig. 2 Geologic map of the study area (box in Fig. 1) showing various litho units (compiled from the geological quadrangle map no-48L, 48P, 49 M, 57D, 58A of the Geological Survey of India)

penetrative foliations observed in the Coorg Block (Fig. 3a, c) are truncated by the later N-S trending S_3 foliation (Fig. 3b). Both the foliations dip steeply (Fig. 3a, b). The fabrics corresponds to the D_1 deformation (S_1) is preserved only in the low strain domains with variable trends and gentle dips (Fig. 3d).

3.2 *Kasargod-Mercara Shear Zone*

The field area is located from latitude E12°25' to E12°45' and longitude N75°00' to N76°00' (Fig. 2). At its western terminus, it is a curvilinear transpressional shear zone that takes a southward turn to converge with the MSZ (Figs. 1 and 2). Human settlements, plantations, and natural dense vegetation make the area difficult for field investigations. The southern stretch of the shear zone is inaccessible because of the reserve forest areas near Virajpet. However, both along and across the shear zone

Fig. 3 Representative field photos from the Coorg Block. **a** Vertical section of gneissic exposure showing E-W fabric with steep dip ~ 80º (photo taken looking N) (Bikash Ranjan Sahoo, 174 cm height for scale). **b** Vertical section of pyroxene granulite showing N-S fabric with steep dip ~ 85º (photo taken looking NE) (Debasis Pal, 165 cm height for scale). **c** and **d** Plan view of gneissic rock showing the penetrative E-W fabric in the Coorg Block. Hammer (45 cm long) head pointing north. Enlarged view of box in **d** shows the earlier fabric preserved within the interfolial domain

traverses were taken to study the lithologies and structure of the area along with the adjacent Coorg Block and WDC. The E-W traverses included the areas in and around Sulya-Ranjha-Kasaragod regions and the other traverses were in the NE-SW direction from Madikeri-Siddapura to further south upto Virajpet cutting across the NE segment of the Coorg Block (Fig. 2). The lithologies encountered during the fieldwork are mostly charnockites, mylonites, mafic granulites, metagabbros, TTG gneisses, kyanite/sillimanite-bearing metapelites, anatectic granitoids, syenites, and quartz-mica schists. The rock types are oriented as per the general regional trend of the shear zone and retained three major deformations represented by S_1, S_2, and S_3 foliations (Figs. 4a-c) among which the third fabric is the most penetrative. The shear zone trends E-W in the western part of the stretch and it curves to the south in the middle segment and terminates at the MSZ in the south near Iritty. The shear zone exhibits dextral slip (Fig. 4d).

3.3 Moyar Shear Zone

The western extension of the MSZ known as the Bavali Shear Zone lies between the latitude N11°35' to N12°00' and longitude E75°30' to E76°50'. Majority of the study area was within human settlements, agricultural lands, and reserve forest areas. The (rarely) exposed rocks in the studied region were found at abandoned quarries, along road cuts and alongside canals. Studies are conducted on inclined, horizontal, sub-horizontal, vertical, and sub-vertical exposures. Traverses were taken around Mananthavady-Kannavam-Sivapuram-Naduvanad-Iritty and Sreekandapuram-Talipparamba-Payyannur (Fig. 2). Exposures of pyroxene granulites, metapyroxenites, quartzites, hornblende-biotite gneisses, deformed/foliated granites, gabbros, and mylonites were noted. In the western, central, and eastern sections of the shear zone, three notable deformation episodes (D_1, D_2, and D_3) were identified (Fig. 5a–d). Mylonitic foliations (S_3) were observed along the high-strain domains (Fig. 5a). In the low strain domains, the early fabric S_1, with its variable trends and gentle dips, is retained. S_2 foliation developed axial planar to the F_1 folds. This foliation trends NNW-SSE to N-S to NE-SW with a steep dip (Fig. 5b). Fabric S_3 trends E-W to WNW-ESE and plunges steeply (Fig. 5a). Orthopyroxene-bearing syn-tectonic felsic melts occurring as anastomosing veins parallel to the shear zone foliation indicate that the last deformation occurred at high-T (> 850 ^{0}C) (Fig. 5c). The western part of the MSZ is a steeply dipping shear zone with dextral kinematics (Fig. 5d).

Fig. 4 Representative field photos from the KMSZ. **a** Vertical section showing the penetrative N-S shear-related foliation in the gneissic rock located along the north-eastern part of the KMSZ (photo taken looking E) (Mr. Ratheesh, 162 cm height for scale). **b** Vertical Exposure showing the NW–SE trending gneissic fabric (photo taken looking NW) (Mr. Karthik, 171 cm height for scale). **c** Plan view showing E-W trending shear zone foliation and folded leucosomes in gneissic rock from the western part of the KMSZ. **d** Plan view of mylonitized foliated granite with S-C fabrics showing dextral shear. Pen (7 cm) head pointing north

Fig. 5 Representative field photos from the MSZ. **a** Plan view of penetrative E-W shear zone foliation in the gneissic rock. **b** Plan view of foliated granite located along the south-eastern part of the MSZ showing the early foliation (S_2); crenulation cleavage, axial plane (S_3) trends NW–SE. **c** Plan view showing orthopyroxene-bearing syn-tectonic felsic melts occurring as anastomosing veins subparallel to the foliation in the eastern part of the MSZ. **d** Plan view of a gneissic rock showing dextral shear. In all the photos pen (7 cm) head points north

Acknowledgements This work forms part of the doctoral dissertation of Saumyaranjan Nanda who has been supported by the DST-SERB through the project (Project code: SR/FTP/ ES-78/2014). I would like to express my gratitude to Prof. Soumyajit Mukherjee (IIT Bombay), for inviting, handling, and reviewing this article in two rounds. Mukherjee (2026) summarized this work.

References

Bhadra, B. K. (2000). Ductile shearing in Attur shear zone and its relation with Moyar shear zone, South India. *Gondwana Research, 3*, 361–369.

Bhadra, S., & Nasipuri, P. (2017). Tectonothermal evolution of a garnet-bearing quartzofeldspathic gneiss from the Moyar shear zone, south India and its bearing on the Neoarchean accretionary tectonics. *Lithos, 274*, 1–18.

Chetty, T. R. K. & Bhaskar Rao, Y. J. (2006). The Cauvery Shear Zone, Southern Granulite Terrain, India: a crustal scale flower structure. *Gondwana Research, 10*(1–2), 77–85.

Chetty, T. R. K., Mohanty, D. P., & Yellappa, T. (2012). Mapping of shear zones in the Western Ghats, Southwestern part of Dharwar Craton. *Journal of the Geological Society of India, 79*, 151–154.

Chetty, T. R. K., Yellappa, T., Nagesh, P., Mohanty, D. P., Venkatasivappa, V., Santosh, M., & Tsunogae, T. (2011). Structural anatomy of a dismembered ophiolite suite from Gondwana: The Manamedu complex, Cauvery suture zone, southern India. *Journal of Asian Earth Sciences, 42*, 176–190.

Drury, S. A., & Holt, R. W. (1980). The tectonic framework of the South Indian craton: A reconnaissance involving LANDSAT imagery. *Tectonophysics, 65*, T1–T15.

Drury, S. A., Harris, N. B. W., Holt, R. W., Reeves-Smith, G. J., & Wightman, R. T. (1984). Precambrian tectonics and crustal evolution in South India. *The Journal of Geology, 92*(1), 3–20.

Ghosh, J. G., de Wit, M. J., & Zartman, R. E. (2004). Age and tectonic evolution of Neoproterozoic ductile shear zones in the Southern Granulite Terrain of India, with implications for Gondwana studies. *Tectonics, 23*(3).

Krishnaraj, N. S., Janardhan, A. S., & Basavalingu, B. (1994). Garnet-kyanite bearing pelitic and associated mafic granulites along Jalsoor-Mercara shear zone: Evidence for deep burial and later uplift. *Geological Society of India, 44*(6), 617–625.

Manjari, A. (1997). Petrogenesis of gneiss to granulite transition: evidence from the Coorg granulite complex, southern India. *Geology in South Asia-II*, 41.

Meißner, B., Deters, P., Srikantappa, C., & Köhler, H. (2002). Geochronological evolution of the Moyar, Bhavani and Palghat shear zones of southern India: Implications for east Gondwana correlations. *Precambrian Research, 114*(1–2), 149–175.

Mukherjee, S. (2026). Introduction to "Structural Geology & Tectonics". In S. Mukherjee (Ed.), *Structural Geology & Tectonics*. Springer. ISBN: 978-981-95-4743-2.

Naha, K., & Srinivasan, R. (1996). Nature of the Moyar and Bhavani shear zones, with a note on its implication on the tectonics of the southern Indian Precambrian shield. *Proceedings of the Indian Academy of Sciences-Earth and Planetary Sciences, 105*, 173–189.

Peucat, J. J., Jayananda, M., Chardon, D., Capdevila, R., Fanning, C. M., & Paquette, J. L. (2013). The lower crust of the Dharwar Craton, Southern India: Patchwork of Archean granulitic domains. *Precambrian Research, 227*, 4–28.

Plavsa, D., Collins, A. S., Foden, J. F., Kropinski, L., Santosh, M., Chetty, T. R. K., & Clark, C. (2012). Delineating crustal domains in Peninsular India: Age and chemistry of orthopyroxene-bearing felsic gneisses in the Madurai Block. *Precambrian Research, 198*, 77–93.

Pratheesh, P., Prasannakumar, V., & Praveen, K. R. (2013). Magnetic fabrics in characterization of magma emplacement and tectonic evolution of the Moyar Shear Zone, South India. *Geoscience Frontiers, 4*(1), 113–122.

Raith, M., Srikantappa, C., Ashamanjari, K. G., & Spiering, B. (1990). The granulite terrane of the Nilgiri Hills (Southern India): characterization of high-grade metamorphism. *Granulites and crustal evolution*, 339–365.

Raith, M. M., Srikantappa, C., Buhl, D., & Koehler, H. (1999). The Nilgirienderbites, South India: Nature and age constraints on protolith formation, high-grade metamorphism and cooling history. *Precambrian Research, 98*(1–2), 129–150.

Rao, B. (2003). Sm–Nd model ages and Rb–Sr isotopic systematics of charnockites and gneisses across the Cauvery shear zone, southern India; implications for the Archaean–Neoproterozoic terrane boundary in the Southern Granulite Terrain. *Memoir—Geological Society of India, 50*, 297.

Ravindra, B. M., & Janardhan, A. S. (1981). Preliminary report on aegerine augite bearing syenite near SuBia Town, Dakshina Kannada District, Karnataka. *Geological Society of India, 22*(8), 399–402.

Rekha, S., Viswanath, T. A., Bhattacharya, A., & Prabhakar, N. (2013). Meso/Neoarchean crustal domains along the north Konkan coast, western India: The Western Dharwar Craton and the Antongil-Masora Block (NE Madagascar) connection. *Precambrian Research, 233*, 316–336.

Rekha, S., Bhattacharya, A., & Chatterjee, N. (2014). Tectonic restoration of the Precambrian crystalline rocks along the west coast of India: Correlation with eastern Madagascar in East Gondwana. *Precambrian Research, 252*, 191–208.

Santosh, M., Yang, Q. Y., Shaji, E., Tsunogae, T., Mohan, M. R., & Satyanarayanan, M. (2015). An exotic Mesoarchean microcontinent: The Coorg block, southern India. *Gondwana Research, 27*(1), 165–195.

Santosh, M., Yang, Q. Y., Shaji, E., Mohan, M. R., Tsunogae, T., & Satyanarayanan, M. (2016). Oldest rocks from Peninsular India: Evidence for Hadean to Neoarchean crustal evolution. *Gondwana Research, 29*(1), 105–135.

Sato, K., Santosh, M., Tsunogae, T., Chetty, T. R. K., & Hirata, T. (2011). Subduction–accretion–collision history along the Gondwana suture in southern India: A laser ablation ICP-MS study of zircon chronology. *Journal of Asian Earth Sciences, 40*(1), 162–171.

Spooner, C. M., & Fairbairn, H. W. (1970). Strontium 87/strontium 86 initial ratios in pyroxene granulite terranes. *Journal of Geophysical Research, 75*(32), 6706–6713.

Srikantappa, C., Raith, M., Ashamanjari, K. G., & Ackermand, D. (1986). Pyroxenites and gabbroic rocks from the Nilgiri granulite terrane, southern India. *Indian Mineralogist, 27*, 62–83.

Srikantappa, C. (1993). High-pressure charnockites of the Nilgiri hills, Southern India. *Memoir Geological Society of India, 25*, 95–110.

Swain, M., Rekha, S., Nanda, S., & Bhattacharya, A. (2023). The anatomy of the 750 Ma Bavali shear zone in South India: Did the integration of India into East Gondwanaland initiate in the mid-Neoproterozoic? *International Journal of Earth Sciences, 112*(8), 2119–2143.

Yang, C. X., Santosh, M., Lloyd, J. C., Glorie, S., Gao, P., Yu, B., Anilkumar, Y., Anoop, K. S., & Kim, S. W. (2023). The Coorg Block, southern India: Insights from felsic and mafic magmatic suites on Mesoarchean plate tectonics and correlation with supercontinent Ur. *Gondwana Research, 118*, 1–36.

Yellappa, T., Santosh, M., Chetty, T. R. K., Kwon, S., Park, C., Nagesh, P., Mohanty, D. P., & Venkatasivappa, V. (2012). A Neoarchean dismembered ophiolite complex from southern India: Geochemical and geochronological constraints on its suprasubduction origin. *Gondwana Research, 21*(1), 246–265.

Geodynamic Evolution of the Indo-Myanmar Ranges, NE India

Thungyani N. Ovung and Bikash Ranjan Nayak

Abstract The Indo-Myanmar Ranges (IMR) situated along the plate boundary of India and Eurasia represent the India-Asia suture zone. These mountain ranges were derived in a trench-accretionary system and later exhumed due to changes in wedge dynamics. In NE India, dismembered Neotethyan ophiolites are exposed in the Indian north-eastern states Nagaland and Manipur. It is spread across Phek and Kiphire districts in Nagaland; while in Manipur, it is observed at Ukhrul and Tengnoupal districts. Here, most of the mantle and crustal units of an ideal ophiolite assemblage are recorded except for the sheeted dykes. Ultramafic rocks have been serpentinized to varying extent indicating the role of high water–rock interaction. The source of the serpentinizing fluids could be seawater or fluids dehydrating from the subducting slab. Plagioclase in basalts is albitic in composition due to alteration. Radiometric dating of microfossil assemblages in the flyschoidal and pelagic sedimentary deposits provided concrete evidences in establishing the timings of formation and emplacement of this ophiolite suite. The association of the Neotethyan oceanic lithosphere with the younger flyschoidal deposits and their structural disposition has been presented in this chapter. Mineralization of chromitites, nickel, and sulfides in ultramafic and mafic rocks of the Naga-Manipur ophiolite belt (NMOB) are also documented in this region.

1 Introduction

The NE India's tectonics (e.g., Gogoi et al., 2022; Goswami et al., 2018, 2020, 2022a, 2022b; Hossain et al., 2021) is still not well understood and has importance for applied aspects such as hydrocarbon exploration. The Indo-Myanmar Ranges are N-S trending mountain belt that is located between Central Myanmar Basin (CMB) to the east and Bengal-Assam Basin to the west. The Indo-Myanmar Range comprises mainly of Mesozoic-Cenozoic flysch deposits with fragments of

T. N. Ovung (✉) · B. R. Nayak
Department of Earth and Atmospheric Sciences, National Institute of Technology Rourkela, Rourkela, Odisha, India
e-mail: ovungtn@nitrkl.ac.in

S. Mukherjee (ed.), *Structural Geology and Tectonics*,
https://doi.org/10.1007/978-981-95-4743-2_4

Neotethyan oceanic lithosphere dispersed along the Naga Hills (NE India) and Chin Hills (western Myanmar) (Naing et al., 2023) (Fig. 1). The Neotethys Ocean basin which existed during Triassic-Cretaceous Periods originated due to the rifting of Lhasa, West Myanmar followed by Sulawesi and Borneo from the northern fringes of Gondwanaland (Metcalfe, 2021; Zhu et al., 2011). The closure of this ocean basin was caused by the northward drifting of India and its subduction beneath the Eurasian plate. The flyschoid deposits and Neotethyan ophiolites of IMR were suggested to have formed initially in a trench-accretionary wedge system, which later evolved to fold-thrust belt due to the oblique collision between India and Sundaland (Morley et al., 2020; Najman et al., 2020; Fareeduddin and Dilek, 2015). The input of thick sediments from the Bengal Basin to the subduction zone led to the exhumation of IMR during the Oligocene–Miocene Periods (Najman et al., 2020). The IMR therefore marks the closure of the Neotethys Ocean basin and represents the India-Asia suture zone.

A late Jurassic age was estimated for the radiolarian cherts from the Naga Hills ophiolite (NHO) (Baxter et al., 2011). Radiolarian fossil assemblage derived from the mud matrix melange beneath the Manipur ophiolite at Ukhrul yielded Paleocene-Eocene age indicating the timing of emplacement for the IMR ophiolites (Aitchison et al., 2019). U–Pb zircon ages of plagiogranites from the NHO yielded an age of 116–118 Ma (Singh et al., 2017). The southern continuation of this ophiolite belt crops out in the Chin Hills (western Myanmar) where the age of rodingite pods intruding the serpentinites was estimated to be 127 Ma (Liu et al., 2016).

From east to west, the IMR is broadly categorized into an eastern IMR core, Inner IMR, and Outer IMR (Maurin & Rangin, 2009). Eastern IMR core is separated from the inner belt by the transpressional Lelon fault (Najman et al. 2020) while, the inner and outer belts are separated by the dextral Kaladan strike-slip fault (Betka et al., 2018). The outer belt is younger and comprises of Neogene sediments, whereas the inner belt is composed of Late Cretaceous-Paleogene sediments. The eastern IMR core exposed at Mt. Victoria region in western Myanmar comprises of Kanpetlet Schist basement, Cretaceous and Triassic turbidites (Sevastjanova et al., 2016), and Jurassic ophiolites (Suzuki et al., 2004).

The IMR in India passes through the north-eastern states of Arunachal Pradesh, Nagaland, Manipur, and Mizoram. The cretaceous Naga-Manipur ophiolite belt (NMOB) is bounded by the Naga metamorphics on the eastern side and Disang Formation to the west (Acharyya, 2010). Naga metamorphic unit is a thick sequence of low-grade metasedimentary rocks such as slates, phyllites, siltstones, quartzites, and quartz-chlorite mica schists (Vidyadharan et al., 1989; Fareeduddin and Dilek, 2015) (Fig. 2a). It consists of Saramati and Nimi Formations. Based on the fossil assemblages, the age of these formations are estimated to be Proterozoic and Maastrichtian, respectively (Acharyya, 2007). The NMOB is a dismembered ophiolite type consisting dominantly of mantle peridotites followed by ultramafic–mafic cumulates such as dunites, chromitites, wehrlites, pyroxenites, and gabbros. The ultramafic rocks have been serpentinized to varying extent. The cumulate section is overlain by upper crustal rocks comprising of plagiogranites and pillowed basalts. The Jurassic Salumi Formation consisting of pelagic sedimentary rocks such as limestones and

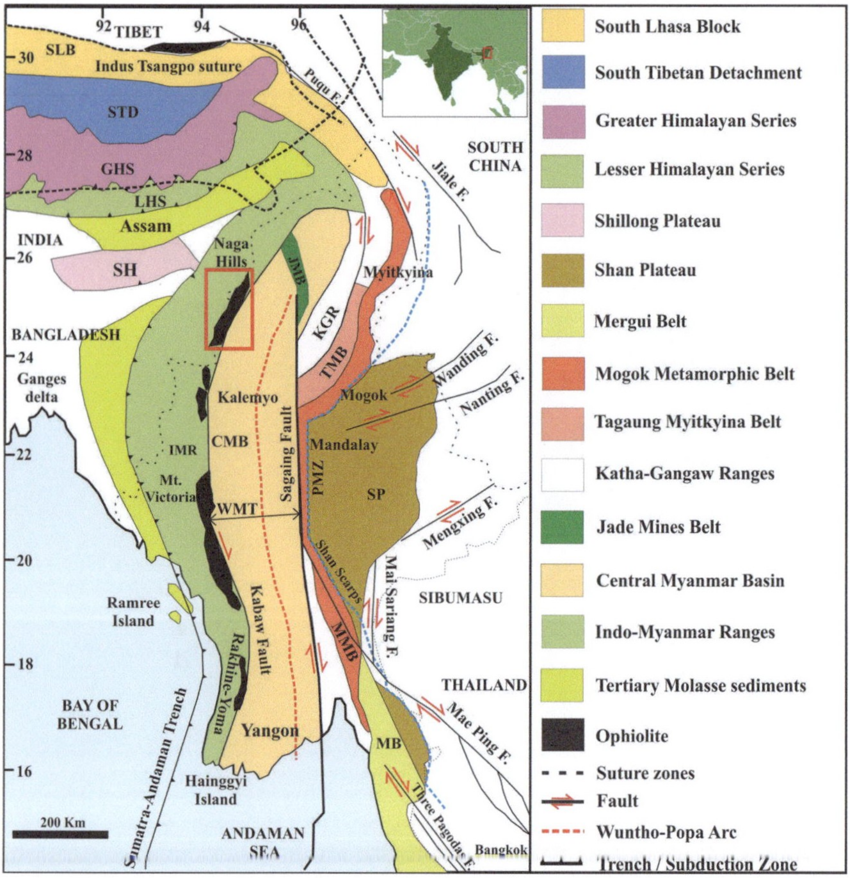

Fig. 1 Simplified geological map showing the Indo-Myanmar Ranges and distribution of Neotethyan ophiolites in NE India and western Myanmar (after Searle et al. 2007). Rectangular red box marks the study area (Lat: 24°15′25.6"-25°41′44.8" N; Long: 94°14′57.57"-94°50′41" E)

radiolarian cherts overlies the upper mantle and crustal units (Fig. 2b). This ophiolite suite is finally overlain by the Eocene-Miocene ophiolite-derived clastics known as the Jopi/Pokphur Formation made of tuffaceous shale, sandstone, polymictic conglomerate containing ophiolitic fragments (Fig. 2c), and greywacke (Acharyya, 2007). To the west of this ophiolite belt is the Disang Formation (Upper Cretaceous-Eocene), which is a shallow marine deposit consisting of finely laminated shales alternating with arenaceous sandstone (Fig. 2d). Further west is the younger Eocene–Oligocene rock known as the Barail Formation made of well-bedded sandstone (Vidyadharan et al., 1989).

The purpose of this chapter is to provide the reader a field guide to understand the complex geology of the Indo-Myanmar Ranges in Nagaland and Manipur and comprehend its formation processes in light of India-Asia collision.

Fig. 2 Field photographs showing: **a** Deformed quartz-chlorite mica schist at the Naga Hills (horizontal exposure view from SE). **b** Folded bedded chert near Phokhungri village (vertical exposure view from E). **c** Polymictic conglomerate made of detrital clasts derived from serpentinites, gabbros, cherts, and basalts (vertical exposure). **d** Disang Formation having well-bedded sandstone interlayered with finely laminated shales near Shangshak village, Manipur (vertical exposure view from S)

2 Discussions

2.1 Evolution of the Neotethyan Oceanic Lithosphere

The Naga-Manipur ophiolite belt represents the remnants of the Neotethyan oceanic lithosphere that was abducted onto the continental margins of India. It is a dismembered type of ophiolite comprising most of the mantle and crustal units except for the sheeted dykes. Here, the mantle peridotites are classified as lherzolites, harzburgites, and cpx-harzburgites (Ovung et al., 2021). In most places, serpentinization has taken place to give a greenish shiny appearance. A huge mantle peridotite outcrop is exposed between Wazeho and Shatiza villages where relict orthopyroxene grains are still intact (Fig. 3a). The rock is finely foliated at some places and shows elongated olivine and pyroxene grains indicative of a tectonite fabric (Fig. 3b). A highly sheared

Fig. 3 Field photographs showing: **a** Outcrop of serpentinized mantle peridotite near Shatiza village, Nagaland (vertical exposure view from NW). **b** Elongated olivine and pyroxene minerals in mantle peridotites exhibiting tectonite fabric (vertical exposure). **c** Highly sheared serpentinites with development of microfolds at Shatiiza village (vertical exposure view from SE). **d** Pseudotachylite within the brecciated serpentinites (vertical exposure view from SE). **e** Needle-shaped tremolites in serpentinized ultramafic rocks at Mokie village (vertical exposure view from E). **f** Ultramafic rock boulders on a hill near Letsami village (inclined exposure view from SW)

and brecciated serpentinite with development of small-scale fold was observed at Shatiza village (Fig. 3c). The axial plane of the folds is parallel to its foliation plane. The presence of pseudotachylite within the brecciated serpentinites is suggestive of a seismic event recorded in the rocks (Fig. 3d). Needle-shaped tremolite minerals are observed within the serpentinites at Mokie village (Fig. 3e). A two km hike from

Mokie village will lead us to a gabbroic exposure where large plagioclase grains are visible within the rock. The metamorphic sole is represented by the presence of high P/T eclogite and blueschist rocks located near Salumi (Ao & Bhowmik, 2014). On a hill near Letsami village, an assortment of ultramafic rock boulders is documented (Fig. 3f). Here, dunites, chromitite pods, and peridotites are altered to varying extent although Cr-spinel and orthopyroxene grains are still preserved.

Along Letsami-Zhipu road, huge outcrop of pillowed basalts is thrust onto the Disang Formation (Fig. 4a). Basalts have undergone greenschist metamorphism locally exhibiting a greenish appearance. Well-preserved chilled margins are also evident (Fig. 4b). Such outcrop is also exposed along Shatiza-Laluri road section. Along Laluri-Mokie traverse, altered basalts are overlain by complex rock types belonging to Phokphur Formation (Fig. 4c). The formation comprises of alternating layers of polymictic conglomerates, shales, and sandstones. Conglomerates consist of well-rounded to angular clasts derived from peridotites, gabbros, cherts, basalts, and sandstones (Fig. 2c). A chert-serpentinite contact is observed near Phokungri village where the contact plane is sheared and brecciated (Fig. 4d) indicating the complex dynamics involved during its emplacement in the subduction zone setting.

Fig. 4 Field photographs showing: **a** Huge outcrop of pillowed basalt overlain by Disang Flysch (vertical exposure). **b** Chilled margins in pillowed basalt at the Naga Hills (vertical exposure). **c** Altered basalts overlain by Phokphur Formation (vertical exposure). **d** Contact between chert and altered ultramafic rock near Phokhungri village (vertical exposure)

2.2 *Mineralization in the Naga-Manipur Ophiolite Belt (NMOB)*

Massive and podiform chromitites are closely associated with peridotites and are observed near Letsami (Nagaland) and Sirohi (Manipur) villages (Fig. 5a). The mineral chemistry of chrome spinel in chromitites of Naga Hills ophiolite exhibits Cr# (Cr/Cr + Al) content between 0.70 and 0.85 indicative of its formation in a forearc region (Fareeduddin & Dilek, 2015). Further south in the Manipur ophiolite, podiform chromitites hosted within harzburgites located near Sirohi village exhibit Cr# content between 0.50 and 0.80, which also supports its formation in subduction setting. According to Arai (1997), chromitites originate in subduction zone setting when a boninitic melt dissolves the orthopyroxenes in harzburgites, which produces a melt where olivine crystals accumulate to form dunitic rock. The residual SiO_2-rich melt then interacts with a primitive melt to form chromitites.

Sulfide mineralization has occurred within the highly altered and ferruginized basalts near Shatiza village, Nagaland (Fig. 5b&c). The common sulfide minerals observed are malachite, azurite, and pyrite. Precipitation of malachite and azurite

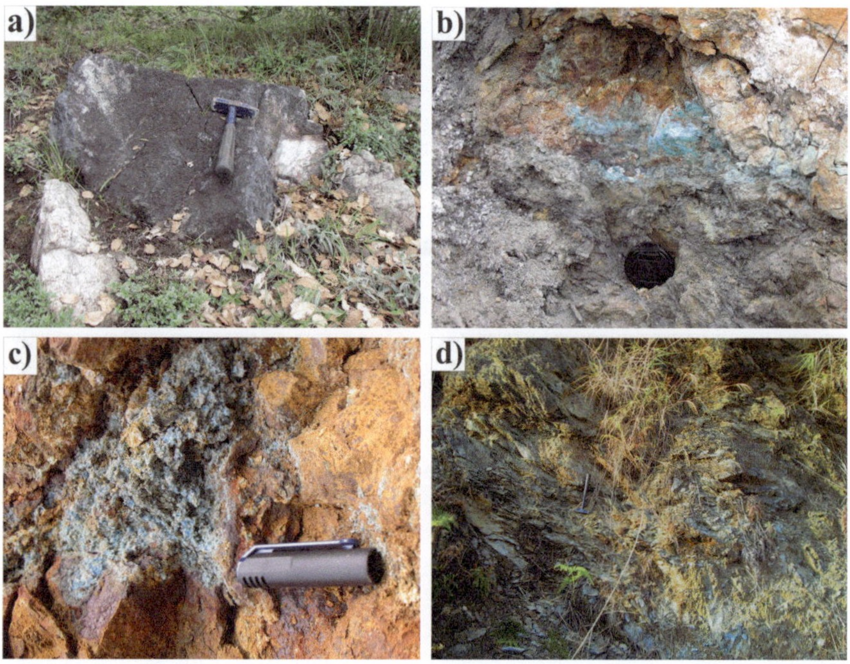

Fig. 5 Filed photographs showing **a** chromitite pod near Letsami village, Nagaland (horizontal exposure view from SW). **b** and **c** sulfide mineralization within altered basalts at Naga Hills ophiolite (vertical exposure). **d** Nickel laterite exposed near Lunghar village, Manipur (vertical exposure view from SW)

might have occurred from the percolating water containing dissolved carbonates and copper that was leached out from the basalt and surrounding carbonate rocks. Such localized mineralization is observed only from that area. Further investigation is required to understand the economic prospect of sulfide mineralization in the NHO.

Nickel laterites are also documented along the Naga-Manipur ophiolite belt. They are exposed along Sirohi-Nungbi, Letsami-Zhipu, Luthur-Salumi, and Salumi-Khongsa roads. It has formed due to the weathering of mantle peridotite rocks on the top surface while the bedrock is still preserved up to some extent. The main source of Ni in such soil profiles are the olivines and orthopyroxenes in ultramafic rocks (Robb, 2005). The presence of Ni-rich serpentine mineral known as népouite has been reported in the weathered peridotites rocks near Lunghar village (Ovung et al., 2017) (Fig. 5d).

3 Conclusions

The Indo-Myanmar Ranges comprising dominantly of flyschoidal deposits (Mesozoic-Cenozoic) of a trench-accretionary system provide a window into the paleo-environmental conditions of this region. Remnants of the Neotethyan oceanic lithosphere, which are now known as the IMR ophiolites, are thrust onto the flyschoidal deposits and are dispersed along the Naga Hills (NE India) and Chin Hills (western Myanmar). Microfossil assemblages present in these trench deposits have provided concrete evidences in establishing the age of its deposition and emplacement of this ophiolite suite. Mantle rocks are the dominant rock type of the IMR ophiolites, which has been serpentinized to varying extent indicating extensive fluid-rock interaction. Serpentine minerals such as lizardite, chrysotile, and antigorite are present in these ultramafic rocks implying its formation in an evolving subduction system at varying P–T conditions. The presence of lizardite and chrysotile indicate low-temperature serpentinization (~200–300 °C), whereas antigorites are suggestive of high-temperature serpentinization (~450–550 °C). Evolving subduction system is also supported by the wide range of Cr# (Cr/Cr + Al) content of chromite (0.12–0.58) in peridotites where the extent of mantle melting is low at the initial stages of subduction initiation and increases at the advanced stages due to the addition of fluids from the dehydrating slab. Deformation tectonite fabric is evident in the elongated olivine and pyroxene grains. Such fabric might have developed due to asthenospheric mantle upwelling or during mountain building events. Crustal unit of the IMR ophiolites is mostly exposed in Nagaland where huge outcrops of pillowed basalts and plagiogranites are observed in Phek and Kiphire districts, respectively. However, alteration in these crustal rocks is pervasive that has altered the primary plagioclase and pyroxene minerals to albite, epidote, sericite, actinolite, and tremolite. Few pyroxene minerals are still intact in some places. Geochemical studies indicate high nickel and chromium concentration in ultramafic rocks of the IMR ophiolites.

Acknowledgements We thank Soumyajit Mukherjee (IIT Bombay) for inviting, handling, and reviewing this chapter. Field-related studies in Nagaland and Manipur were conducted through the professional development allowances (PDA) of TNO at NIT Rourkela and partly through the Department of Science and Technology (DST)-funded project (Grant No: SR/S4/ES-389/2008), where she was involved with the project as a research fellow. Mukherjee (2026) summarized this work.

Abbreviations

CMB	Central Myanmar Basin
IMR	The Indo-Myanmar Ranges
NHO	Naga Hills ophiolite
NMOB	Naga-Manipur ophiolite belt

References

Acharyya, S. K. (2007). Collisional emplacement history of the Naga-Andaman ophiolites and the position of the eastern Indian suture. *Journal of Asian Earth Sciences, 29*, 229–242.

Acharyya, S. K. (2010). Tectonic evolution of Indo-Burma Range with special reference to Naga-Manipur Hills. In S. Ibotombi (Ed.), *Indo-Myanmar Ranges in the Tectonic Framework of the Himalaya and Southeast Asia* (Vol. 75, pp. 25–43). Memoir Geological Society of India.

Aitchison, J. C., Ao, A., Bhowmick, S., Clarke, G. L., Ireleand, T. R., Kachovich, S., Roeder, T., Truscott, N., Zhen, Y., & Zhou, R. (2019). Tectonic evolution of the western margin of the Burma microplate based on new fossil and radiometric age constraints. *Tectonics, 38*(5), 1718–1741.

Ao, A. & Bhowmik, S. K. (2014). Cold subduction of the Neotethys: the metamorphic record from finely banded lawsonite and epidote blueschists and associated metabasalts of the Nagaland Ophiolite Complex, India. *Journal of Metamorphic Geology, 32*(8), 829–672: 860.

Arai, S. (1997). Origin of podiform chromitites. *Journal of Asian Earth Sciences, 15*, 303–310.

Baxter, A. T., Aitchison, J. C., Zyabrev, S. V., & Ali, J. R. (2011). Upper Jurassic radiolarians from the Naga Ophiolite, Nagaland, northeast India. *Gondwana Research, 20*, 638–644.

Betka, P. M., Seeber, L., Thomson, S. N., Steckler, M. S., Sincavage, R., & Zoramthara, C. (2018). Slip-partitioning above a shallow, weak decollement beneath the Indo-Burman accretionary prism. *Earth & Planetary Science Letters, 503*, 17–28.

Fareeduddin & Dilek, Y. (2015). Structure and petrology of the Nagaland-Manipur Hill Ophiolitic Mélange zone, NE India: A Fossil Tethyan Subduction Channel at the India-Burma Plate Boundary. *Episodes, 38*(4), 298–314.

Gogoi, M. P., Gogoi, B., & Mukherjee, S. (2022). Tectonic instability of the petroliferous upper Assam valley (NE India): A geomorphic approach. *Journal of Earth System Science, 131*, 18.

Goswami, T. K., Bezbaruah, D., Mukherjee, S., Sharmah, R. K., & Jabeed, S. (2018). Structures and morphotectonic evolution of the frontal fold-thrust belt, Kameng river section, Arunachal Himalaya, India. *Journal of Earth System Science, 127*, 88.

Goswami, T. K., Mahanta, B. N., Mukherjee, S., Syngari, B. R., & Sarmah, R. K. (2020). Orogen-transverse structures in the eastern Himalaya: Dextral Riedel shear along the Main Boundary Thrust in the Garu-Gensi area (Arunachal Pradesh, India), implication in hydrocarbon geoscience. *Marine and Petroleum Geology, 114*, Article 104242.

Goswami, T., Gogoi, M., Mahanta, B. N., Mukherjee, S., Saikia, H., Shaikh, M., Kalita, P., Baral, U., & Sarmah, R. (2022a). Brittle tectonics in the western Arunachal frontal fold belt, India: Change in stress regime from pre-collisional extension to collisional compression. *Geological Journal, 57*, 5019–5038.

Goswami, T. K., Kalita, P., Mukherjee, S., Mahanta, B. N., Sarmah, R. K., Laishram, R., Saikia, H., Gogoi, M., Macharari, R., & Oza, B. (2022b). Basement cross-strike Bomdila fault beneath Arunachal Himalaya: Deformation along curved thrust traces, seismicity and implications in hydrocarbon prospect of the Gondwana sediments. *Geological Journal, 57*, 4974–4999.

Hossain, Md. S., Khan, Md. S. H., Abdullah, R., & Mukherjee, S. (2021). Late Cenozoic transpression at the plate boundary: Kinematics of the eastern segment of the Dauki Fault Zone (Bangladesh), and tectonic evolution of the petroliferous NE Bengal Basin. *Marine and Petroleum Geology, 131*, Article 106133.

Liu, C.-Z., Zhang, C., Xu, Y., Wang, J.-G., Chen, Y., Guo, S., Wu, F.-Y., & Sein, K. (2016). Petrology and geochemistry of mantle peridotites from the Kalaymyo and Myitkyina ophiolite (Myanmar): Implications for tectonic settings. *Lithos, 264*, 495–508.

Maurin, T., & Rangin, C. (2009). Structure and kinematics of the Indo-Burmese wedge: Recent and fast growth of the outer wedge. *Tectonics 28*(2): TC 2010.

Metcalfe, I. (2021). Multiple Tethyan Ocean basins and orogenic belts in Asia. *Gondwana Research, 100*, 87–130.

Morley, C. K., Naing, T. T., Searle, M., & Robinson, S. A. (2020). Structural and tectonic development of the Indo-Burma ranges. *Earth Science Reviews, 200*, Article 102992.

Mukherjee, S. (2026). Introduction to "Structural Geology & Tectonics". In S. Mukherjee (Ed.), *Structural Geology & Tectonics*. Springer. ISBN: 978-981-95-4743-2.

Naing, T. T., Robinson, S. A., Searle, M. P., Morley, C. K., Millar, I., Green, O. R., Bown, P. R., Danelian, T., Petrizzo, M. R., & Henderson, G. M. (2023). Age, depositional history and tectonics of the Indo-Myanmar Ranges, Myanmar. *Journal of the Geological Society of London, 180*, 1–27.

Najman, Y., Sobel, E. R., Millar, I., Stockli, D. F., Govin, G., Lisker, F., Garzanti, E., Limonta, M., Vezzoli, G., Copley, A., Zhang, P., Szymanski, E., & Kahn, A. (2020). The exhumation of the Indo-Burman Ranges, Myanmar. *Earth and Planetary Science Letters, 530*, Article 115948.

Ovung, T. N., Ray, J., Ghosh, B., Mandal, D., Dasgupta, P., & Paul, M. (2017). Occurrence of népouite in the serpentinite of the Manipur Ophiolite Belt, Northeastern India: Implication for melt rock interaction in a supra-subduction zone. *Journal Geological Society of India, 90*, 154–158.

Ovung, T. N., Ghosh, B., & Ray, J. (2021). Petrogenesis of the Neotethyan ophiolites from the Indo-Myanmar Ranges: A review. *International Geology Review, 63*(12), 1437–1449.

Robb, L. (2005). *Introduction to ore-forming processes* (p. 373). Hoboken, New Jersey: Blackwell Science Ltd., John Wiley & Sons, Inc.

Searle, M. P., Noble, S. R., Cottle, J. M.,Waters, D. J., Mitchell, A. H. G., Hlaing, T. & Horstwood, M. S. A. (2007). Tectonic evolution of the Mogok metamorpohic belt, Burma (Myanmar) constrained by U–Th–Pb dating of metamorphic and magmatic rocks. *Tectonics, 26*, TC3014.

Sevastjanova, I., Hall, R., Rittner, M., Paw, S., Naing, T. T., Alderton, D. H., & Comfort, G. (2016). Myanmar and Asia United, Australia left behind long ago. *Gondwana Research, 32*, 24–40.

Singh, A. K., Chung, S.-L., Bikramaditya, R. K., & Lee, H. Y. (2017). New U-Pb zircon ages of plagiogranites from the Nagaland-Manipur Ophiolites, Indo-Myanmar Orogenic Belt, NE India. *Journal of the Geological Society, 174*(1), 170–179.

Suzuki, H., Maung, M., Aye, K. A., & Takai, M. (2004). Jurassic radiolarian from chert peb-bles of the Eocene Pondaung Formation, Central Myanmar. *Neues Jahrburg Geology Paläontology, 231*, 369–393.

Vidyadharan, K. T., Joshi, A., Ghose, S., Gaur, M. P. & Sukla, R. (1989). Manipur Ophiolites. Its geology, tectonic setting and matallogeny. In N.C. Ghose (Ed.), *Phanerozoic Ophiolites of India* (pp. 197–212). Patna: Sumna Publication.
Zhu, D. C., Zhao, Z. D., Niu, Y., Dilek, Y., & Mo, X. X. (2011). Lhasa terrane in southern Tibet came from Australia. *Geology, 39*, 727–730.

Innovative Technology to Harness Geothermal Energy from Active Ridge Flanks and Oceanic Islands: Advantage for the Oil Industries to Bring Energy Transition

Dornadula Chandrasekharam

Abstract An innovative technology to harness geothermal energy from the active Red Sea ridge flanks is presented. The oil industries have the expertise and capabilities to install off-shore platforms to harness this energy. Each geothermal power plant over the off-shore platform can generate 40 MWe minimum considering the available heat over the flanks. Instead of abandoning and/or decommissioning non-producing and non-profitable wells, the oil industries can utilize the entire heat available from these wells using geothermal technology. This innovative technology can be implemented over the ocean islands that are depending completely on imported fossil fuels. While electricity can be supplied to these oceanic islands, the energy from the ocean ridge flanks can be utilized for generating freshwater using desalination technology and also for powering the Middle East and countries around the Red Sea for sustainable development with food-water-energy security. Each off-shore platform can generate 40 MWe, equivalent to 315 billion kWh which can generate 19 billion m^3 of freshwater. In the future, oil industries can play a major role to achieve or move towards creating a Net-Zero Emissions (NZE) status in the world.

Keywords EGS · Geothermal energy · Off-shore oil platform · Red Sea · Oceanic islands · Mid-ocean ridges · Abandoned oil wells

1 Introduction

In the past, the oil industry's business starts with exploration, drilling, and recovery of oil/gas and abandoned the non-performing/non-profitable wells. But now the scenario has changed completely, where oil industries go beyond oil/gas recovery and the wells are never abandoned but become profitable for another half century after the production dwindles. Such wells become geothermal wells generating power and

D. Chandrasekharam (✉)
Izmir Institute of Technology, Izmir, Türkiye
e-mail: dchandra50@gmail.com

S. Mukherjee (ed.), *Structural Geology and Tectonics*,
https://doi.org/10.1007/978-981-95-4743-2_5

Table 1 Major oil and gas producing countries

Producer	Mt	bcm
USA	706	949
Russia	512	722
Saudi Arabia	511	99
Canada	255	184
China	195	191
Iraq	201	No gas
Iran	130	235
WORLD	4141	4014

Mt: Million tons, bcm: billion cubic meters
Iraq does not produce gas (IEA, 2021)

increasing the profit of the industry. Thus oil industry will dominate in the energy sector even when the hydrocarbon reservoirs become dry. This is true concerning onshore and off-shore wells. Besides onshore oil and gas production platforms, off-shore environment is often considered the domain of oil and gas fields in the energy sector. In the present scenario, more than 85% of the energy requirement is mitigated through the utilization of oil and gas. List of major oil and gas producing countries is listed in Table 1. Production of both oil and gas is nearly equal.

These fossil fuels are responsible for ever-increasing CO_2 emissions and responsible for the increasing global temperature (Figs. 1 and 2).

If the current emissions trend continues, the world will have to face both physical and economic crises in the future. Hence, it is recommended that countries increase renewable energy sources in their primary energy mix so that the carbon intensity curve could be maintained at or above 2D level if not 4D level (Fig. 3) (IEA, 2021).

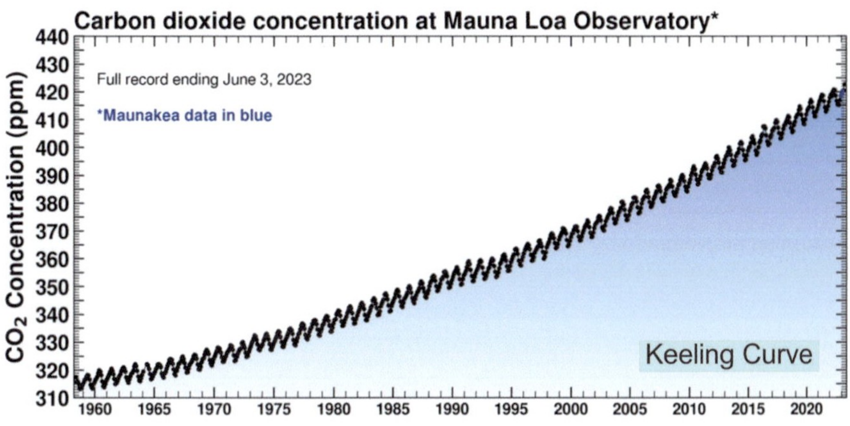

Fig. 1 Historic and modern fossil fuel-related CO_2 emissions (*Source* https://gml.noaa.gov/ccgg/trends/)

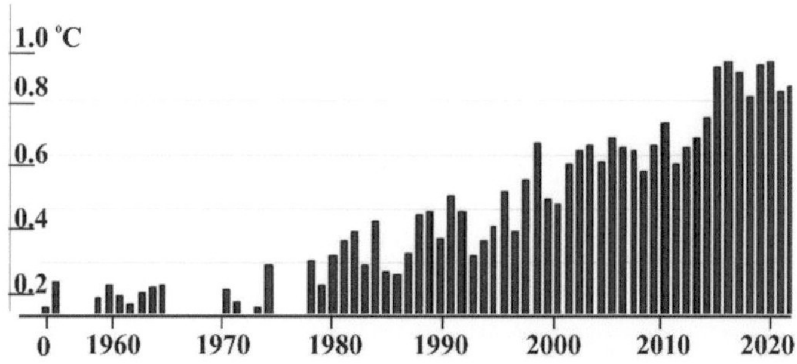

Fig. 2 Rise in global temperature with time (*Source* https://www.weforum.org/agenda/2022/11/global-temperatures-have-fluctuated-since-1880-climate-change)

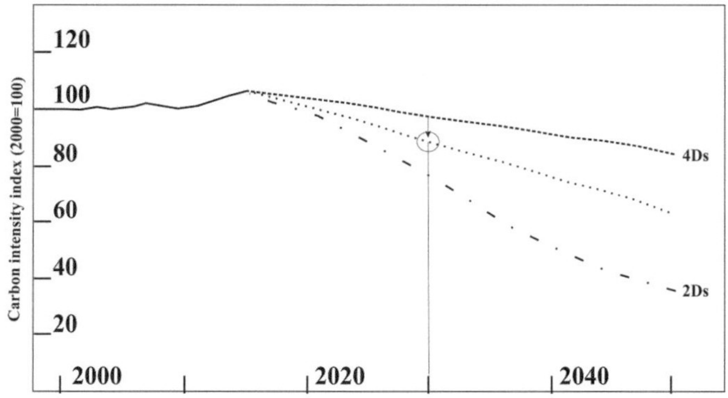

Fig. 3 Status of the carbon intensity curve in the 4D and 2D scenarios (Chandrasekharam et al., 2022; IEA, 2021)

Reducing emissions (Fig. 3) and controlling the temperature rise below 1.5 °C (Fig. 2) are the primary concerns of the countries and the oil industry. Driving the energy sector to the Net-Zero Emissions scenario (NZE) (IEA, 2021) needs a tremendous effort by the energy sector. Although the renewable energy sector is making efforts to increase the primary energy mix using off-shore wind and floating solar PV panels, the major contribution by these sources does not exceed greater than 20% of the total primary energy supply (Fig. 4) (IEA, 2021). The wind and solar PVs have very low efficiency and can support baseload supply. Manufacturing of solar PV cells is an energy-intensive process that depends on fossil fuels. The amount of CO_2 emissions during this process is enormous (2020).

In the case of solar PV, quantity of quartz required to manufacture 1 MWe of electricity is about 10 tons. The quartz is transformed to metallurgical grade first and then to electrical grade before cells are made. The total amount of CO_2 emissions during these two stages is 33129×10^6 kg. This amount is far higher compared to emissions from geothermal source which is about 450 g/kWh (2020). Hence, this source is not considered in this paper.

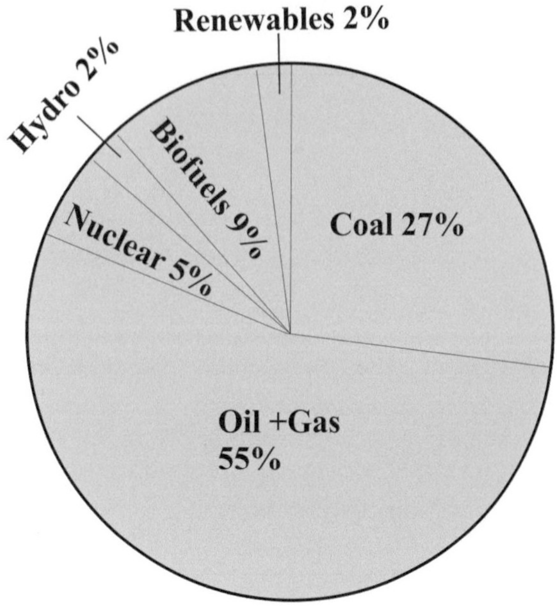

Fig. 4 Energy sources from IEA (2021) report

The contributions of conventional energy and renewable energy (geothermal) in the world development scenario are shown in Table 1 (Adapted from Shanmugam, 2023). It is apparent from Table 2 that geothermal energy is the best renewable energy option available to follow fossil fuels in the world's energy source supply compared to wind and solar PV as these two sources have low efficiency and have no major impact on the global CO_2 emissions (Shanmugam, 2023).

Global CO_2 emissions have no significant impact from such small contributions from low carbon energy sources like wind and solar PV (Shanmugam, 2023).

The current impression about CO_2 emissions, global temperature rise, and climate-related issues is that these events are being caused by using fossil fuels and solutions are being sought to substitute the usage of fossil fuels by low carbon-free energy source like wind and solar PV. But as seen from Table 2, these sources cannot support large demand and their efficiencies are too low and involve large investments. For example, the other mitigation strategy to reduce CO_2 emissions is to store CO_2 underground in oil reservoir or induce water–rock interaction between CO_2 and the rock-forming minerals and trap the CO_2 as calcite. This technology is known as carbon capture and storage or carbon sequestration. These techniques are feasible at megaton scale, the current projects storing 30 Mt of CO_2 per year. But to store CO_2 at a scale to control climate change issues (gigatons of storage) is economically and commercially not developed (Krevor et al., 2023). This involves a clear understanding of fluid dynamics associated with underground geological strata and storage management, seismicity associated with large quantity of storage (Vilarrasa & Carrera, 2015) and development of multi-stage and multi-basin scale models. In order to increase the contribution by renewable energy, especially geothermal,

Table 2 Energy sources and their contributions (although wind and solar PVs are also contributing their efficiency, percent contribution is low and they are not the focus of this paper) (adapted from Shanmugam, 2023)

No	Property types	Fossil fuel	Geothermal
1		Oil Natural gas	
2	Percentage of world energy	80	20
3	Supports infrastructure and agriculture	Yes	Yes
4	Petroleum products	Byproducts critical for modern living	Critical energy transition minerals like Li are byproducts
5	Energy density	Very high	Very high
6	Occurrence	Subsurface	Subsurface
7	Reliability	Very high and continuous supply	Very high and continuous supply
8	Energy storage	Cheap and efficient	No need to store-99% online with 98% efficiency
9	Emission of CO_2	Oil: 817 kg/MWh	0.893 kg/MWh
		Gas: 193 kg/MWh	
10	Damage to environment	Yes	None
11	Research funding	Low	Low
12	Availability	Unlimited	Greater than fossil fuels
13	Ultimate effect on NZE policy	Not compatible	Compatible

only the oil industry has the capacity to increase the geothermal output as well as maintain the fossil fuels' contribution to the primary energy source. This process to some extent contributes to the NZE concept since the CO_2 emissions from the fossil fuels are balanced by geothermal energy by the same industry and the unused wells are brought back to use generating low carbon energy.

The only source that has high efficiency and supports baseload electricity supply is the earth's heat available in the form of geothermal energy sources, available on land as hydrothermal sources or as radiogenic heat locked up in subsurface rocks such as granites and basalts. Oil and gas wells do encounter such thermal regime in the reservoirs, in the form of hot water and heat. These heat sources are ignored in the past by the oil industries due to the false impression that these thermal sources are not profitable to explore and exploit. Thus, many oil and gas wells that are dry and/or non-producing are abandoned. Even millions of such wells lie unutilized, and the number is ever-increasing.

However, exploration and exploitation of geothermal energy are not new to the petroleum industry. Several abandoned and non-productive oil and gas wells are being converted to geothermal wells, and considerable clean energy is being supplied to the consumers in the recent years (Bu et al., 2014; Caulk & Tomac, 2017; Davis & Michaelides, 2009).

In this paper, the mechanism of extracting geothermal energy from abandoned and non-productive productive wells is discussed.

Besides this routine business of converting the abandoned wells to profitable geothermal wells, oil industries have much bigger role to play in the world energy transition scenario and as leader in providing low carbon energy source for the next few decades.

This is possible by tapping the thermal energy from the mid-ocean ridges and oceanic islands. In this regard, a new innovative technology to harness geothermal energy by the oil industries from active mid-ocean ridge flanks and ocean islands is presented. The location of sites for such exploration and exploitation of this clean and CO_2-free energy are highlighted in this review. The amount of power that can be generated from ridge flanks and islands is higher by several orders compared to regular geothermal power generation plants operating in the world today, with efficiency greater than 90%. These geothermal sites can provide supercritical temperatures (375 °C) for power generation. The drilling technology being adopted by the oil industries can very well be utilized to harness this source of clean energy for sustainable development.

2 Existing Technology to Develop Abandoned or Low-Producing Oil Wells

The majority of oil and gas reservoirs, distributed over wide geographic locations (Fig. 5), host compressed high-temperature fluids that are responsible for oil to migrate to the surface through the production well. Since these reservoirs are not being recharged, over some time due to the extraction of oil and gas, the pressure in the reservoir decreases, leaving considerable oil in the reservoir. This becomes a problem for the industry due to the lowering of profit and increasing cost to maintain the well. Thus, due to reservoir management issues non-productive and low-productive wells get abandoned.

The present number of abandoned wells in the world is over 30 million (Caulk & Tomac, 2017; Nian & Cheng, 2018) and in the US alone this number is exceeding 3 million (Fig. 6). Abandoned wells, in general, are owned by a registered oil industry while orphan wells are drilled by private owners, and these wells have no official records. When such wells are non-productive or non-profitable, the owners abandon them.

The most serious problem with abandoned and orphaned wells, in the present context of climate change, is methane emissions (Brandt et al., 2014; Kang et al., 2019a, 2019b; Saunois et al., 2016; Townsend-Small et al., 2016). While CO_2 and methane are greenhouse gases (GHG), methane has 34 times more Global Warming Potential (GWP) compared to carbon dioxide (IPCC, 2013; Saunois et al., 2016;

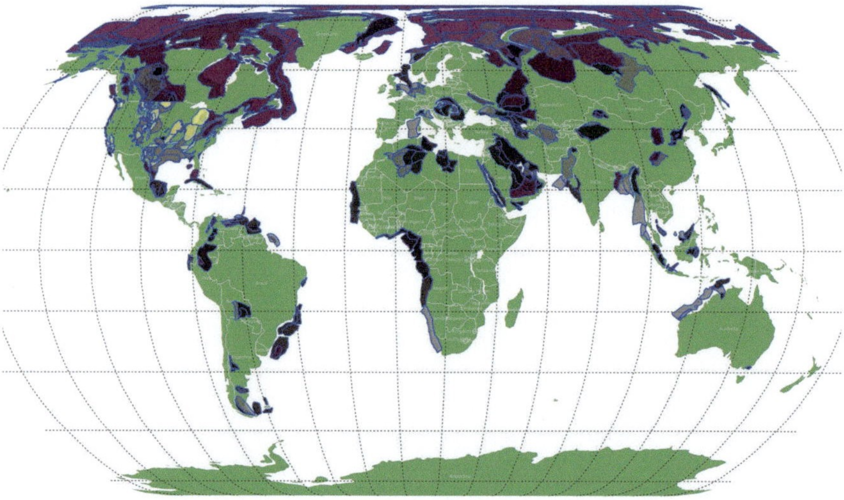

Fig. 5 Distribution of oil and gas fields in the world (Black: Oil fields; Purple: Gas fields. Adapted from Noorollahi et al., 2022, https://databayou.com/oil/andgas.html)

Fig. 6 Number of abandoned and non-production oil and gas wells in USA (adapted from Boutot et al., 2022)

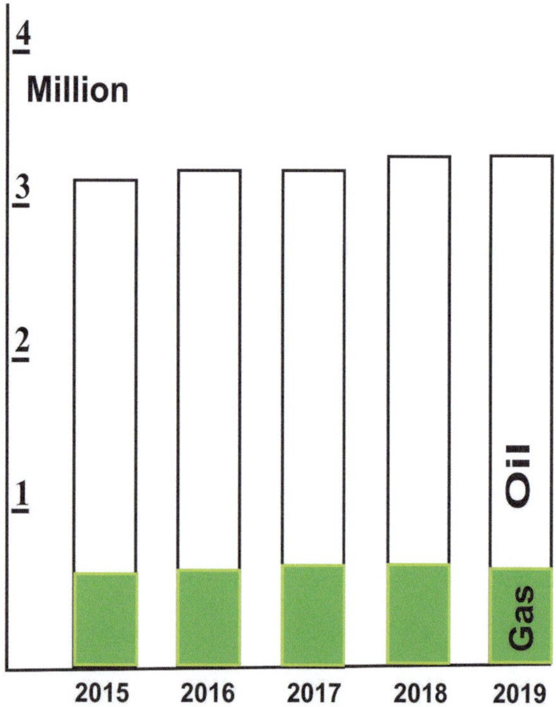

Shindell et al., 2012). The amount of methane emissions varies from 90 to 3000 kg/y/well as shown in Table 3. The cost incurred to rectify the environmental impact caused due to these uncontrolled emissions, also known as the social cost of methane emissions (SCM), is considerable and is US$ 5/kg of methane emissions (Kang et al., 2019a, 2019b). Since these emissions are continuous for decades, they can be used as a source of clean energy. This is the cost incurred to plug the well, and the controlled emission has been estimated using the following equation:

$$c_p(\text{well}) = c_{p,t} \cdot m_t$$

where $m_t =$ is the ton of methane mitigated over a time (t) period per well, C_p is the cost of plugging the well (Kang et al., 2019a, 2019b).

The quantity of power that can be generated from an average 200 kg/y/well of methane (Table 3) is about 6 kWh/m^3 of methane.

There are two ways to avoid methane emissions. One is to flare the gas or use it as direct energy. Flaring cost is about US$ 219/ton of gas flared while the gas can be supplied to domestic and industrial establishments for heating and for other applications, and the supply cost can be recovered (piping, installation, etc.) and the wells can be made as profit-making wells. This is much better than plugging the wells to arrest methane leakage since part of the capital cost incurred in drilling the well can be recovered by using it as a source of thermal energy for domestic and industrial purpose.

Plugging is not cost-effective. Plugging these wells costs between US$ 10,000 and 1,000,000 per well (Karakurt et al., 2012; Muehlenbachs, 2015; Yusuf et al., 2012) and this cost will be much higher for off-shore wells. The technology and mitigation strategies are not viable for implementing agencies in the past.

There are, however, other wells that are abandoned due to poor yield and/or the well are dry. The wells become non-profitable when the yield declines. Unrecoverable oil left in the reservoir is always of the order of 20% of the reserve which is quite a substantial quantity. The yield declines due to a drop in the reservoir pressure. The

Table 3 Methane emissions from abandoned oil and gas wells (average 200 kg/y/well excluding 3000 anomaly)

	Average methane emissions kg/y/well	Region	References
1	190	Pennsylvania	Kang et al. (2015a)
2	30	West Virginia	Riddick et al. (2019)
3	360	UK	Boothroyd et al. (2016)
4	90	Colorado	Townsend-Small et al. (2016)
5	150	Pennsylvania	Etiope et al. (2013)
6	258	Pennsylvania	Kang et al. (2016b)
7	3000	Pennsylvania	Kang et al. (2016b)
8	300	Pennsylvania	Kang et al. (2016b)

bottom hole temperature of the dry wells is sufficiently high that they can be used for extracting the heat for power generation or for direct application.

Over a period of time, due to the technological development between both geothermal and oil industries, these abandoned wells are now being utilized for power generation. This is a win–win situation for both oil industries and the green energy sector. Since the cost of drilling can be partly recovered by the oil industry, geothermal sector will save more than 50% of its drilling cost (Anna et al., 2020; Bu et al., 2012, 2014; Nian & Cheng, 2018).

Extensive investigations have been carried out to evaluate the cost–benefit ratio of converting the abandoned wells to generate geothermal power as well as extraction of heat for other direct applications using numerical models (Kharseh et al., 2015; Kujawa et al., 2005, 2006; Mokhtari et al., 2016; Nian & Cheng, 2018). Technology to convert abandoned oil wells to geothermal wells, design of the wells, configuration, temperature and pressure conditions and the amount of electricity and heat that can be generated have been discussed, and several numerical models have been published by several workers.

Thus, there are several options available to convert abandoned oil and gas wells into active geothermal wells. Extensive thermo-numerical models have been published to evaluate the performance of such converted wells (Noorollahi et al., 2022). A few of the technologies being adopted are shown in Table 4 and Fig. 7.

Thermal, economic, and exergy analyses of power generation from abandoned oil and gas wells, using geothermal Organic Rankine Cycle (ORC) technology, indicate that 3 KW to 6 MWe and greater than 4 MWt can be obtained from a single well (Table 4). The average electricity cost from such power plants varies from 3 to 5 US cents/kWh (Cheng et al., 2014; Kharseh et al., 2015; Mokhtari et al., 2016) with a payback period of 4 to 5 years (Kharseh et al., 2015). One of the best uses of electricity generated from abandoned oil and gas wells, using ORC technology, is to generate freshwater through the desalination of seawater to support food and water sectors, especially in regions, such as the Gulf countries where water and food imports are very high (Chandrasekharam et al., 2018b; Noorollahi et al., 2022). Desalination using geothermal energy is very cost-effective and avoids climate change issues since the CO_2 emissions are far less (Chandrasekharam, 2018, 2021; Chandrasekharam et al., 2018a, 2018b; Chandrasekharam et al., 2016a, 2016b). Even if 50% of the 30 million abandoned wells are retrofitted, the minimum power that can be generated is on the order of 150 million MWe. One can calculate the amount of CO_2 that can be saved and the cost that can be recovered from these wells (see the CO_2 emissions by oil-fired power plants in Table 2).

Now that the technology and economics are well established in converting abandoned and orphaned wells to geothermal wells, off-shore oil companies can look for high-potential geothermal sources and sites over the ridge flanks of the spreading mid-ocean ridges. The amount of energy from geothermal systems associated with the spreading mid-ocean ridge thermal regime is enormous as evident from the Iceland geothermal systems, that is generating 754 MWe (Hutter, 2021).

Although there are logistic and technological issues to access the mid-ocean ridges like the Mid-Atlantic, Mid Indian, and East Pacific rise, there are other younger

Table 4 Technologies are being adopted to extract geothermal power from abandoned wells

	Country	Field	Well depth m	Extraction type	Method of analyses	Bottom hole temp °C	Power output KW	References
1	Poland	Jachówka K2	2870	Double pipe heat exchanger	Field and experimental	80	255	Kujawa et al. (2005, 2006)
2	China	Shuguang	600	In situ combustion	Field and experimental	200	6000	Zhang et al. (2008)
3	USA	Texas	3000	Double Pipe heat exchanger	Computational	140	3	Davis and Michaelides (2009)
4	China	Huaibei	4000	Heat exchanger	Numerical	180	5	Bu et al. (2012, 2014)
5	China		5000	Double pipe heat exchanger	Transient heat numerical model	150	135	Cheng et al. (2014)
6	Italy	All fields		Double pipe heat exchanger	Statistical analyses	135	800	Alimonti and Soldo (2016)
7	China	Shengli	3000	Annulus tubing	Mathematical model	125	180	Nian and Cheng (2018)
8	World	Review		Double heat exchanger	Numerical modelling	140	3.4	Anna et al. (2020)
9	India	4 fields	2400–5500	Heat exchanger	Statistical analyses	150–285	100–500	Singh (2020), Singh et al. (2018)

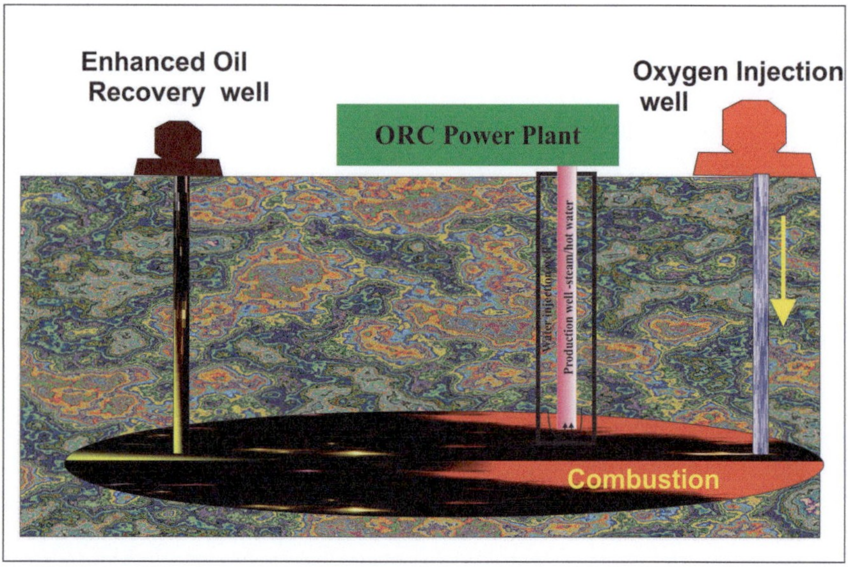

Fig. 7 Schematic diagram showing the technology to recover oil (enhanced oil recovery) and heat (geothermal) from non-productive oil wells. Heat exchanger and combustion methods are shown

spreading ridges, like the Red Sea spreading ridge, that are close to the continental margins and are easily accessible by the ocean drilling oil rigs. Other sites that the oil rigs have access are the volcanic islands. The volcanic island with high heat flow and thin oceanic crust has high geothermal potential that can be harnessed by the oil rigs.

2.1 ORC Plants

ORC is a vapor-based closed-loop thermodynamic cycle that can generate electricity using hot water. The feed in sources to the system are geothermal waste heat from thermal and cement manufacturing plants, waste incinerators. The heat from the liquid is absorbed by a heat exchanger (H) delivered to a secondary fluid that has very low boiling point, like any organic liquid (Fig. 8). In the past, Freon was in use as a secondary fluid but due to environmental concern this has been replaced by a low boiling point fluid that is not harmful to the climate system. Here heat is converted into mechanical form (turbine T) that generates electricity (G). After the work cycle, the organic vapor condenses in the condenser and returns to the heat exchanger. Thus, it works on closed-loop technology. The geothermal fluid is returned back into the reservoir using an injection well (Di Pippo, 2012). Rankine cycle is a well-known technology used in steam engine that works on steam but in ORC an organic fluid is used to convert heat into mechanical and then to electrical energy. The ORC turbines,

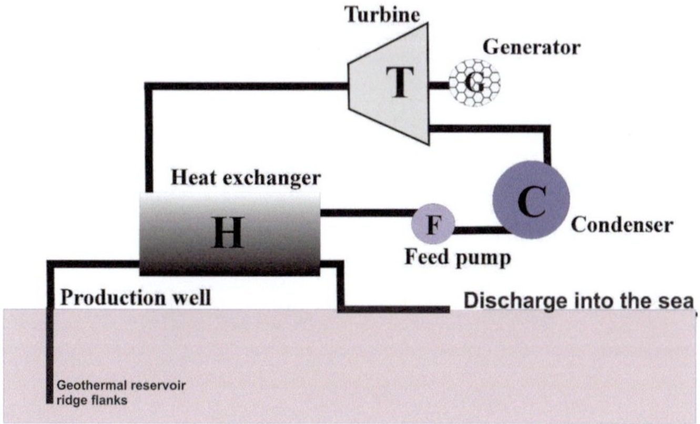

Fig. 8 Schematic diagram showing a simple binary geothermal power plant

as they are popularly known, are best suited for stand-alone power generation, for distributing electricity using microgrids. A majority of geothermal power plants in the world use ORC technology (Di Pippo, 2012). These are best suited for remote places with remote operations and are capable of generating 1 to 5 MWe with 95% efficiency in all weather conditions (Di Pippo, 2012).

3 New Innovative Technology to Harness Geothermal Energy from Ridge Flanks

Ocean environment is a repository of geothermal systems with high heat flow values over the mid-ocean spreading ridges and oceanic islands. The heat flow values are high in the ocean environment relative to the continents (Fig. 9). This is due to the shallow disposition of the upper mantle, where melting of basalts occurs, in the ocean environment that varies from 1200 to 1450 °C. Thus, the 1200 °C isotherm is located at shallow regions in many younger volcanic islands, and hence, the heat flow values are higher with younger ocean crust (Fig. 10).

Most of the islands in the Pacific Ocean and young spreading ridges like the Red Sea register very high heat flow values (see Fig. 10). The presence of high-temperature hydrothermal systems over the flanks of Canary Islands has been reported.

These are the regions suitable for oil industries to tap this carbon-free energy and be the leader in energy supply and energy transition in the world.

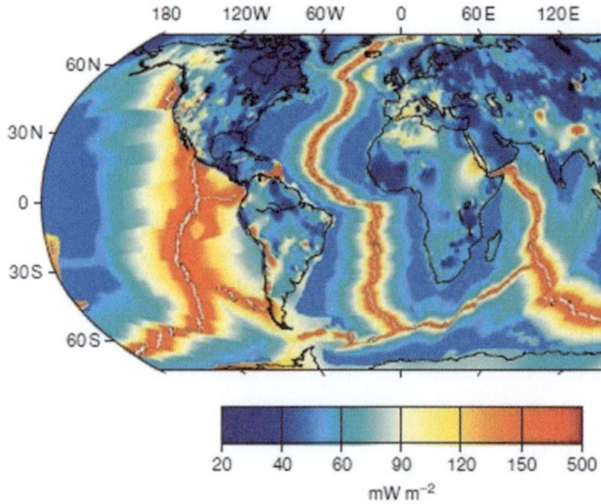

Fig. 9 Heat flow values over the world. The heat flow values over the ocean ridges and oceanic islands are much higher compared to the rest of the land and ocean masses (Juapart et al., 2015)

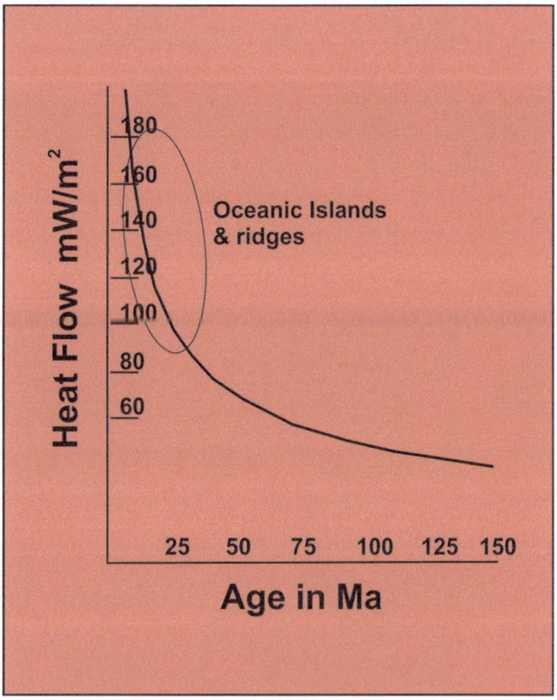

Fig. 10 Age versus heat flow in the ocean environment. Younger volcanic islands have higher heat flow values relative to the older islands (Juapart et al., 2015)

3.1 Red Sea Geothermal Regime

The idea of targeting the Red Sea flanks emerged during the world debate on how to support the carbon-free energy demand for the future dream city/world, the NEOM. The doctrine of NEOM states that energy being supplied to this dream city should be renewable energy free from a carbon footprint. Only geothermal energy qualifies for this condition, and hence the Red Sea flanks are targeted (Chandrasekhar & Chandrasekharam, 2023). Mid-ocean ridges (Atlantic, Indian Ocean, and East Pacific Rise) are loci of extremely high geothermal regime, with surface temperatures rising to 1300 °C (Fig. 11). Actual observation and measurements were made in Iceland which is located over the northern part of the Mid-Atlantic Ridge (MAR). The 1300 °C isotherm (melting temperature of basalts at atmospheric pressure) is located at a depth of about 20 km over the flanks of MAR (Birch & Halunen, 1966, Flovenz & Saemundsson, 1993; Lucazeau et al., 2006). This thermal regime helped Iceland to establish a geothermal-based country. Currently, Iceland is utilizing geothermal energy to generate 755 MWe and 2367 MWt for direct applications like greenhouse cultivation, dehydration, space heating and cooling, snow melting, and swimming pools (Ragnarsson et al., 2021) (Fig. 11).

But not all countries are located over the flanks of the spreading ridge like Iceland. However, many countries, located near an active spreading ocean ridge, have access to such geothermal sources. One such young, active spreading ridge is the Red Sea Ridge (Fig. 4).

The Red Sea rift is the youngest spreading rift basin in the world, active since 30 Ma (Almalki et al., 2015). Its magmatic activity is represented by sheet dike

Fig. 11 Geothermal power plant on the flanks of the Mid-Atlantic Ridge Iceland (modified after Flovenz & Saemundsson, 1993)

intrusions (G in Fig. 4) that are 40,000 years old and hot brine pools in the valleys adjacent to the flanks. The minimum distance between the ridge flanks and the continental margins (sea coast) from Saudi Arabia and Djibouti, Eritrea, is 150 km (Chandrasekharam et al., 2016a, 2016b). The length of the ridge is 2000 km from the Gulf of Aden to the southern tip of the Sanai Peninsula, and the depth of the ridge flank from the sea surface is about 500 m (Fig. 12) (Allan, 1970; Bosworth & Stockli, 2016; Bosworth et al., 2005; Rasul & Stewart, 2019).

The heat flow values over the entire ridge flanks vary from 86 to 205 mW/m^2 (the global average heat flow is 60 mW/m^2) (Girdler, 1970; Girdler & Evans, 1977; Makris et al., 1991; Sclater, 1966; Sclater et al., 1980) (Fig. 12). The entire Red Sea (from the flanks to the ridge) is much hotter than the world average heat flow of 60 mW/m^2 (Girdler, 1970; Sclater, 1966). This results due to upwelling of magma along the rift axis, represented by dike swarms, sheet dikes, and disposition of Moho at shallow depth below the flanks (8 to 14 km), as deciphered from geophysical investigations along the ridge flanks (Bosworth et al. 2019; Cochran et al., 1986, Salem et al., 2014a, 2014b; Schettino et al., 2019), and circulation of supercritical hydrothermal fluids (Fig. 12). Thus the thermal regime indicates temperatures of 350–370 °C over the ridge flanks at depth varying from 2.5 to 4 km (Chandrasekharm, 2023; Chandrasekhar & Chandrasekharam, 2023). Thus, borewells can tap the hydrothermal fluids and/or heat from the dry hot rocks at such depths (Degens & Ross, 1969). The temperature can be harnessed by circulating either water or carbon dioxide using a double-pipe extraction configuration (Fig. 13).

3.2 Technology to Extract Heat from the Ridge Flanks

Extracting heat from high heat-generating granites, known as Enhanced Geothermal Systems (EGS), involves hydrofracturing granites at depths varying from 3 to 5 km and circulating fluids to capture the heat from these granites for power generation (Tester et al., 2006). Induced seismicity is a common phenomenon during hydrofracturing granites in EGS. This problem has been solved in recent years by using loop technology that connects an injection well with a production well in the hot rock as shown in Fig. 14. Thermo-hydro-numerical computer model using COMSOL V6 software indicates that with an injection flow rate of 500/s and with a distance of 5 km between the injection and production wells (single loop), with a bottom hole temperature of 375 °C, the temperature at the production wellhead would be 275 °C (Chandrasekharam et al., 2023). This will be able to generate 58 MWe using Organic Rankine Cycle (ORC) technology.

In the case of ridge flanks, the rock type being basalt, it is easier to establish loop technology using oil drilling rigs. Over the flanks, the temperature at 2 km would be >200 °C. Borewells over the flanks either can tap the supercritical geothermal brines (Degens & Ross, 1969) or the temperatures can be extracted using loop technology. Each borewell tapping the supercritical temperature regime can generate 40 MWe

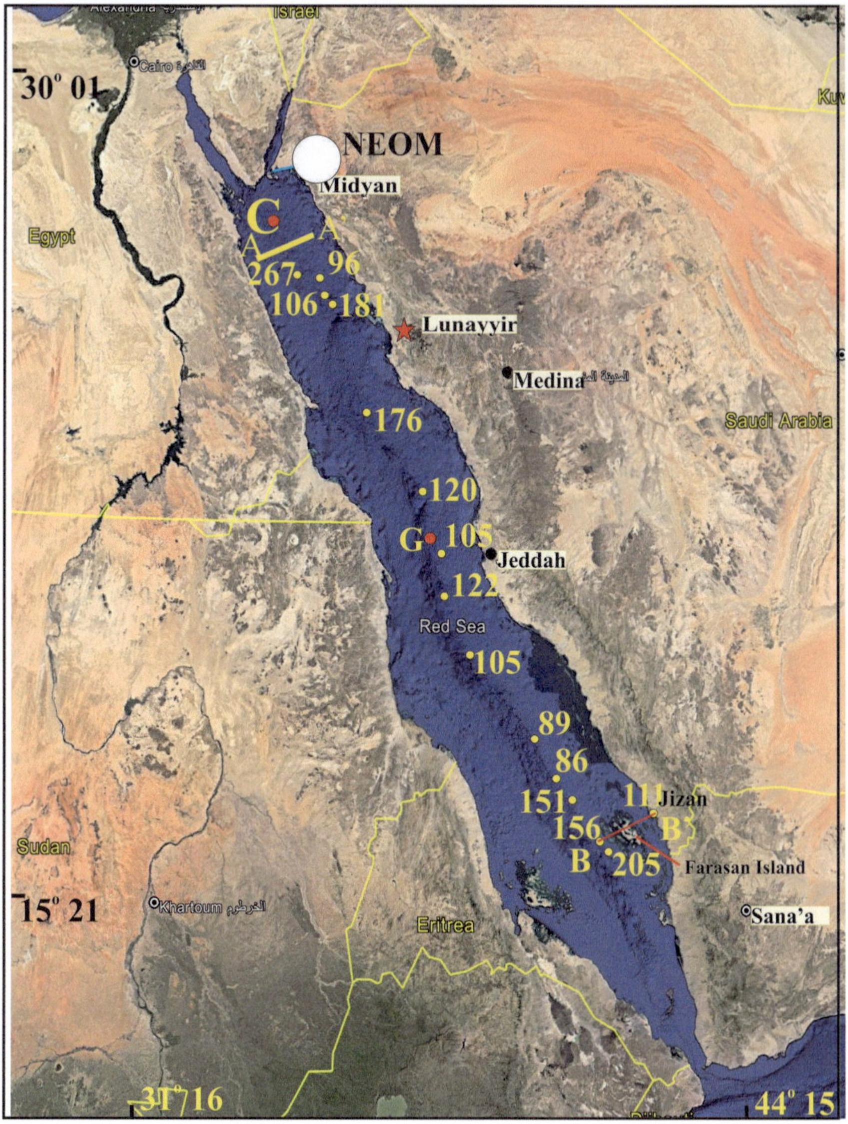

Fig. 12 Red Sea spreading ridge showing the measured heat flow values over the ridge flanks. B-B' cross-sectional profile is shown in Fig. 4. Numbers indicate heat flow values in mW/m^2. Lunayyir is an active volcano on the Arabian Shield adjacent to the Red Sea. C: Deepest part of the ridge axis-the Conrad deep; G: Gabbroic sheet intrusion site (adapted from Chandrasekhar & Chandrasekharam, 2023)

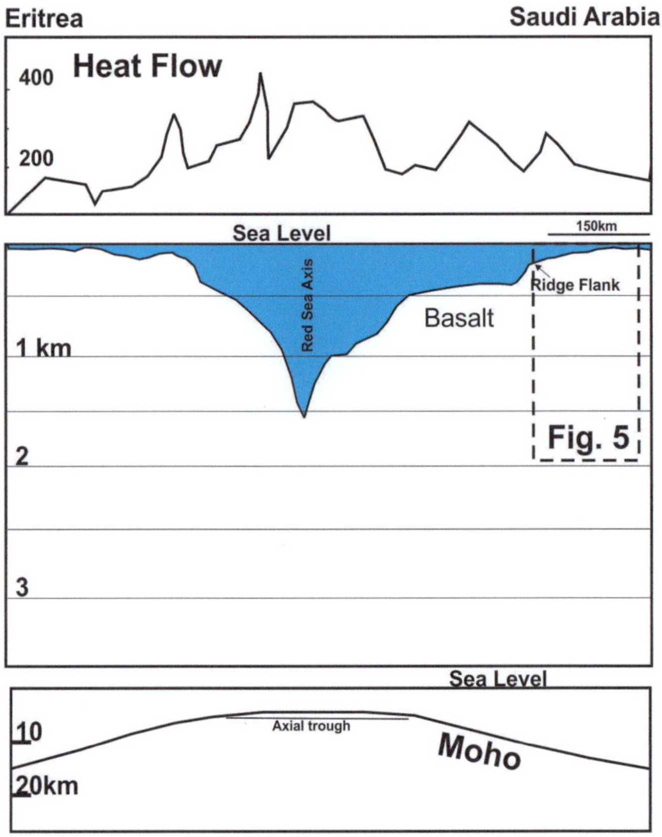

Fig. 13 Heat flow, ridge profile, and Moho position of the Red Sea rift basin (source adapted from Chandrasekhar & Chandrasekharam, 2023). Heat Flow units: mW/m^2; Scale on the left axis indicates depth below sea level. See Fig. 14 for dotted rectangle. The Profile is drawn across lines B-B' in Fig. 12. B' is over the Farsan Islands on the western coast of Saudi Arabia. The section along A-A' is similar to B-B' but the depth of the rift valley is >2 km

by circulating 60 kg/s of fluids using the loop technology (Cladouhos et al., 2016, 2018).

Oil rigs can extract this geothermal energy over the ridge flanks, as shown in Fig. 5. Since these rigs generate electricity, they are self-sustained and do not need an external energy support system. They can be stand-alone rigs.

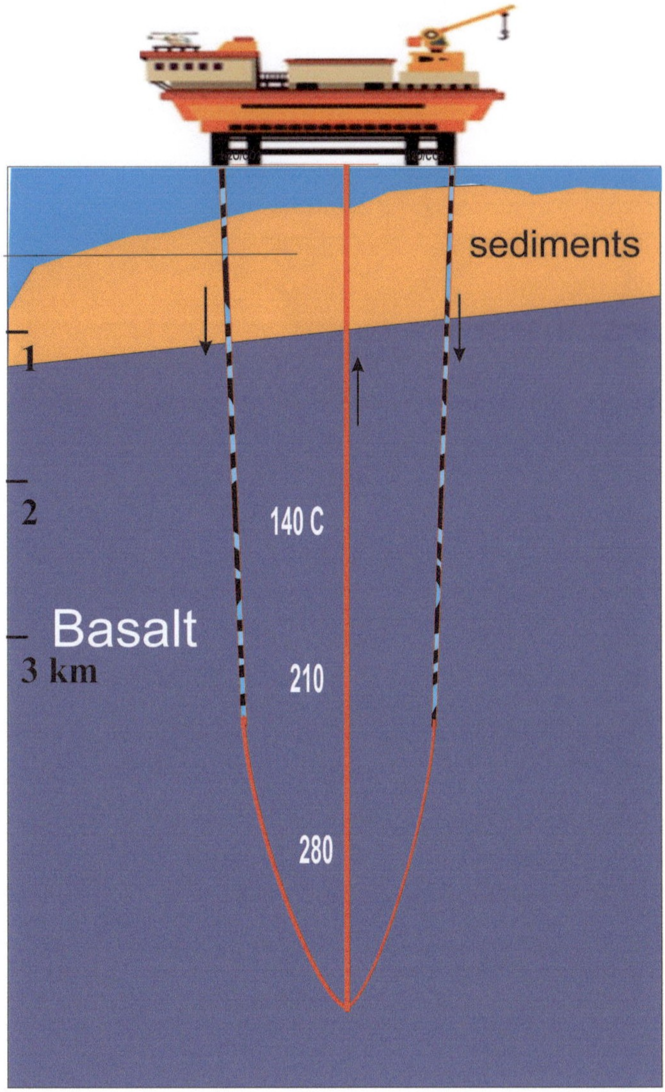

Fig. 14 Loop technology to tap geothermal energy from ridge flanks by off-shore oil platform. Axis scale: depth below the sea level. Oil drilling rigs can be anchored over the basalts (hard rocks) over the flanks and tap the thermal energy and convert it to electrical energy using ORC technology, described in Sect. 2.1

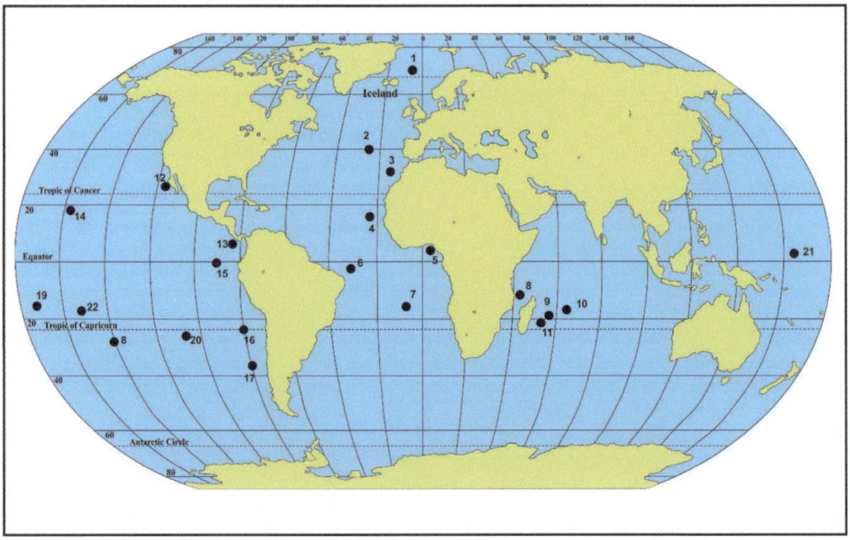

Fig. 15 Oceanic Islands of the world (modified after Humphreys & Niu, 2009). Each red circle includes several islets given in Table 5. All the islands, although associated with volcanic activity, are habitable

3.3 Technology to Tap Geothermal Energy from Ocean Islands

Oceanic Islands are formed due to submarine volcanic eruptions and hence are loci of high geothermal regimes. There are 108 islands in the Pacific Ocean, 56 in the Atlantic Ocean, and 25 in the Indian Ocean (Fig. 15) (Humphreys & Niu, 2009). A few of the most potential islands that are listed in Table 2).

Since these are volcanic islands, the heat flow is 100–250 mWm^2 (Indian and Atlantic Ocean Islands), 100–150 mWm^2, for islands in the East Pacific Rise (Harris & McNutt, 2007; Sclater et al., 1980). The technology to harness geothermal energy from these islands is similar to that described in Sect. 3.2. As an example, a few ocean islands that are prospective hot spot islands, concerning geothermal energy are described in the following section. These islands are associated with mantle plumes, where there is constant upwelling of molten lava from the mantle at the rate of 10^3 kg/s (Harvey & Harvey, 2010; Zhao, 2007). These plume locations are shown in Fig. 16. The important islands associated with such plumes are listed in Table 5.

In the Atlantic Ocean, extensive fumarolic activity over the Tenerife and Gran Canary Islands gives evidence of shallow geothermal systems over the flanks of the volcanic islands (Harvey & Harvey, 2010; Schmincke & Sumita, 1988). Considering the age of these islands (12 Ma, Schmincke & Sumita, 1988), the Curie depth of the

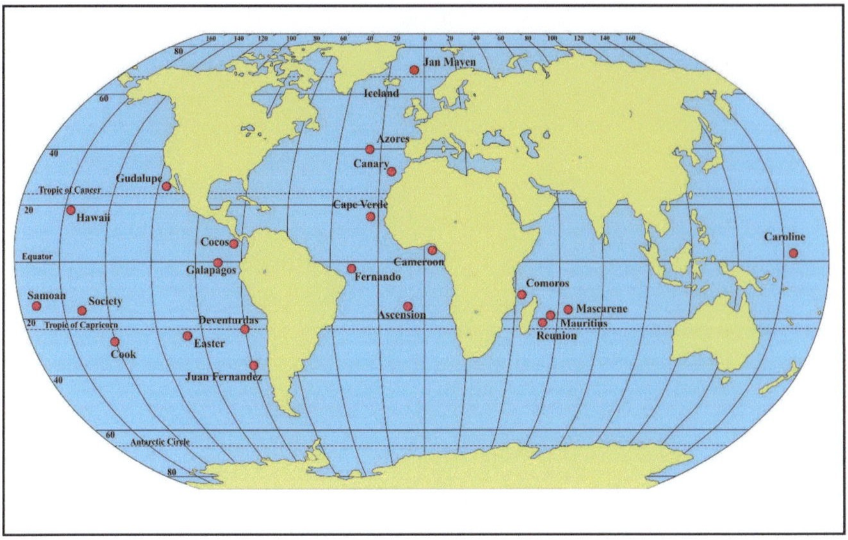

Fig. 16 Oceanic Islands associated with mantle plumes/hot spots (adapted from Zhao, 2007). The location of the plumes is associated with high heat flow as seen from Fig. 9. The area of high heat flow is larger in the Pacific Ocean compared to the Atlantic and Indian Oceans. This resulted in the evolution of several oceanic islands in the Pacific Ocean. Similarly, the Canary group of Islands in the Atlantic Ocean is associated with a regional high heat flow regime between the Atlantic Ridge and the continental rifts over NW Africa and SW of Spain and Portugal (Fig. 9)

Table 5 A few important oceanic islands are associated with mantle plumes Humphreys & Niu, 2009 (*Source*, R: Reference Island in Fig. 6. Number: Indicates the islets in the Island)

No	Atlantic Ocean	Number of islets	No	Indian Ocean	Number of islets	No	Pacific Ocean	Number of islets
1	Jan Mayen		8	Comoros	5	12	Guadalupe	
2	Azores	10	9	Mauritius		13	Cocos	
3	Canary	8	10	Mascarene		14	Hawaii	12
4	Cape Verde	10	11	Reunion		15	Galapagos	23
5	Cameroon	5				16	Desventuradas	3
6	Fernando de Noronha					17	Juan Fernandez	3
7	Ascension					18	Cook	12
						19	Samoan	4
						20	Easter	
						21	Caroline	10
						22	Society	10

Canary Islands is around 10 km (Klügel et al., 2020; Li & Wang, 2018) Richter et al. (1991), Saleh et al. (2012), Sui et al. (2018), Yang and Li (2014).

Global aeromagnetic analyses show that the Curie depth below all the islands in the Pacific Ocean falls within the range of 10 to 14 km and Curie depth of volcanic islands and seamounts younger than 10 Ma and those that are associated with active volcanism varies from 5 to 10 km (Li & Wang, 2018). Unlike the Canary Islands, the Curie depth below the older volcanic islands chain in the Atlantic Ocean, like the Cameroon Islands (30–65 Ma,) varies from 16 to 41 km (Njeudjang et al., 2020).

Similarly, the Pacific islands have enormous geothermal energy as evident from the presence of plumes and the shallow Curie depth (Li & Wang, 2018). These are potential sites that the oil industries can explore. The oil industries have the expertise and infrastructure to install off-shore platforms and have the drilling technological expertise. The depth of drilling is not large as shown in Fig. 17. As an example, as shown in Fig. 17, the drilling depths, below the water level, are shallow compared to the depth where DSDP deep drilling sites were located.

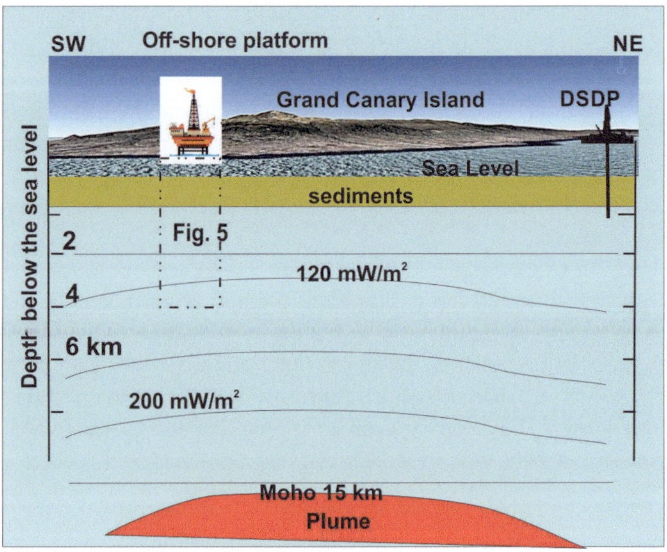

Fig. 17 Grand Canary Island, in the Atlantic Ocean, lies over a plume (see Fig. 16) and ideally located to tap the geothermal energy from the flanks (see Fig. 9 for heat flow value). The DSDP site was established to sample the ocean sediments from this island. Oil industry can establish drilling rigs over the flanks of the Island to extract the geothermal energy. The depth and heat available below the flanks are similar to that shown in Fig. 14. Geothermal site over the Grand Canary Island, Atlantic Ocean (adapted from Alonso et al., 2019, Heat flow values from Juapart et al., 2015)

4 Discussion

Oil exploration and production industries have the expertise in establishing off-shore drilling platforms and in directional drilling which is very essential to establish loop-EGS technology. The most suitable site to implement this technology by the oil industry is the Farasan Islands, located between the Red Sea flank and the Saudi Arabian coast and Conrad deep (267 mW/m^2) located on the northern part of the Red Sea rift axis (Fig. 4). The heat flow value below the island is 156 mW/m^2 with Moho placed at a depth of 15 km. Critical temperature is expected at 2 km and 40 MWe (equivalent to 316 billion kWh/y) can be generated from this site by circulating 60 kg/s fluid (Chandrasekhar and Chandrasekharam, 2023; Cladouhos et al., 2018). Initially, three wells at these two sites can generate over 240 MWe (3 wells at each site, about 2 billion kWh) from 6 wells. This can be extended to the entire 2000 km long flank area subsequently. While CO_2 emissions from fossil fuels are 0.439 kg/kWh of power generated, the CO_2 emissions from these EGS geothermal sites are negligible. The ORC technology is very useful in generating power from the abandoned oil and gas wells and from the million tons of methane that is being flared. The oil industries thus can flip the coin and become net-zero carbon emitters by implementing EGS geothermal technology in abandoned oil well sites and from the ridge flanks and oceanic islands. The energy from these sites can be utilized for generating electricity, or the heat can be utilized for direct applications like desalination, greenhouse cultivation, and dehydration of agricultural produce. Since water security is needed for many islands, the electricity is thus generated from the geothermal source.

About 16 kWh of electricity (sourced from fossil fuels) is needed to generate 1 m^3 of freshwater through the desalination method (Ghaffour et al., 2014), and the desalinated water is currently sold at US$ 6/m^3 (Chandrasekharam et al., 2022; Ghaffour et al., 2014). At each off-shore platform, about 19 billion m^3 of desalinated water can be generated. These off-shore platforms will be able to generate electricity as well as freshwater. The electricity can be supplied to energy starved Middle East countries (those countries that are dependent on imported fossil fuels, e.g., Jordan, Syria) through the geothermal corridor (Chandrasekharam, 2023) or this low-carbon energy-generated freshwater can be supplied water-starved countries like the GCC and Middle East countries. World over fossil fuel-based desalination plants are in operation, the maximum being in Saudi Arabia (Chandrasekharam, 2018, 2022). Although the cost of fossil fuels and geothermal sourced desalinated water is similar (Table 6), there are no hidden costs in geothermal sourced desalinated water.

In the case of islands, this EGS power technology in conjunction with the generation of desalinated water will be a great support for sustainable development for all the islands shown in Fig. 15 and will increase their GDP. The Pacific Island countries have always been dependent on Asia, Europe, and the Americas over the last several decades. However, in the last two decades support from Europe and the Americas dwindled and the PIC had to depend on the economy of Asia. Papua New Guinea,

Table 6 Cost of desalinated water using fossil fuel and renewable sources of energy (adopted from Chandrasekharam, 2018, 2022)

Energy source	Cost/m^3 in US$
Fossil fuels	1.6
Solar PV	6.6
Geothermal	1.6

Fiji, and Solomon Islands are mostly populated with a total population of about 9 million (Juswanto & Ali, 2016).

Agriculture and tourism constitute major revenue for the governments of the PIC but over a period of time, these countries are not able to increase their GDP for several reasons. The main reason is the import of fossil fuels. The cost of imports accounts for 15% of the countries' GDP (ESCAP, 2022). Because of high energy costs, the economic growth of the population is highly restricted. There are no cold storage facilities, for example, to preserve food, to use modern electronic equipment to access wide information, and access to wider financial services (ADB, 2020; Dornan, 2014; Lal & Kumar, 2022).

Even though the Asian Development Bank provides financial support to the electricity sector (604 million US$ between 2007 and 2020, ADB, 2020), most of the support is spent on imports of energy. Financial aid of 700 million US$ is available for the energy sector in the current year, unless this amount is spent on building self-sustained energy sector, this amount will again go into fuel imports.

Over the last several decades, fossil fuels were the primary energy sources for all the Ocean Islands (for the world as well), especially for the Pacific Ocean islands (Lal & Kumar, 2022). Due to the global climate change issues being experienced by the countries, renewable energy sources are developed intensely over the last decade and these sources are able to contribute to the primary energy source mix. However, due to low efficiency and issues related to baseload supply of electricity and power storage issues, especially the wind and solar PV sources (Chandrasekharam & Ranjith, (2020); Lal & Kumar, 2022), the contribution of the above energy sources to the primary energy source is limited. However, geothermal energy, due to its ability to support baseload electricity supply, with high efficiency, as discussed in the previous section, is being considered as a potential source to replace a large quantity of fossil fuels as primary energy source, especially in the island countries. Adopting the technology described in this paper, the oil industry can bring a huge energy transition, especially in PIC and make significant contribution to their development and substantially enhance their GDP (Lal & Kumar, 2022). Thus the oil industries can bring in an energy transition by promoting geothermal energy in the PIC by the year 2024, as shown in Fig. 18.

The Pacific Island Countries (PIC), in conjunction with the oil industries, should build a sustainable energy source by tapping the geothermal energy available around the countries as discussed above by using this financial aid. The volcanic islands can support the generation of a minimum of 40 MWe geothermal energy on the

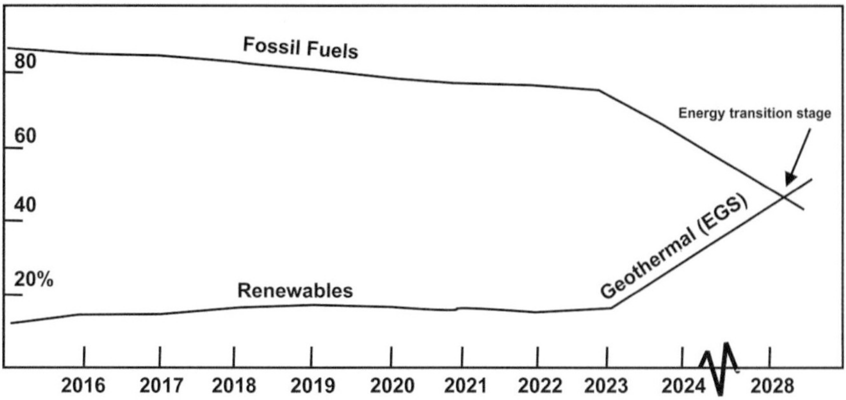

Fig. 18 Primary energy supply. Energy transition period using EGS (adapted from Lal & Kumar, 2022)

islands listed in Table 2 initially with the financial support from ADB and increase the contribution of renewable energy in the primary source energy supply and reduce the imports of fossil fuels. This will potentially close the renewable energy gap in the next decade (Fig. 18).

The Pacific Island Countries (PIC), though are not major GHG emitters, are the victims of climate change calamities caused due to emissions from other economically developed countries. The per-capita carbon dioxide emissions from the PIC (including those in the Indian Ocean) were about 3 tons, in 2018, and the ratio of the world to PIC emissions ratio is 0.03, which is negligible. The oil industries can contribute to their economy by providing geothermal energy and supporting their GDP which will help the population manage climate-related issues caused by the developed economies (Lal & Kumar, 2022; WB, 2021; Brandt et al. (2016).

The contribution of oil industries to energy transition is immense. These industries can become carbon savers from the current carbon emitters. Each off-shore platform over the ridges and the islands can generate a minimum of 40 MWe, as discussed above, and this amount in the future can be increased making geothermal as the main primary energy supplier to the world. The greatest beneficiaries are the island countries, located in all three oceans, with abundant food-energy water for sustainable development and enhancing their GDP.

Over decades, the fuel imports-GDP-of these countries have not changed. Although investments in renewable energy relative to fossil fuels exceeded over the last five years (Fig. 19), the percentage of primary energy source is always been the fossil fuels (Fig. 18). With the implementation of this new innovative technology of harnessing energy from the ocean ridge flanks and oceanic islands, described in this paper, there will not be a divide between fossil fuels and geothermal energy and oil industry will become the sole energy contributor to the world and supply will be balanced between the geothermal and fossil fuels.

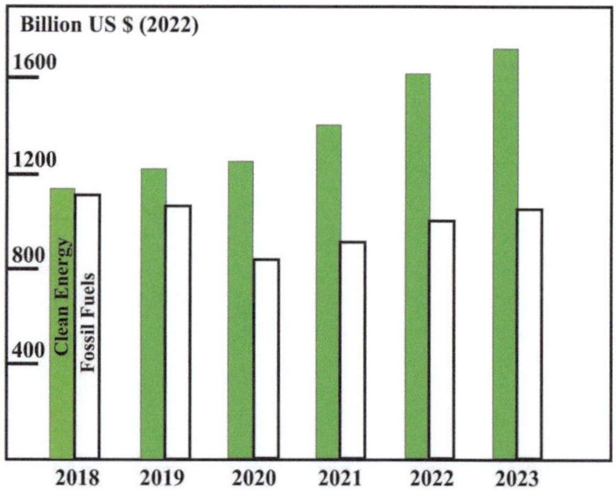

Fig. 19 Global investment in renewable and fossil fuels energy (IEA, 2023)

Thus, with this new technological innovation, oil industries can reverse this scenario in the coming decade. This is a win–win situation to both the parties. Oil industries can expect their dominance over energy supply forever. Off-shore oil and gas platforms thus play an important role in the global energy sector. Recent technological developments, design structure, hydrodynamic analyses using theoretical models, numerical simulations, and experimental tests have provided solution to several technical issues faced by the off-shore platforms. In addition, improved flexible pipelines and cable system for remote monitoring of the platforms have been added to this technological development (Wen-hua et al., 2022).

5 Conclusions

The new innovative technology to tap geothermal energy from the ridge flanks and island volcanic will bring a revolution to the oil sector in energy supply and security. The off-shore platforms along the flanks of the Red Sea axis can generate both desalinated water and provide water security to all the Red Sea countries as well as to the countries along the geothermal corridor. A minimum of 100 to 150 off-shore platforms each with generating capacity of 40 MWe can be installed, providing about 1900 to 2500 billion m^3 of desalinated water to the water-starved countries on the eastern side of the ridge flank. In the case of PIC and other islands in the Indian and Atlantic Oceans, electricity and freshwater (if needed) can be provided to reduce oil imports. This will provide food-energy-water to the countries around the Red Sea as well as Middle East countries. Similarly, the PIC as well as other islands in the Indian and Atlantic oceans can become energy independent, reduce oil dependency, and enhance their GDP. Thus, in the future oil industries can play a great role in climate change issues over the entire work. The oil industries, by adopting this new technology, can extend their operation of the non-productive wells, without decommissioning them and add new dimension to the energy scenario by opening several off-shore platform to harness the geothermal energy both from the ridge flanks and the ocean islands. All the 77 islands in the Pacific Ocean can support generation of about 3080 MWe, assuming 40 MWe from each off-shore platform located on the island flanks. This energy can be utilized to support food-energy-water security to all the islands, and in addition, the energy can also be utilized to generate other green energy sources like hydrogen.

Acknowledgements The author thanks TUBITAK for its support (under project No: 120C079) in the preparation of this paper.

Abbreviations

ADB	Asian Development Bank
CO_2	Carbon dioxide
°C	Temperature in degrees Celsius
COMSOL	Computer program for multi-physics simulation of reservoir
DSDP	Deep Sea Drilling Project. A multinational ocean research drilling program to study the oceans
EGS	Enhanced Geothermal Systems. Artificially made geothermal reservoir in hot rocks devoid of water/fluid circulation
ESCAP	Economic and Social Commission for Asia and the Pacific
GCC	Gulf Cooperation Council
GDP	Gross Domestic Product: monetary measure of the market value of all the final goods and services produced in a specific time period by a country or countries
GHG	Greenhouse Gases
GWP	Global Warming Potential (global warming impact between two greenhouse gases)
IEA	International Energy Agency
KWe	Kilo Watt electricity
kWh	Kilo Watt-hours
km	Kilometer
MAR	Mid-Atlantic Ridge (plat boundary where new crust is generated). Region of highest heat flow in the ocean
Ma	Million years
MWe	Mega Watt electricity
MWt	Mega Watt of heat from geothermal systems
mW/m^2	Amount of heat transmitted from the earth over an area of 1 m^2
NZE	Net-Zero Emissions Amount of CO_2 consumed is equal to CO_2 emitted
NEOM	Saudi Arabia's conceived Dream project proposed to be built on the NW margin of the Arabian Shield, near Red Sea coast

ORC	Organic Rankine Cycle (technology of generating power using organic fluid as heat carrier)
PIC	Pacific Island Countries: Island countries in the Pacific Ocean
SCM	Social cost of methane emissions (climate damage cost due to ton of methane emitted)
U S A	United States of America
2D	Rise in global temperature by 2 °C

References

ADB. (2020). *Pacific energy update Asian Development Bank, Philippines* (51p). Publication Stock No. ARM210008-2.

Alimonti, C., & Soldo, E. (2016). Study of geothermal power generation from a very deep oil well with a wellbore heat exchanger. *Renewable Energy, 86*, 292–301.

Allan, T. D. (1970). Magnetic and gravity fields over the Red Sea. *Philosophical Transactions of the Royal Society of London, 267*, 153–180.

Almalki, K. A., Betts, P. G., & Ailleres, L. (2015). The Red Sea-50 years of geological and geophysical research. *Earth Science Reviews, 147*, 109–140.

Alonso, M. Padrón, E., Sumino, H., Hernández, P. A., Melián, G. V., Ramos, M. A., Rodríguez, F., Padilla, G., Merino, M. G., Amonte, C., & Pérez, N. M. (2019). Heat and Helium-3 fluxes from Teide Volcano, Canary Islands, Spain. *Geofluids*, ID 3983864, 12. https://doi.org/10.1155/2019/3983864

Anna, C., Barbara, T., & Anna, S. (2020). The utilization of abandoned petroleum wells in geothermal energy sector. Worldwide trends and experience. In *E3S web of conferences, ICoRES 2019* (Vol. 154, p. 05004). https://doi.org/10.1051/e3sconf/202015405004

Birch, F. S., & Halunen, A. J., Jr. (1966). Heat-flow measurements in the Atlantic Ocean, Indian Ocean, Mediterranean Sea, and Red Sea. *Journal of Geophysical Research, 71*, 583–586.

Bosworth, W., & Stockli, D. F. (2016). Early magmatism in the greater Red Sea rift: Timing and significance. *Canadian Journal of Earth Science, 53*, 1–72.

Bosworth, W., Huchon, P., & McClay, K. (2005). The Red Sea and Gulf of Aden Basins. *Journal of the African Earth Sciences, 43*, 334–378.

Boutot, J., Peltz, A. S., McVay, R., & Kang, M. (2022). Documented orphaned oil and gas wells across the United States. *Environmental Science and Technology, 56*, 14228–14236.

Brandt, A. R., Heath, G. A., Kort, E. A., O'Sullivan, F., Pétron, G., Jordaan, S. M., Tans, P., Wilcox, J., Gopstein, A. M., Arent, D., Wofsy, S., Brown, N. J., Bradley, R., Stucky, G. D., Eardley, D., & Harriss, R. (2014). Methane leaks from North American natural gas systems. *Science, 343*, 733–735. https://doi.org/10.1126/science.1247045

Brandt, A. R., Heath, G. A., & Cooley, D. (2016). Methane leaks from natural gas systems follow extreme distributions. *Environmental Science and Technology, 50*, 12512–12520. https://doi.org/10.1021/acs.est.6b04303

Bu, X., Ma, W., & Li, H. (2012). Geothermal energy production utilizing abandoned oil and gas wells. *Renewable Energy, 41*, 80–85. https://doi.org/10.1016/j.renene.2011.10.009

Bu, X., Ma, W., & Gong, Y. (2014). Electricity generation from abandoned oil and gas wells. *Energy Sources, Part A, 36*, 999–1006. https://doi.org/10.1080/15567036.2010.551255

Caulk, R., & Tomac, I. (2017). Reuse of abandoned oil and gas wells for geothermal energy production. *Renewable Energy*. https://doi.org/10.1016/j.renene.2017.05.042

Chandrasekhar, V., & Chandrasekharam, D. (2023). Red Sea geothermal belt: Potential clean energy source to power NEOM and nearby countries, Chap. 9. In N. Rasul & I. Stewart (Eds.), *Rifting and sediments in the Red Sea and Arabian Gulf regions* (356p). CRC Press.

Chandrasekharam, D. (2018). Water for the millions: Focus Saudi Arabia. *Water Energy Nexus, 1*, 142–144.

Chandrasekharam, D. (2021). Geothermal energy for food and water security for Yemen: A review. *Arabian Journal of Geosciences, 14*, 253.

Chandrasekharam, D., Baba, A., Ayzit, A., & Singh, H. K. (2022). Geothermal potential of Kaymaz and Sivrihisar granites, Eskisehir region, western Anatolia. *Renewable Energy, 196*, 870–882. https://doi.org/10.1016/j.renene.2022.07.035

Chandrasekharam, D., Lashin, A., Al Arifi, N., Al Bassam, A., & Varun, C. (2016a). Desalination of seawater using geothermal energy to meet future fresh water demand of Saudi Arabia. *Water Resource Management*. https://doi.org/10.1007/s11269-016-1419-2

Chandrasekharam, D., Lashin, A., Al Arifi, N., & Al-Bassam, M. (2016b). *Red Sea geothermal provinces* (1st ed., 220p). CRC Press.

Chandrasekharam, D., Lashin, A., Al Arifi, N., & Al-Bassam, A. M. (2018a). Energy and Food security through desalination using geothermal energy: Eritrea. *Arabian Journal of Geosciences, 11*, 523. https://doi.org/10.1007/s12517-018-3892-9

Chandrasekharam, D., Lashin, A., Al Arifi, N., & Al-Bassam, A. M. (2018b). Desalination of seawater using geothermal energy for food and water security: Arab and Sub-Sahara countries, Chap. 4. In G. Gnaneswar (Ed.), *Handbook on sustainable desalination handbook—Process design and implementation strategies* (54p, 590p). Elsevier Pub.

Chandrasekharam, D., & Ranjith Pathegama, G. (2020). CO2 emissions from renewables: solar pv, hydrothermal and EGS sources. *J. Geomecha. Geophys. Geoenergy Georesour*. https://doi.org/10.1007/s40948-019-00135-y

Chandrasekharam, D., Singh, M., & Baba, A. (2023). Sahinkalesi massif, a resurgent dome and super-hot EGS province: Hasandag Stratovolcanic province, Central Anatolia (Renewable Energy—under review).

Chandrasekharam, D. (2023). Geothermal power corridor—Connecting the middle east countries. *Bulletin of the Mineral Research and Exploration* (in press).

Chandrasekharam, D. (2022). Enhanced geothermal systems (EGS) for UNSustainable development goals. *Discover Energy*. https://doi.org/10.1007/s43937-022-00009-7

Cheng, W.-L., Li, T.-T., Nian, Y.-L., & Xie, K. (2014). Evaluation of working fluids for geothermal power generation from abandoned oil wells. *Applied Energy, 118*, 238–245.

Cladouhos, T. T., Petty, S., Swyer, M., Uddenberg, M., Grasso, K., & Nordin, Y. (2016). Results from Newberry Volcano EGS Demonstration, 2010–2014. *Geothermics, 63*, 44–61. https://doi.org/10.1016/j.geothermics.2015.08.009

Cladouhos, T., Petty, S., Bonneville, A., Schults, A., & Sorlie, C. F. (2018). Super-Hot EGS and the Newberry Deep Drilling project. In *Proceedings, 43rd workshop on geothermal reservoir engineering* Stanford University, Stanford, California, February 12–14, 2018. SGP-TR-213.

Cochran, J. R., Martinez, F., Steckler, M. S., & Hobart, M. A. (1986). Conrad Deep: A new northern Red Sea deep. Origin and implications for continental rifting. *Earth and Planetary Science Letters, 78*, 18–32.

Davis, A. P., & Michaelides, E. E. (2009). Geothermal power production from abandoned oil wells. *Energy, 34*, 866–872.

Degens, E. T., & Ross, D. A. (1969). *Hot brines and recent heavy metal deposits in the Red Sea* (p. 609). Springer.

Delhotal, K. C., de la Chesnaye, F. C., Gardiner, A., Bates, J., & Sankovski, A. (2006). Mitigation of methane and nitrous oxide emissions from waste, energy and industry. *The Energy Journal, 27*, 45–62. http://www.jstor.org/stable/23297075

Di Pippo, R. (2012). *Geothermal power plants* (3rd ed., pp. 463–474). Elsevier.

Dornan, M. (2014). Access to electricity in small island developing states of the pacific: Issues and challenges. *Renewable and Sustainable Energy Reviews, 31*, 726–735.

ESCAP. (2022). *Pacific perspective 2022. Accelerating climate action* (66p). U. N. Economic and Social Commission for Asia and Pacific, France, Report.

Flovenz, O. G., & Saemundsson, K. (1993). Heat flow and geothermal processes in Iceland. *Tectonophysics, 225*, 123–138.

Ghaffour, N., Lattemann, S., Missimer, T., Ng, K. C., Singa, S., & Amy, G. (2014). Renewable energy driven innovative energy efficient desalination technologies. *Applied Energy, 136*, 1155–1165.

Girdler, R. W. (1970). A review of Red Sea heat flow. *Philosophical Transaction of the Royal Society of London, 267*, 191–203.

Girdler, R. W., & Evans, T. R. (1977). Red Sea heat flow. *Geophysical Journal of the Royal Astronomical Society, 51*, 245–251.

Harris, R. N., & McNutt, M. K. (2007). Heat flow on hot spot swells: Evidence for fluid flow. *Journal of Geophysical Research, 112*, B03407. https://doi.org/10.1029/2006JB004299

Harvey, C. C., & Harvey, M. C. (2010). The prospectivity of hotspot volcanic islands for geothermal exploration. In *Proceedings world geothermal congress*.

Humphreys, E. R., & Niu, Y. (2009). On the composition of ocean island basalts (OIB): The effect of lithospheric thickness variation and mantle metasomatism. *Lithos, 112*, 118–136.

Hutter, G. W. (2021). Geothermal power generation in the world 2015–2020 update report. In *Proceedings world geothermal congress 2020+1*, Reykjavik, Iceland, April–October 2021.

IEA. (2021). *Net zero by 2050. A road map for the global energy secto*r (224p). International Energy Agency, France, Report.

IEA. (2023). *World energy investment 2023* (181p). International Energy Agency (France) Report.

IPCC. (2013). *Climate change 2013: The physical science basis. contribution of Working Group I to the fifth assessment report of the Intergovernmental Panel on Climate Change* (pp. 1535). Cambridge University Press.

Juapart, C., Labrosse, S., Lucaazeau, F., & Mareschal, J.-C. (2015). In S. Gerald (Ed.), *Treatise on geophysics* (2nd ed., Vol. 7, pp. 223–270). University of California.

Juswanto, W., & Ali, Z. I. (2016). Economic growth and sustainable development in the Pacific Island countries. In *2016 Pacific update conference: Inclusive growth for enhanced resilience 18–19 July 2016* (8p) University of the South Pacific. Asian Development Bank Brief 2016-6.

Kang, M., Kanno, C. M., Reid, M. C., Zhang, X., Mauzerall, D. L., Celia, M. A., Chen, Y., & Onstott, T. C. (2019a). Direct measurements of methane emissions from abandoned oil and gas wells in Pennsylvania. *Proceedings of the National Academy of Sciences, United States of America, 111*, 18173–18177. https://doi.org/10.1073/pnas.1408315111

Kang, M., Mauzerall, D. L., Ma, D. Z., & Celia, M. A. (2019b). Reducing methane emissions from abandoned oil and gas wells: Strategies and costs. *Energy Policy, 132*, 594–601.

Karakurt, I., Aydin, G., & Aydiner, K. (2012). Sources and mitigation of methane emissions by sectors: A critical review. *Renewable Energy, 39*, 40–48. https://doi.org/10.1016/j.renene.2011.09.006

Kharseh, M., Al-Khawaja, M., & Hassani, F. (2015). Utilization of oil wells for electricity generation: Performance and economics. *Energy, 90*, 910–916.

Klügel, A., Villinger, H., Römer, M., Kaul, N., Krastel, S., Lenz, K.-F., & Wintersteller, P. (2020). Hydrothermal activity at a cretaceous seamount, canary archipelago, caused by rejuvenated volcanism. *Frontiers in Marine Science, 7*, Article 584571. https://doi.org/10.3389/fmars.2020.584571

Krevor, S., Coninck, H., Gasda, S. E., Ghaleigh, N. S., Gooyert, V., Hajibeygi, H., Juanes, R., Neufeld, J., Jennifer J. Roberts, J. J., & Swennenhuis, F. (2023). Subsurface carbon dioxide and hydrogen storage for a sustainable energy future. *Nature Reviews Earth & Environment, 4*, 102–118. https://doi.org/10.1038/s43017-022-00376-8

Kujawa, T., Nowak, W., & Stachel, A. A. (2005). Analysis of the exploitation of existing deep production wells for acquiring geothermal energy. *Journal of Engineering Physics and Thermophysics, 78*, 127–135.

Kujawa, T., Nowak, W., & Stachel, A. A. (2006). Utilization of existing deep geological wells for acquisitions of geothermal energy. *Energy, 31*, 650–664.

Lal, R., & Kumar, S. (2022). Energy security assessment of small Pacific Island countries—Sustaining the call for renewable energy proliferation. *Energy Strategy Reviews, 41*,100866. https://doi.org/10.1016/j.esr.2022.100866

Li, C. F., & Wang, J. (2018). Thermal structures of the Pacific lithosphere from magnetic anomaly inversion. *Solid Earth: Geodynamics Earth and Planetary Physics, 2*, 52–66. https://doi.org/10.26464/epp2018005

Lucazeau, F., Bonneville, A., & Escartin, J. (2006). Heat flow variations on a slowly accreting ridge: Constraints on the hydrothermal and conductive cooling for the Lucky Strike segment (Mid-Atlantic Ridge, 37°N). *Geochemistry Geophysics Geosystems, 7*(7), Q07011. https://doi.org/10.1029/2005GC001178; ISSN: 1525-2027.

Makris, J., Tsironidis, J., & Richter, H. (1991). Heat flow density distribution in the Red Sea. *Tectonophysics, 198*, 383–393.

Martinez, F., & Cochran, J. R. (1988). Structure and tectonics of the northern Red Sea: Catching a continental margin between rifting and drifting. *Tectonophysics, 150*, 1–32.

Mokhtari, H., Hadiannasab, H., Mostafavi, M., Ahmadibeni, A., & Shahriari, B. (2016). Determination of optimum geothermal Rankine cycle parameters utilizing coaxial heat exchanger. *Energy, 102*, 260–275.

Muehlenbachs, L. (2015). A dynamic model of cleanup: Estimating sunk costs in oil and gas production. *International Economic Review, 56*, 155–185. https://doi.org/10.1111/iere.12098

Nian, Y. L., & Cheng, W. L. (2018). Insights into geothermal utilization of abandoned oil and gas wells. *Renewable and Sustainable Energy Reviews, 87*, 44–60.

Njeudjang, K., Kana, J. D., Tom, A., Essi, J. M. A., Djongyang, N., & Tchinda, R. (2020). Curie point depth and heat flow deduced from spectral analysis of magnetic data over Adamawa volcanic region (Northern Cameroon): Geothermal implications. *SN Applied Sciences, 2*, 1330. https://doi.org/10.1007/s42452-020-3099-z

Noorollahi, Y., Naseer, M. N., & Siddiqi, M. M. (2022). *Utilization of thermal potential of abandoned wells: Fundamentals, applications, and research* (p. 484p). Academy Press.

Ragnarsson, A., Steingrímsson, B., & Thorhallsson, S. (2021). Geothermal development in Iceland 2015–2019. In *Proceedings world geothermal congress 2020+1*, Reykjavik, Iceland, April–October 2021.

Rasul, N. M. A., & Stewart, I. C. F. (Eds.) (2019). *Geological setting, palaeoenvironment and archaeology of the Red Sea* (788p). Springer Nature.

Richter, H., Makris, J., & Rhim, R. (1991). Geophysical observations offshore Saudi Arabia: Seismic and magnetic measurements. *Tectonophysics, 198*, 297–310.

Riddick, S., Mauzerall, D. L., Celia, M. A., Kang, M., Bressler, K., Chu, C., & Gum, C. (2019). Measuring methane emissions from abandoned and active oil and gas wells in West Virginia. *Science of the Total Environment, 651*, 1849–1856. https://doi.org/10.1016/j.scitotenv.2018.10.082

Saleh, S., Salik, M., & Pamukcu, O. (2012). Estimating Curie point depth and heat flow map of northern Red Sea Rift of Egypt and its surroundings from aeromagnetic data. *Pure and Applied Geophysics, 170*, 863–885. https://doi.org/10.1007/s00024-012-0461-0

Salem, A., Green, C., Ravat, D., Singh, K. H., East, P., Fairhead, J. D., Mogren, S., & Biegert, E. (2014a). Depth to Curie temperature across the central Red Sea from magnetic data using the de-fractal method. *Tectonophys, 624*(625), 75–86.

Salem, A., Blakely, R., Green, C., Fairhead, D., & Ravat, D. (2014b). Estimation of depth to top magnetic sources using the local wavenumber approach in an area of shallow Moho and Vurie depth—The Red Sea. *Interpretation (Society of Exploration), 2*, 10–19. https://doi.org/10.1190/INT-2013-0196.1

Saunois, M., Jackson, R. B., Bousquet, P., Poulter, B., & Canadell, J. G. (2016). The growing role of methane in anthropogenic climate change. *Environmental Research Letters, 11*, 120207.https://doi.org/10.1088/1748-9326/11/12/120207

Schettino, A., Macchiavelli, C., & Rasul, N. M. A. (2019). Plate motions around the Red Sea since the early Oligocene. In N. M. Rasul & C. F. Stewart (Eds.), *Geological setting, palaeoenvironment and archaeology of the Red Sea* (pp. 203–272), 788p.

Schmincke, H. U. & Sumita, M. (1988). Volcanic evolution of Gran Canaria reconstructed from apron sediments: Synthesis of VICAP project drilling. In P. P. E. Weaver, H. U. Schmincke, J. V. Firth, & W. Duffield (Eds.), *Proceedings of the ocean drilling program, Scientific Results* (Vol. 157).

Sclater, J. G. (1966). Heat flow in the northwest Indian Ocean and Red Sea. *Philosophical Transactions of the Royal Society London, A, 259*, 1099, 271–278.

Sclater, J. G., Jaupart, C., & Galson, D. (1980). The heat flow through oceanic and continental crust and the heat loss of the Earth. *Reviews of the Geophysics and Space Physics, 18*, 269–311.

Shanmugam, G. (2023). Fuels, climate change, and the vital role of CO2 plays in thriving people and plants on planet earth. *Bulletin of the Mineral Research and Exploration, (MTA)* (in press).

Shindell, D., Kuylenstierna, J. C. I., Vignati, E., van Dingenen, R., Amann, M., Klimont, Z., Anenberg, S. C., Muller, N., Janssens-Maenhout, G., Raes, F., Schwartz, J., Faluvegi, G., Pozzoli, L., Kupiainen, K., Höglund-Isaksson, L., Emberson, L., Streets, D., Ramanathan, V., Hicks, K., … Fowler, D. (2012). Simultaneously mitigating near-term climate change and improving human health and food security. *Science, 335*, 183–189. https://doi.org/10.1126/science.1210026

Singh, H. K. (2020). Geothermal energy potential of Indian oilfields. *Geomechanics and Geophysics for Geo-energy and Geo-resources, 6*, 19. https://doi.org/10.1007/s40948-020-00148-y

Singh, H. K., Sinha, S. K., & Chandrasekharam, D. (2018). Geothermal energy potential of Indian oilfields. *ONGC Bulletin*, 12–22.

Sui, D., Wiktorski, E., Roksland, M., & Basmoen, T. A. (2018). Review and investigations on geothermal energy extraction from abandoned petroleum wells. *Journal of Petroleum Exploration and Production Technology*. https://doi.org/10.1007/s13202-018-0535-3

Tester, J. W., Anderson, B. J., Batchelor, A. S., Blackwell, D. D., DiPippo, R., Drake, E. M., Garnish, J., Livesay, B., Moore, M. C., Nichols, K., Petty, S., Toksöz, M. N., & Veatch Jr., R. W. (2006). *The future of geothermal energy-impact of enhanced geothermal systems on the United States in the 21st century*. MIT (Massachusetts Institute of Technology), 372s.

Townsend-Small, A., Ferrara, T. W., Lyon, D. R., Fries, A. E., & Lamb, B. K. (2016). Emissions of coalbed and natural gas methane from abandoned oil and gas wells in the United States. *Geophysical Research Letters, 43*, 2283–2290. https://doi.org/10.1002/2015GL067623.2015GL067623

Vilarrasa, V., & Carrera, J. (2015). Geologic carbon storage is unlikely to trigger large earthquakes and reactivate faults through which CO_2 could leak. *Proceedings of the National Academy of Sciences of the United States of America, 112*, 5938–5943.

WB. (2021). *World development indicators—Countries and economies: The World Bank Group*. Retrieved March 28, 2023, from http://data.worldbank.org/country

Wen-hua, W. U., Yan, Z., Ying, G., Bai Cheng, L., Qing Zhen, L. U., Zhaokuan, L. U., & Jun, Y. (2022). An over view of structural design, analysis and common monitoring technologies for floating platform and flexible cable and riser. *China Ocean Engineering, 36*, 511–531. https://doi.org/10.1007/s13344-022-0044-5

Yang, Y., & Li, M. (2014). Short-time performance of composite-medium line-source model for predicting responses of ground heat exchangers with single U shaped. *International Journal of Thermal Sciences, 82*, 130–137.

Yusuf, R. O., Noor, Z. Z., Abba, A. H., Hassan, M. A. A., & Din, M. F. M. (2012). Methane emission by sectors: A comprehensive review of emission sources and mitigation methods. *Renewable and Sustainable Energy Reviews, 16*, 5059–5070. https://doi.org/10.1016/j.rser.2012.04.008

Zhang, L., Peking, U., & Yuan, J. (2008). Energy from abandoned oil and gas reservoir. SPE 115055. In *Proceedings 2008 SPE Asia Pacific oil & gas conference and exhibition*, Perth, Australia.

Zhao, D. (2007). Seismic images under 60 hotspots: Search for mantle plumes. *Gondwana Research, 12*, 335–355.

Significance of Oriented-Core in Sub-surface Geological Exploration

Sukanta Goswami

Abstract Oriented-core facilities are crucial in exploratory drilling for several reasons. These include structural information, understanding formation stress, maximizing resource recovery, fluid flow prediction, porosity and permeability analysis, directional drilling support and reducing non-productive time with cost efficiency. Different steps such as (*i*) detail mapping and (*ii*) sketching are followed before commencing any borehole for sub-surface drilling. After getting the solid cylindrical drill cores from sub-surface geologic units, the depth-wise variation in rock characters are studied to match the predicted analogy. The study of litho-structural attributes from the cylindrical cores can give some information on geologic settings of different rock units vis-a-vis deformation structures, alterations, slips, compression-extension directions etc. These in turn can reveal (mineralized) fluid movements and zones of weakness to demarcate possible loci of mineralization. Determination of actual attitude of different structural features, e.g., micro-cracks/fractures/joints, faults, hinges and axial planes of folds, intersection of different planes, unconformity, alteration zone require study of oriented drill cores. Although we are accustomed with the general appearance of rock structures in the exposed outcrops in terms of plan or section views, the same structures can often create ambiguity when observed in cylindrical drill cores. From a single drill core, large-scale features cannot be interpreted. However, cumulative study of small-scale features can give clues to interpret regional features.

Keywords Core orientation · Structural control · Mineral exploration · Core axis · Structural geology

---

S. Goswami (✉)
Sakariya Mines and Minerals, Bidhannagar, Kolkata, West Bengal, India
e-mail: sukantoyahoo@gmail.com

S. Mukherjee (ed.), *Structural Geology and Tectonics*,
https://doi.org/10.1007/978-981-95-4743-2_6

1 Introduction

Core drilling is an important aspect in exploration and depending on the objectives, borehole can be drilled vertically or at any angle after considering the expected geometry of the target to be explored. Drift and deviation must be zero or close to zero to maintain the accuracy. Systematic core drilling practice starts from borehole collar survey with Differential Global Positioning System (DGPS) followed by different stages like magnetic declination correction in location fixing, deviation survey and orientated core study and control measurement of the borehole closure depth at the end. In general, vertical boreholes are drilled in undeformed, sub-horizontal/low-dipping terrains. However, structures such as faults, fractures, and foliations usually occur as inclined to highly dipping planes. Therefore, inclined boreholes are more common in such cases of deformed terrains for exploration.

For vertical borehole, drift concept does not apply unlike the inclined borehole. In case of inclined boreholes, the difference between the original azimuth and the expected/planned azimuth is called drift, which is measured on horizontal plane. Deviation is a measurement of inclination difference between the actual path and expected/planned path of the borehole. Deviation is measured on a vertical plane containing the two paths (Fig. 1). During each run of core drilling (where the recovered core length can be ~3 m), random rotation of recovered core about its long axis (i.e., core axis) within barrel can disqualify them for structural analysis. However, the depth data can be measured and accurately interpreted in litho-logging along with true thickness of litho-units. But, the actual attitude of structural features cannot be directly determined even if the solid cores are acquired from ideal inclined borehole without any drift-deviation.

Marjoribanks (2010) discussed various exploration techniques, including oriented-core analysis, and highlighted its significance in understanding sub-surface structures. The same subject in the context of exploratory drilling has also been discussed subsequently (e.g., Laubach et al., 2004; Zhang et al., 2013a, 2013b). Unrolled images of drill cores are of great use to interpret structural geology deciphered from drill cores (e.g., Chakraborty & Mukherjee, 2020). Another important aspect of studying drill cores is to identify drilling-induced fractures (Chatterjee & Mukherjee, 2023), consider the natural fractures only, and then interpret the stress regime.

As far as structurally controlled mineral exploration is concerned, accuracy is very much essential in deciphering structural data from cores. Especially, uranium, gold or metallic sulfide-bearing minerals are highly sensitive to deformation, changes in Eh-pH condition, fluid interaction along veinlets, micro-cracks, and hydrothermal fluid flow along fractures (Goswami et al., 2023). Identification of mineralized plane is crucial.

This article provides means to reduce errors in structural analysis from non-oriented-cores. I will discuss core orientation process and few techniques used in interpreting the structural data.

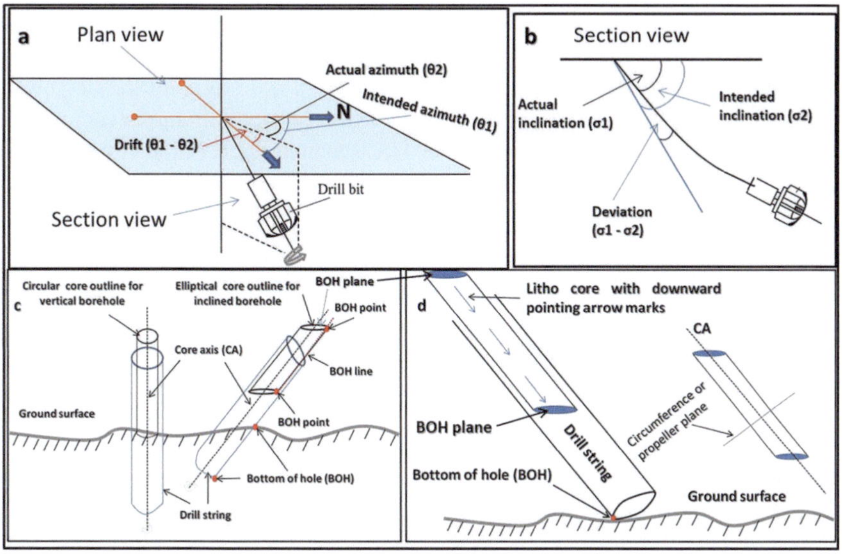

Fig. 1 **a** Three dimensional representation of an imaginary borehole depicting the horizontal plane of measuring azimuth and drift. **b** Section view of imaginary borehole to depict the inclination and deviation concept. **c** Vertical and inclined drill pipe disposition on flat sub-horizontal ground surface with characteristic core recovery. **d** Inclined borehole on flat ground surface give bottom of hole point and the elliptical section of the cylindrical core

2 How to Get Oriented-Cores

Few key updated references covering a broad spectrum of methods and best practices for obtaining oriented drill cores are Chap. 4 of Jahn et al. (2008), Laubach et al. (2009), Bacon et al. (2010), Chap. 6 of Marjoribanks (2010), Zhang et al. (2013a, 2013b), Schlumberger (2015, 2017), Baker Hughes (2017), etc.

1. Well defined precise azimuth and inclination of the core axis along the borehole path length are needed. Drift and deviation must be accurately checked and should be close to zero (Fig. 1a, b).
2. Demarcation of bottom of the hole (BOH) point and line are to be marked properly on core. The touching point of inclined drill string with the surface is designated as the BOH (Fig. 1c, d). It is clear that for vertical borehole on flat topography BOH is difficult to assign because entire rod circumference will touch the ground at a time instead of a point. However, in case of vertical borehole on sloping surface BOH can be assigned. Moreover, for inclined borehole drilled through any sub-horizontal surface, the elliptical outline of the BOH plane indicate the intersection of horizontal plane and the inclined drill pipe. Since, the generalized topographic expression of any terrain is a manifestation of exhumation, the overall tendency sub-horizontal unloading joints, and in case of an initial undeformed sedimentary terrains bedding parallel fissility is common. Therefore, the intersection between

such sub-horizontal weak planes and inclined drill pipes often give elliptical outlines.

3. Arrow markings must be placed on each piece of core to point the downward direction (Fig. 1d). The core axis (CA) is considered to pass along the center of the core parallel to such arrow marks. The CA is actually pole of the circumference plane.

In the undeformed/less deformed well exposed sedimentary terrain the rocks being drilled exhibit well measured constant and standard structural orientation of bedding, foliation and even fractures. These well-defined structures can be projected sub-surface to provide useful means of orienting the core (Fig. 2a). However, in these methods there are several limitations as follows:

I. In case of deformed terrain attitude of beds or any structural features are less likely to be constant so that they can be assigned as standard (Fig. 2b).

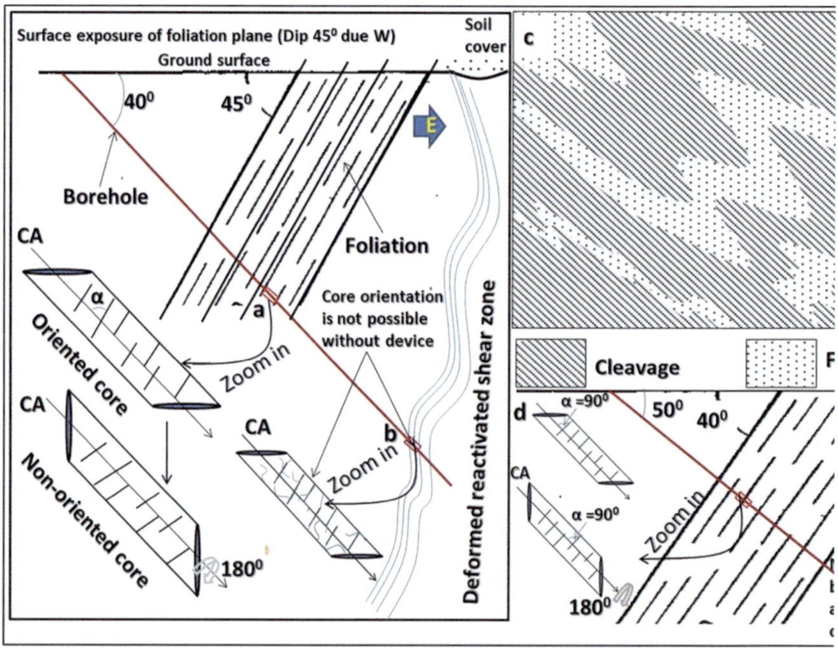

Fig. 2 Inclined borehole cut across two different lithounits. **a** The subsurface foliation planes which continued up to the surface and correlatable, hence the recovered core from this part is easier to orient unlike the wavy folded unit. **b** The intensely sheared unit show wavy irregular pattern without any surface exposure. Cores from this portion cannot be oriented manually. **c** Geological unit with folded beds and associated axial planar cleavage development. Note that cleavage planes are more reliable than bedding for orientating any core of this segment. **d** A typical case of inclined borehole normal to a foliation plane just like vertical borehole on horizontal ground. The intersection plane will always give circular outline normal to the CA, thus orientation is not possible (modified after Marjoribanks, 2010)

II. In case of folded terrain cleavage may be better representative than bedding and can be used with caution after confirming the consistency in attitude (Fig. 2c). But, when more than one set of cleavage present at low angle to each other there are chances of error in identifying the correct set and avoid ambiguity. Although youngest cleavage usually gives most consistent and reliable information (Annels & Hellewell, 1988), it is a difficult task without experienced eyes and sound structural geological understanding, especially in multiply deformed terrain.
III. Practically, the standard surface outcrop features chosen for orienting the core, must cut the core at low angle ($\alpha < 65°$) where α is the angle between core axis and the feature. For high angle between the structural feature of interest and CA, orientation is not possible because when α is close to 90° trace will remain unchanged irrespective of core rotation (Fig. 2d). The details in this regard are elaborated in the next section (i.e., Sect. 3, Concept of planes, lines and angles) of this paper.
IV. Moreover, in case of virgin areas where sparse outcrops, it may not be possible to assign any planar features with constant attitudes in orienting core.

Therefore, mechanical instrumentation (core orientation device assembly) is required to resolve the problems in such cases, which can be expensive and time-consuming processes. Qualitative lithologs can give valuable information on different lithology, depth of occurrence, thickness, presence of deformation structures, broad chronological information etc. However, accurate attitude measurements are not always possible from qualitative approach. To avoid limited opportunity of qualitative study, one must have clear idea regarding the purpose of measuring structures (i.e., certainly not to gather numerous eye-catching data tables without implication). Hence, until we utilize the quantitative data in calculating attitude/orientation of important features and plot those in section and map, oriented-core remains underutilized.

In general, the measurement of the angle (α) of planar structure with the CA is taken into consideration without the angle (β) between the same planar structure and bottom of hole (BOH) line during litho-logging of borehole core. In fact, the combination of α and β is meaningful and complementary just like dip and dip direction of any plane. Therefore, it is clear that the main challenge in structural interpretation from core is related to the BOH identification. If by any method BOH points and lines (as discussed above) can be demarcated, structural analysis of the cores would be possible.

3 Concept of Planes, Lines and Angles

Oriented-core can be useful in measuring attitude of structures by two prime methods: 1. Core frame and compass, 2. Internal core angle method. In the first method, the oriented-core piece is fixed in a simple device as per the actual ground position.

Structural data are measured simply with a compass. Marjoribanks (2010) described different types of core frames. The internal angle method is relatively faster but need clear concept on prerequisites such as internal reference lines and planes such as core axis (CA), bottom of hole line (BOH), intersection ellipse, circumference plane.

Since the cores are solid cylindrical part of drilled lithounit, trace of any planar features (e.g., fault/fracture, foliation) on the core surface appears with curved/sub-circular to elliptical outline depending upon the orientation of the concerned plane. The planes at 90° to CA are known as the circumference plane or the propeller plane. The term intersection ellipse is used for the planes at low angle to CA. With increasing angle between the CA and the plane, intersection outline changes gradually from elliptical to circular (Fig. 3a, b). On rotation of core at different angles (clockwise) the view is sketched (Fig. 3a) to make out about the CA, BOH, intersection plane and circumference plane. Thus, the BOH plane and the intersection plane need to be visualized first. If the intersection ellipse plane is observed the sloping E-E′ line can be projected into a circular horizontal plane to define the β angle measured clockwise from the BOH point to the E point (Fig. 3b, c). Thus, intersection plane at low angle gives inflection points but if the plane is parallel to the CA, no inflection points are formed (Fig. 3d). After measuring the intersection planes and internal core angles (α and β) the data are to be converted into the actual strike, dip amount and direction using simple trigonometric equations (Fig. 4a, b) or by stereonet plots. True thickness of lithounit is calculated for both oriented and non-oriented-cores by using only α and the apparent thickness along the CA. Lineation measurement is not an easy task from core. If penetrative lineation, which is clearly visible on the core surface and BOH plane (i.e., the plane containing BOH), pass through the core axis then only lineation can be measured from the oriented-core (Fig. 4c). Lineation along CA can be marked as L-L′, in which L is conventionally marked as the lower end, which makes angle γ with the down-direction of CA. Similar to β measurement (in case of planer intersection along circumference plane), from BOH to L in clockwise sense the angle δ is measured (Fig. 4d). Therefore, γ and δ can provide complete information on lineation's attitude (plunge and trend).

To measure the β and the δ angles, simple 180° plastic protectors are to be cut with crescent-shaped outlines to fit the half of core circumference (Fig. 5a). For α and γ angle measurements, plastic scale and protector are fixed with screw so that the device can be rotated to different angular positions (Fig. 5a). The β and the δ angles are to be measured clockwise after demarcating the planes. Figure 5b shows two different planes (1. vein parallel to CA and 2. fault plane at low angle to CA) (Fig. 5b). However, in case of intersection plane measurement, the α angle should be <65° because with increasing this angle between intersection plane and CA intersection ellipse gradually becomes circular and E-E′ line demarcation become difficult (Fig. 5c). Figure 5d explains how to measure the α angle from a calcite vein.

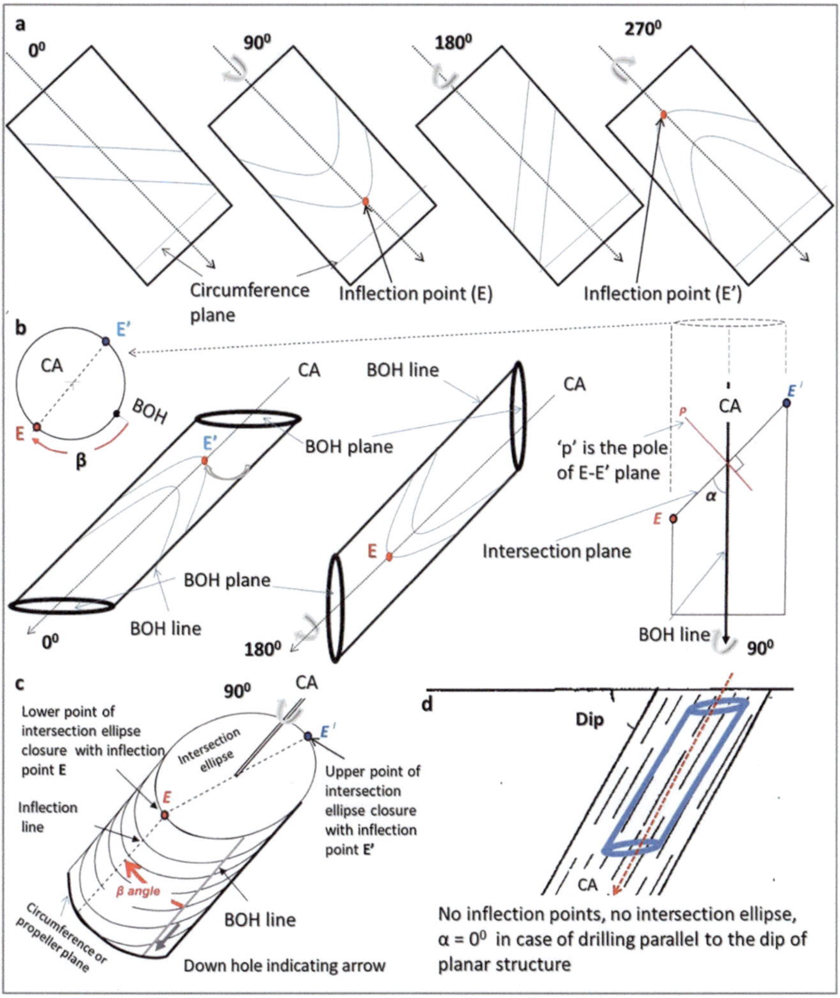

Fig. 3 **a** An example of intersection pattern of any plane/vein cut at acute angle to the core axis. With progressive rotation of the core with CA, the intersection pattern varies. Hence the α angle must be carefully determined after connecting the two inflection points with an imaginary line. The angle between this imaginary line and CA will give α angle. **b** Projection of intersection ellipse on horizontal surface give circular outline. This circumference plane is used to measure β angle from BOH. **c** 3D visualization of elliptical intersection plane with E-E′ major axis. **d** No inflection point if CA is parallel to the plane (modified after Marjoribanks, 2010)

4 Qualitative and Quantitative Approaches

Small-scale faults are often well observed in cores with measurable slips. Smaller-scale faults can mimic larger-scale faults (Pumpelly's rule: Pumpelly et al., 1894). Regional fault usually manifested in cores as brecciated zones of broken core and

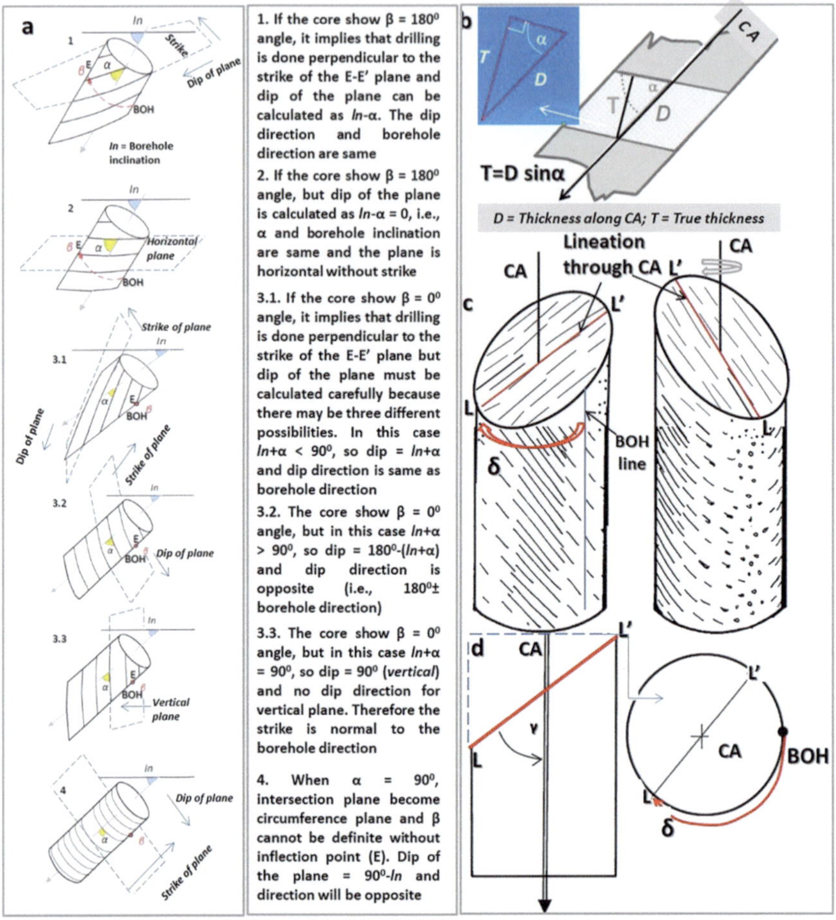

Fig. 4 **a** 3D visualization of intersection plane on core with trigonometric implication to construct geometrically possible orientation and determine the attitude of planes from combination of α and β angle. **b** True thickness determination from apparent thickness along CA. **c** Broad idea on appearance of penetrative lineation from core. **d** Lineation measurement from core (modified after Marjoribanks, 2010)

clay development with significant core loss (Fig. 6a). Finding out attitude of faults may be difficult from the brecciated zones even if the cores are oriented correctly. However, displacement of older veins or fracture planes across the minor small-scale fault can give clues to regional deformation styles. Oriented-cores are needed in case of minor fault characterization because locus of the α angle forms a cone around the CA (Fig. 6b). Therefore, after pinpointing the exact attitude of small fault from oriented-core, it is better to look for if any correlation of the fault zone from the adjacent boreholes (Fig. 6c). Similarly, closely spaced axial plane cleavage of folded terrain may be distinct in core but folded beds themselves become obscure. Even if

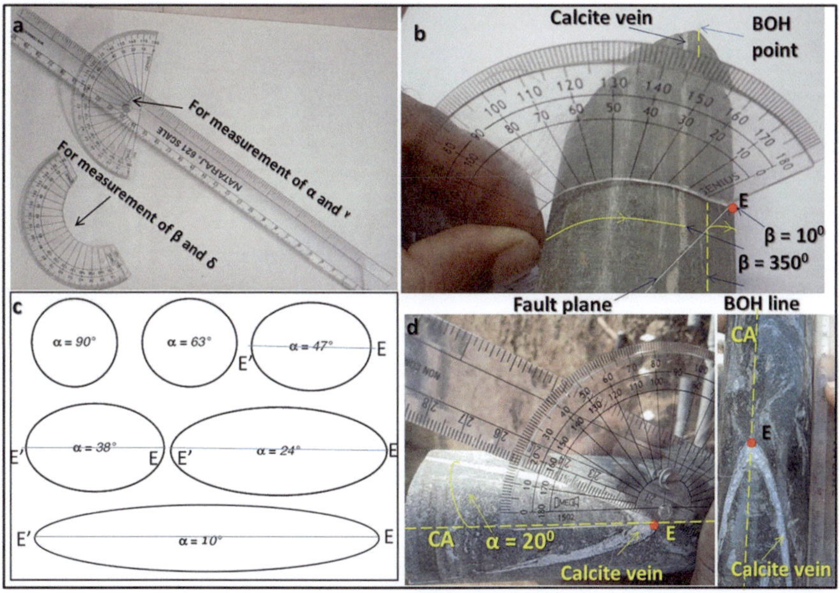

Fig. 5 **a** Use of plastic protector and scale in measuring α, β, γ and δ angle. **b** β angle measurement process (same process for δ angle also). **c** Different intersection ellipse according to the corresponding α angles. **d** α angle measurement of intersecting vein

the small-scale folded planes are cut by drill cores, careful observations would be required to explain the complicated appearance due to three sets of inflection points (i.e., for two limbs E1-E1′ and E2-E2′ plus the hinge/axial parts FA (Fig. 7a–d). A simple case is depicted in which fold axes are oriented almost at right angle to the CA (Fig. 7a-d). When CA is not normal to the FA, complicated appearance is the outcome for the asymmetric fold shape on the core surface. Thus, fold study is mostly confined to qualitative approaches because mostly these structures are larger than the core diameters (e.g., A-size = 27 mm, B-size = 36.5 mm, N-size = 47.6 mm, H-size = 63.5 mm), which make it difficult to quantify the attitude of axial planes, limbs, axes and hinges.

5 Problems and Solutions for the Structurally Controlled Mineralization

For practical purposes use of stereonet plot is one of the most basic prerequisite for understanding planar and linear features. The azimuth/direction and inclination of the borehole can give direct attitude of CA. From CA plot circumference plane can be constructed as great circle at 90°. CA is the pole of circumference plane or the propeller plane. BOH line determination is best possible in stereonet because, it

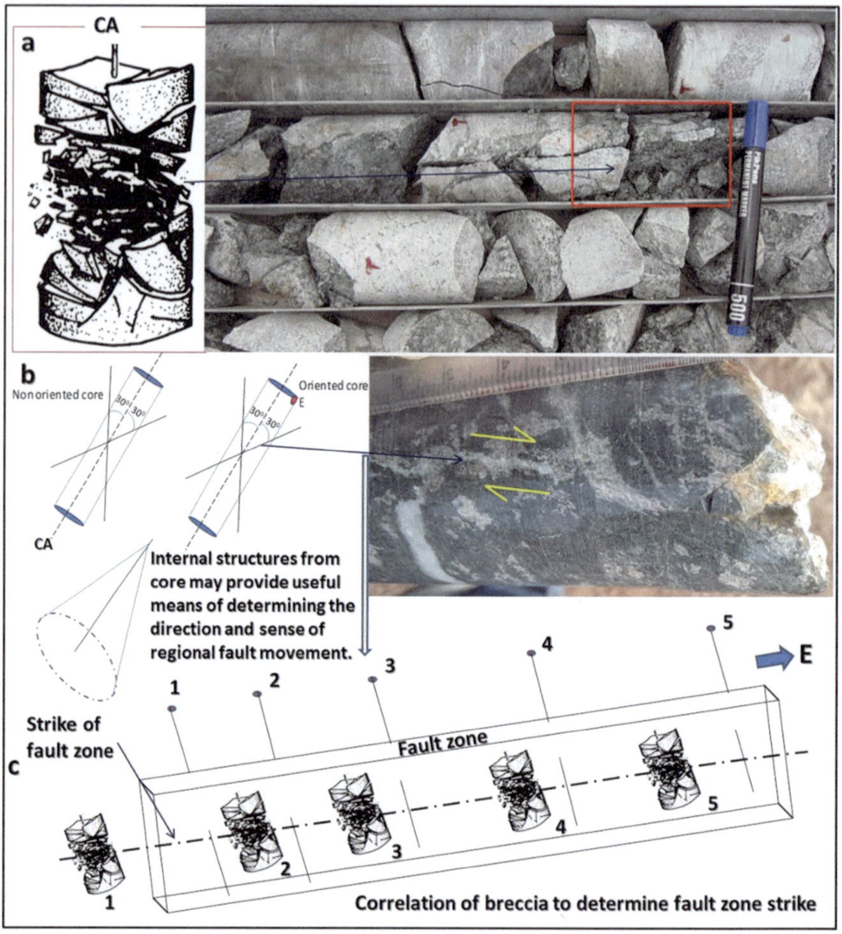

Fig. 6 **a** Brecciated zone in core with sketch to. **b** Importance of oriented core with α and β combination as complementary to each other, i.e., without β angle α has no meaning and vice versa. Only α angle with CA give conical loci of all possible orientation of planes. After having β angle from BOH plane attitude can be fixed. **c** Correlation of brecciated zone without oriented core from a combination of different array of boreholes and corresponding depths from cores (modified after Marjoribanks, 2010)

is the intersection line between the vertical plane containing the CA and the lower half of circumference plane. Borehole can be represented as a line (AB) oriented in space and the projection of the line AB can be observed as core axis point CA (Fig. 8a, b). Similarly, the other planes and lines (e.g., circumference plane, imaginary vertical plane containing CA and BOH line, BOH line) can be constructed from initial azimuth and inclination data. For example, in case of an inclined SW directed borehole circumference plane will exhibit NE directed inclination and vertical plane

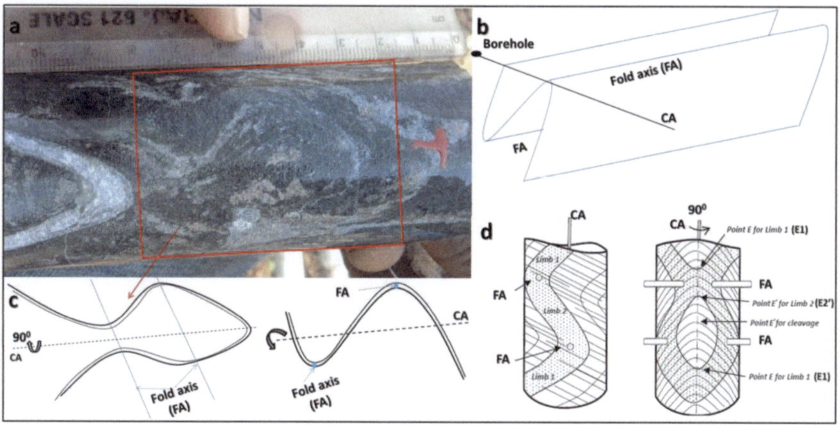

Fig. 7 **a** Appearance of folded unit in core. Note that the wavy ellipse without proper closure indicate changes in orientation of a single plane. **b** Three dimensional view of folded unit and borehole orientation. Depending up on the scale of folding the limbs of opposite dip direction can be manifested. **c** Intersection planes of folded unit at different faces of core. **d** Sketch depicting the limbs and hinge portion of folded unit at different faces of core (modified after Marjoribanks, 2010)

containing the CA and BOH line will show NE-SW strike (Fig. 8a–c). The steps are already well described (Marjoribanks, 2010). An overview of the steps is:

(i) From the calculated α and β angles from CA and BOH on stereonet it is possible to find directly the attitude of planes. However, the plane of measurement has to be defined such a way that CA, intersecting plane being measured (represented by the long axis of the intersection ellipse i.e., E-E′) and the pole (P) of the intersection plane represented by the ellipse lie on it (Fig. 8d). The main target is to plot the pole (P), from which the attitude of the plane can directly be calculated.

(ii) Now, the already measured value of β from core study can be counted along the great circle of circumference plane with clockwise direction from BOH point on stereonet (Fig. 8e). If the measured value of β is less than 90° the point E can be assigned on the circumference plane but for 90°–270° angle of β, both E and E′ will be visible along the opposite points of the circumference (Fig. 8f, i). For other angles, either E or E′ will be visible at a time on the stereonet. Several instances are represented for β angle (45°, 90°, 135°, 225°, 270° and 315°) for better visualization (Fig. 8e–j).

(iii) After plotting β angle (E and E′ cases), already available α angle can be used to plot the depicted plane containing CA and E and/or E′ (i.e., plane of observation in Fig. 8d). Therefore, three possible conditions may be observed when E or E′ or both E and E′ appear on the stereonet. For the $\beta = 0°–89°$ and 271°–360° ranges, E point occupies the space on stereonet unlike the $\beta = 91°–269°$ case when E′ plots on the net. In both the cases the great circle needs to be constructed through CA and either E or E′. When both E and E′ occur together

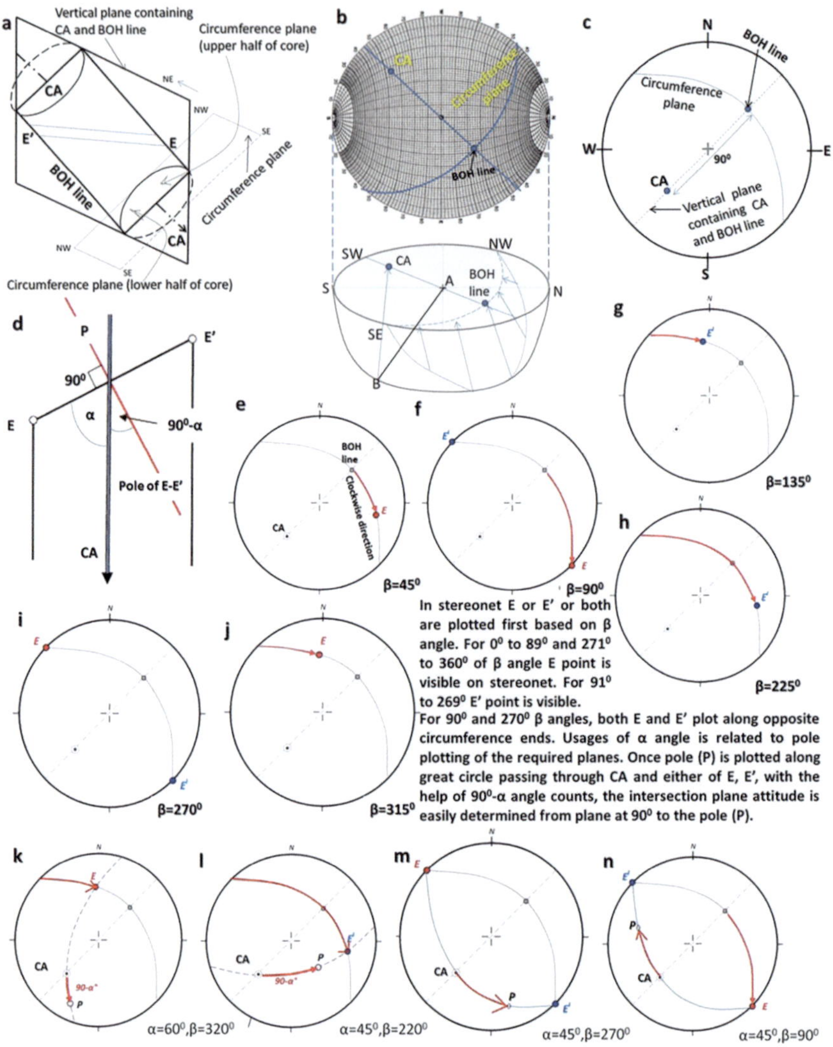

Fig. 8 **a** Vertical plane containing the CA and the BOH line shows intersection plane as E-E′ line and **b** corresponding stereogram depict BOH line as point. **c** Infinite numbers of parallel circumference planes may intersect the vertical plane containing CA give infinite points, which can be connected as BOH line. Therefore BOH line can give a point on a particular circumference plane. Thus the loci of circumference planes must be imagined. **d** A diagrammatic representation of intersection plane with its pole and CA. **e**–**n** Different combination of α and β angles with corresponding stereogram for determining the pole and finally the required attitudes of intersection planes (modified after Marjoribanks, 2010)

there are two option of great circle construction and any one option of great circle construction can be chosen. However, in all cases the main concept of pole (P) demarcation is to count 90°-α angle along great circle from CA toward E′ and away from E. These three possibilities must be analyzed always before determining pole as point P (Fig. 8k–n). Now, from the pole the required plane at 90° can be easily constructed. The dip and strike of the plane normal to the pole is directly measured from stereonet.

6 Oriented-Core in Resource Exploration

After having the detail ideas on mechanical aspects of drill core studies, the prime focus must be made on understanding of economic mineralization along weak deformed planes. Generally, metalliferous hydrothermal fluid migration and ore mineralization along fractures depend on several factors, e.g., pressure, temperature, pH, and Eh. Usually narrow linear fracture zones are often explored for sub-surface information by drilling boreholes at right angle to the strike with different inclinations (Emmons, 1924, 1926; Spurr, 1907). In this context, a basic knowledge on distribution pattern and sequence of element specific zonation of hydrothermal vein at different depth, pressure and temperature range is important. In order of decreasing temperature, the sequence of element concentration may be conceptualized as W, Mo, Cu, Zn, Pb, Ag, Au, Hg, and Sb (Chen & Grapes, 2007; Chen et al., 1996; Zhuang et al., 2006), during the upward and outward transport of hydrothermal fluid from the usual granitic pluton source. Thus, depending up on paleogeothermal gradient, hydrothermal materials get deposited at different depths.

Interestingly, the position of uranium (U) is not specified as such because of uncertainty and its distinct geochemical attributes. For vein type uranium exploration, pitchblende, coffinite and uraninite minerals along vein need to be identified first. Uranium mineralization is influenced by different factors like pressure, temperature, fluid pH, Eh, host rock geochemistry, wall rock alteration, and fluid interaction vide electron gain/loss criteria. However, for structural controlled or litho-controlled uranium mineralization age and spatial geotectonic distribution must be evaluated along with paragenesis. Thus, after narrowing down the target use of oriented-core only give true picture of actual ore body geometry at preliminary level. For the principal 16 type settings (Dahlkamp, 1993) of uranium exploration, oriented-core can give best possible sub-surface information in most of the cases.

A systematic analysis is made to make out about the significant usages of oriented-core as follows:

1. For demarcation of unconformity plane and associated fracture attitudes in case of Unconformity contact and sub unconformity epi-metamorphic type mineralization.
2. To calculate dimension of mineralized veins in case of both granite related and non-granite related fracture fillings.

3. To visualize the exact ore body geometry of all types of sandstone associated sub-types, e.g., tabular, palaeo channel, and roll front curvature determination.
4. Collapse breccias and associated deformed zone complex can also form pipe like pattern, which can be caught by oriented-core.
5. Quartz pebble conglomerate (QPC) types are relatively less sensitive with reference to oriented-core, rather they are more dependent up on age and suitable geological settings. However, to demarcate such basal sub-surface conglomerate bodies in deformed terrains oriented-core may be useful.
6. Use of oriented-core is not much relevant in exploration of lignite, black shale or phosphorite, coal-graphite, metasomatite, and volcanic type of mineralization. However, whenever some specific sub-surface plane or trend demarcation is required in certain cases of mineralization settings oriented-core may be useful.
7. Different types of surficial mineralization like peat-bog, karst-cavern, calcrete related mineralization are not much dependent up on oriented-core.
8. Finally it is clearly understood that for sub-surface exploration of any strata controlled or structural plane or vein bound settings, best output is given by oriented-core.

7 Discussions and Conclusions

It is already clarified that BOH demarcation is the main task to orient a core. If the azimuth/direction and inclination of CA are known and attitude of any significant plane (e.g., dip amount and direction of foliation) is persistent in surface and sub-surface core, then without any BOH indicating device core may be oriented successfully. In such cases careful repetitive measurements of large number of data have to be cross checked and then the measured data of plane from outcrop on ground surface are to be converted into α angle. Now, the directly measured α angle from CA in core for the same plane are to be compared (Fig. 9a). Once proper correlation is established, the core may be oriented based on ground surface data of the particular marker plane which is continuing in the core, azimuth, and inclination of CA. After orienting the core by this indirect process, BOH can be possible to demarcate if the core is placed with exact inclination and azimuth. From BOH point BOH line can be constructed parallel to the CA along the core surface and then β angle can be measured. However, in case of β angle the authenticity can be evaluated by crosschecking the same from mechanical instrument results (Fig. 9b). Another way of BOH determination is very much challenging and cumbersome job, in which the BOH surface has to be examined. Although the chance of error is high in this strategy, examination of huge set of cores can give fair ideas on BOH surface for inclined boreholes which always tend to produce elliptical core end surface. This is due to the fact that almost all the plutonic rocks or any sedimentary rocks found on the Earth's surface today was buried, compacted and came across the diagenetic stages and then finally got exhumed to the present surface. During this exhumation erosion and pressure release is main mechanism of exhumation, which lead to develop unloading

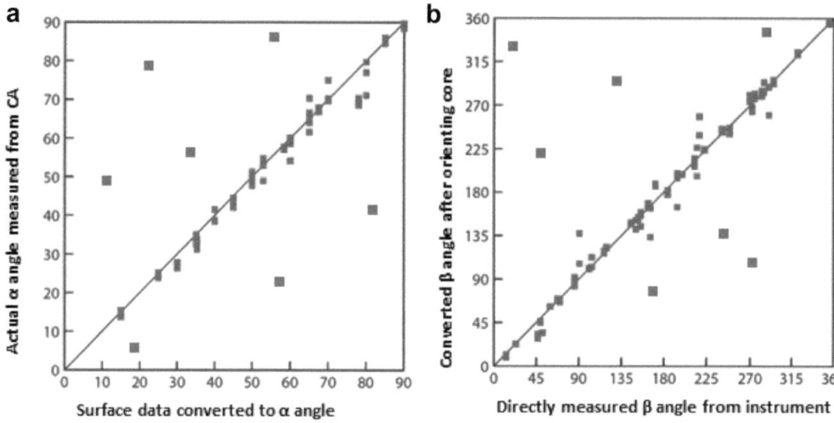

Fig. 9 Hypothetical representation. **a** Cross checking of measured α values from CA and converted surface data of same plane. **b** Converted β angle check from manually oriented core and mechanical device data

joints and associated horizontal slab developments. For sedimentary rocks bedding joints and associated weal planes and partings are very common feature. Similarly, sheeted slabby layers are not uncommon feature in granitic terrain. The presence of such unloading joints and associated weak zones often give us opportunity to get elliptical surface (i.e., intersection between inclined cylindrical core barrel and sub-horizontal weak planes) for inclined borehole cores. In such elliptical BOH plane, the BOH point can be marked after having down hole arrow marks (Figs. 1c–d and 3b). However, the criteria is often become restricted when such weak unloading/exhumation associated pressure release features are not well developed.

Irrespective of the limitations of normal drill core, it is clear that oriented-core do not always imply about instrumentation and huge expenditure. Simple non-oriented-core may be oriented without any instrumentation if by any method it is possible to make the core as it was inside the ground before removal. In fact, available mechanical instrumental devices to determine the BOH in inclined boreholes can also have possibilities of potential errors. The BOH point producing spear rod can often be damaged by operational mistakes, e.g., distortion of the spear rod due to bending, fast dropping of the spear rod over the rock being drilled may show ambiguity and numerous BOH dots (non-optimum speed lead to bouncing and creation of several impact marks). The opposite scenario can also be possible when spear head could not touch the rock and no BOH mark is generated. Therefore, it can be concluded that after broad demarcation of any ore body infilling boreholes are necessary to validate the result and increase the degree of confidence because the boreholes provide broadly one dimensional linear sample. In this regard oriented-core analysis give the additional opportunity to visualize the 3D within the cylindrical core.

Acknowledgements Soumyajit Mukherjee (IIT Bombay) invited, reviewed and edited this chapter. Mukherjee (2026) summarized this work.

References

Annels, A. E., & Hellewell, E. G. (1988). The orientation of bedding, veins and joints in core; a new method and case history. *International Journal of Mining and Geological Engineering, 5*, 307–320.

Bacon, C., Simm, R., & Redfern, J. (2010). *3D seismic interpretation: A primer for geologists* (2nd ed.). Cambridge University Press.

Baker Hughes. (2016). *Oriented-core services for reservoir characterization*. Baker Hughes Technical Report. https://www.bakerhughes.com

Chakraborty, M., & Mukherjee, S. (2020). Structural geological interpretations from unrolled images of drill cores. *Marine and Petroleum Geology, 115*, Article 104241.

Chatterjee, S., & Mukherjee, S. (2023). Review on drilling-induced fractures in drill cores. *Marine and Petroleum Geology, 151*, Article 106089.

Chen, L., Booker, J. R., Jones, A. G., Wu, N., Unsworth, M. J., Wei, W., & Tan, H. (1996). Electrically conductive crust in southern Tibet from INDEPTH magnetotelluric surveying. *Science, 274*, 1694–1696.

Chen, G. N., & Grapes, R. (2007). *Granite genesis: In-situ melting and crustal evolution* (278 p). Springer Science & Business Media, Science. ISBN 978-1-4020-5891-2 (e-book).

Dahlkamp, F. J. (1993). *Uranium ore deposits* (p. 460). Springer.

Emmons, W. H. (1924). Primary downward changes in ore deposits. *Transactions of the American Institute of Mining and Metallurgical Engineers, 70*, 964–992.

Emmons, W. H. (1926). Relations of metallifeous load systems to igneous intrusions. *Transactions of the American Institute of Mining and Metallurgical Engineers, 74*, 29–70.

Goswami, S., Tiwari, R. P., Choudhury, D. K., Saravanan, B., & Sinha, D. K. (2023). Reactivated fracture-controlled uranium mineralization: an example from NNE-SSW Kamaguttapalle-Kammapalle tract, Kadapa district, Andhra Pradesh, India. *Ore and Energy Resource Geology, 15*, Article 100034.

Jahn, F., Cook, M., & Graham, M. (2008). *Hydrocarbon exploration and production* (2nd ed.). Elsevier.

Laubach, S. E., Lander, R. H., & Austin, T. E. (2004). Fracture porosity creation and persistence in a basement-involved Laramide Fold, Upper Cretaceous Frontier Formation, Green River Basin, USA. *AAPG Bulletin, 88*, 901–923.

Laubach, S. E., Olson, J. E., & Gross, M. R. (2009). Mechanical and fracture stratigraphy. *American Association of Petroleum Geologists Bulletin, 93*, 1413–1426.

Marjoribanks, R. (2010). *Geological methods in mineral exploration and mining* (2nd ed). Springer. ISBN 978-3-540-74370-5.

Mukherjee, S. (2026). Introduction to "Structural Geology & Tectonics". In S. Mukherjee (Ed.), *Structural Geology & Tectonics*. Springer. ISBN: 978-981-95-4743-2.

Pumpelly, R., Wolff, J. E., & Dale, T. N. (1894). *Geology of the green mountains in Massachusetts*. US Geological Survey, Monographs, 23.

Schlumberger. (2015). Techniques for obtaining oriented drill cores in complex geologies. *Schlumberger Oilfield Review, 27*, 45–58.

Schlumberger. (2017). *Advanced core orientation techniques for unconventional reservoirs*. Schlumberger technical report. https://www.slb.com

Spurr, J. E. (1907). A theory of ore deposition. *Economic Geology, 2*, 781–785.

Zhang, L., Yin, S., & Li, Z. (2013). Oriented-core analysis and its application in fracture evaluation. *Journal of Petroleum Exploration and Production Technology, 3*(2), 113–120.

Zhuang, W. M., Chen, G. N., Lin, X. M., Peng, Z. L., & Ma, H. M. (2006). Composition of oxygen isotopes and discussion on the formation of the Changkeng Au-Ag deposit, Guangdong Province. *Journal of Jilin University Earth Science, 36*, 521–526.

4D Monitoring of Analogue Models in CT Scanners Using Linear Markers of X-ray Highly Absorbent Materials

E. L. Pueyo, T. Román-Berdiel, E. Willingshofer, L. H. Ros, F. Beekman, A. Peiro, R. Soto, L. Rincón, M. J. Ramón, and A. M. Casas

Abstract Physical analogue models, commonly known as "sand boxes," help understanding structural geological deformational processes. Medical-type CT X-ray scanning is a proven method employed for monitoring the internal deformation of sand boxes in space and time. However, standard 4D reconstruction approaches of acquired CT scans (including digital volume correlation techniques) do not always enable an accurate monitoring and analysis of the strain patterns when deformation is highly non-cylindrical, non-coaxial or rotational. In this contribution, we focus on the use of highly absorbent X-ray markers as novel materials to obtain more accurate 4D monitoring in analogue models. We show how to optimize the type and proportion of linear and dispersed markers with respect to standard sand. We demonstrate that red lead or iron powders produce sufficient brightness to be used as markers to image internal deformation in the radiograms and enabling to be detected in standard imaging files, which will fundamental for the application of digital image correlation techniques in the aforementioned settings. Few preliminary experiments are shown to assess the applicability of the methodology in sand boxes together with the statement of some future challenges and developments.

Keywords Sand boxes · Analogue modelling · 3D reconstruction · Red-lead and iron markers

E. L. Pueyo (✉) · R. Soto · M. J. Ramón
IGME (CSIC), Unidad de Zaragoza, Zaragoza, Spain
e-mail: unaim@igme.es

E. L. Pueyo · T. Román-Berdiel · A. Peiro · R. Soto · M. J. Ramón · A. M. Casas
Unidad Asociada en Ciencias de la Tierra IGME (CSIC)/Universidad de Zaragoza, Zaragoza, Spain

T. Román-Berdiel · A. Peiro · A. M. Casas
Geotransfer-IUCA, Dpto. Ciencias de la Tierra, Universidad de Zaragoza, Zaragoza, Spain

E. Willingshofer · F. Beekman
Department of Earth Sciences, Utrecht University, Utrecht, Netherlands

L. H. Ros · L. Rincón
Servicio de Radiología, Hospital Universitario Miguel Servet, SALUD Aragón, Zaragoza, Spain

S. Mukherjee (ed.), *Structural Geology and Tectonics*,
https://doi.org/10.1007/978-981-95-4743-2_7

1 Introduction

Classic procedures for the study of analogue models consist of taking top-view images as well as photogrammetric characterization of the relief at different times during the development of the model. Once the model is finished, serial cross sections (slices) allow investigating the interior of the model to perform the 3D reconstruction of that final stage. However, in the intermediate ones, this operation, which implies the destruction of the model, cannot be carried out. For more than thirty years, computed tomography (CT) scanning techniques have help overcoming this problem and understanding deformation patterns in the interior of physical analogue models (sand boxes) in a non-destructive means and without altering the development of the model (e.g., Colletta et al., 1991; Fedorik et al., 2019; Harris et al., 2012; Schreurs et al., 2003; Yakymchuk et al., 2012; Yang et al., 2016). A fully and accurate 4D monitoring of strain patterns in the model interior is still challenging when modeling geological structures with significant along-strike variations that usually entail for non-coaxial, non-cylindrical deformation or have undergone vertical axis rotations (Soto et al., 2006). Increasing the reliability of model reconstruction in 4D must balance the use of materials with enough density (and radiologic) contrast together with the suitability of their rheological properties. In this sense, several materials have been proposed to enhance the identification of the stratigraphic horizons of the model in the CT images without compromising the model's integrity; layers of plastic markers (Willingshofer et al., 2018a), mixtures of quartz and feldspar sands with other denser materials as Zirshot ceramic microbeads (Zwaan et al., 2018a), corundum (Panien et al., 2005) and Polydimethylsiloxane (PDMS; Zwaan et al., 2018b), Liaver glass (Warsitzka et al., 2019), etc. See recent review by Zwaan and Schreurs (2023). All these materials give the necessary density contrasts in the radiograms and allow characterizing the evolution of the geometry of the stratigraphic horizons during the modeling. Therefore, the combined use of standard techniques (as topography derived from photogrammetric methods and temporally spaced series of CT sections of the model interior) together with image correlation (digital volume) ones have been successfully applied to quantify strain during model evolution (Adam et al., 2013; Schmid et al., 2022; Zwaan et al., 2018a, 2018b, 2020). However, uncertainty increases when trying to relate the temporal series of images in models displaying important obliquity (i.e., non-coaxial, non-cylindrical, and rotational deformation) because of the significant out-of-plane movement of the model particles across the DICOM radiographic sections (Digital Imaging and Communication in Medicine). This is a well-known issue from the 2D cross-section balancing techniques (Apotria, 1995; Strayer & Suppe, 2002 and references therein) with relevant implications in shortening estimations (Pueyo et al., 2004, 2016; Pedrera et al., 2023; Sussman et al., 2012), among others.

Lead, due to its high density and high atomic number, is an excellent absorber of X-rays. The use of highly absorbent X-ray linear markers as the lead tetraoxide (commonly known as red lead or minium) in static models was first proposed (Ramón, 2013; Ramón et al., 2013) to monitor deformation patterns in 3D simulations of non-coaxial and rotational structures scanned with CT devices and photogrammetric techniques aiming to overcome the aforementioned restrictions. These models were inspired in actual bent (Rodríguez-Pintó et al., 2016) and conical anticlines (Pueyo et al., 2021, 2024) from the south Pyrenean front (External Sierras).

The experiments performed by Ramón et al. (2013) modeled non-cylindrical and non-coaxial fold geometries and used linear markers aiding for the 3D structural restoration. In these models, in addition to the layer modeling (EVA foam plates), a supplementary reference system (initial X and Y axes; Fig. 1), needed to monitor internal deformation in 3D, was accurately defined as two sets of orthogonal grids screen-painted. The amount of ink (red lead dissolved with turpentine) was varied by changing the line width allowing for the main (X axis of 0.5 mm) and secondary sets (Y axis of 2 mm) to be differentiated due to the different X-ray absorption and the derived differential radiologic contrast of both axes. This genuine 3D reference system allows to relate the deformed and non-deformed stages. The rationale behind these CT scanned models was to test the performance and reliability of a novel 3D restoration method designed to use paleomagnetic vectors in natural scenarios (e.g., fold and thrust belts with a complex 4D development) (Ramón et al., 2012, 2016a, 2016b).

In this chapter, we introduce lead tetraoxide (LT) powder as a linear marker in the interior of the physical analogue models allowing for tracking deformation in 4D. We report on optimal mixtures of lead tetra oxide with standard quartz sand to be detected by medical CT scanning techniques. Comparisons with iron powder (IP), recently proposed with a similar purpose (Beekman et al., 2019) are also carried out. For this purpose, we highlight the differences of X-ray absorption of red lead and iron sand in a series of radiograms of experiments and physical analogue models, and explore their applicability, limitations and future challenges.

2 Materials and Methods

Our first goal was to search for an optimized mixture between red-lead powder and standard quartz sands used in analogue modeling. This is because pure red lead powder is very dense (8300 kg m^{-3}) and highly absorbs X-rays producing the dazzling of the radiograms as previously demonstrated in static simulations (Ramón, 2013; Ramón et al., 2013). Therefore, a first set of experiments (1 and 2) was conducted to find the best proportions in weight of pure red lead powder with respect to the total weight of the marker regularly used in the laboratory of Zaragoza University; quartz sand with densities varying between 1650–1800 kg m^{-3} (Román-Berdiel et al., 2019). Differences lie, therefore, in the proportion of the two components.

The first checking test was performed in a plastic box of dm dimensions (Figs. 2 and 3) in the analogue modelling lab of the University of Zaragoza (UZ), where the samples consisting of layers of pure quartz sand (Román-Berdiel et al., 2019) intercalated with linear makers with different proportions of red lead were prepared.

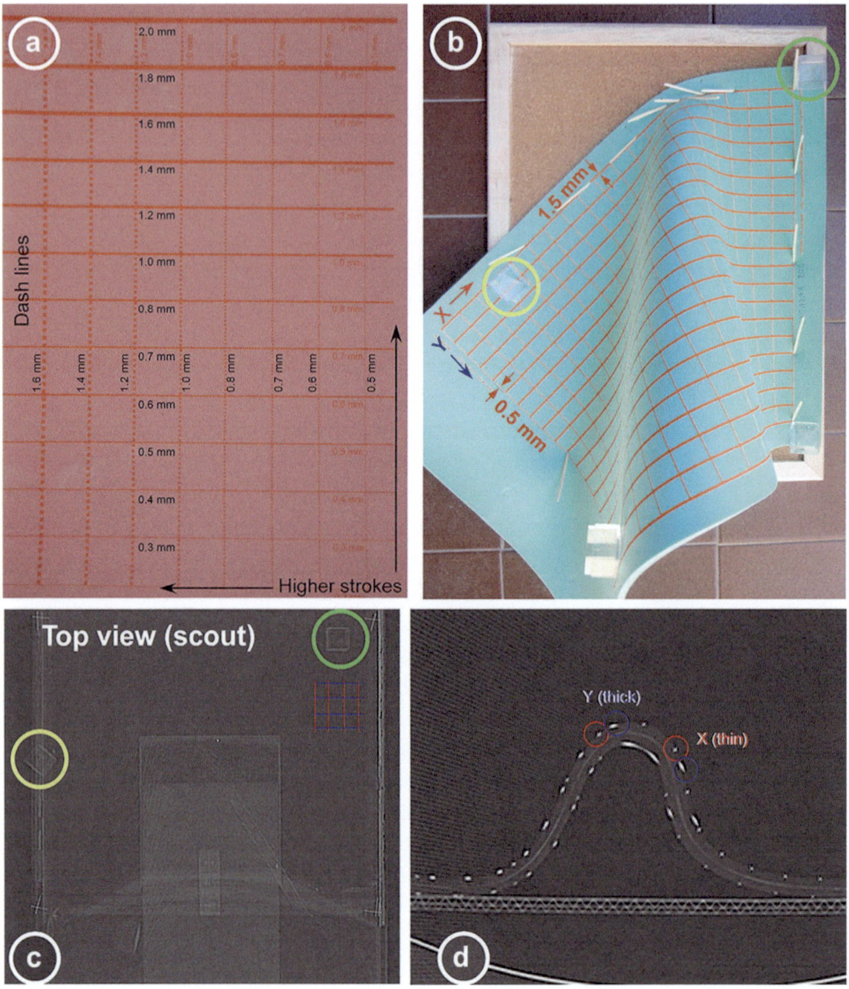

Fig. 1 Screen-painting techniques simulating a radio-opaque grid reference system. **a** Trials of different thicknesses and dashed lines aiming to generate a differential X-ray absorption. **b** Simple model of the San Marzal anticline (plastic boxes highlighted in green and yellow circles are used for reference in the DICOM images). **c** CT top-view (scout) imaging of the model in **b** and **d**. One representative section showing how the X (in red) and Y (in blue) references are identified in the radiograms (taken from Ramón, 2013)

The CT scanning of this box (experiment 1) was performed at the "Miguel Servet" University Hospital (HUMS) from the Aragonian public health system (SALUD) in Zaragoza (Fig. 4a). Additionally, another checking box (experiment 2) with both mixtures (red lead and iron sand [Beekman et al., 2019]), was later performed to show the suitability of our approach in the Utrecht TecLab.

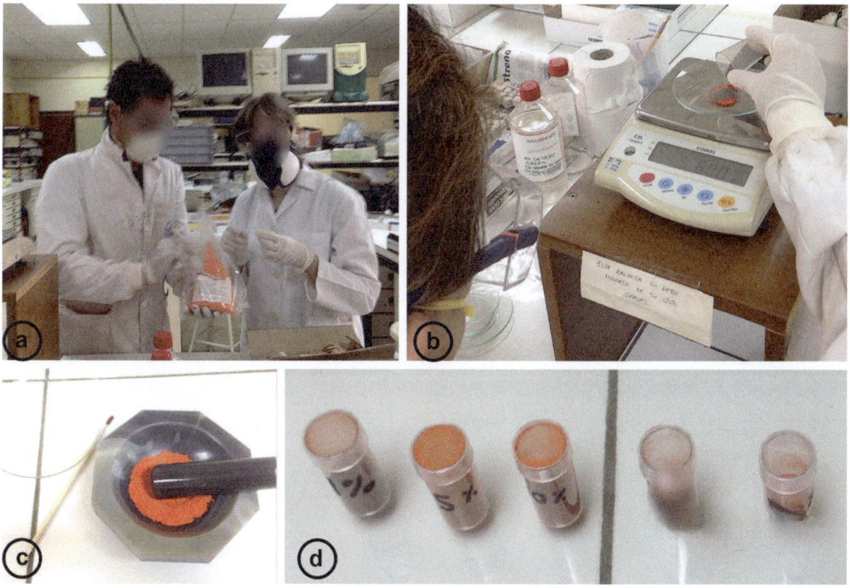

Fig. 2 Experiment 1: Searching the optimal proportion of red lead and quartz sand as linear markers in the analogue modeling laboratory (University of Zaragoza). **a** Marker mixture preparation. Special safety measurements were taken in order to prevent inhalation or skin contact with LT. **b** Weighting the mixtures to estimate the density. **c** Mortar to homogenize the sand-red lead mixture. **d** final mixtures 2, 5, 10, 25 and 50%

Additional experiments performed (standard physical analogue models) are a modification of the thrust wedge models with along-strike tapered, silicone-floored multilayers performed by Storti et al. (2007) (Fig. 5a). The experimental setting of these models is similar to the one described by the aforementioned authors (Fig. 5) but we adapted it to the TecLab of the University of Utrecht (UU) conditions, particularly to the CT scanner dimensions. Besides, we re-scaled some parameters to optimize the lab time and the gantry diameter (see Table 1 and Figs. 5 and 7). The basic configuration of these experiments consists of along-strike tapered, pre-deformational sediments (the sandpack) overlain or not by a basal viscous layer (silicone polymer; Table 2 and Fig. 5a).

The experiments were conducted in a deformation box with a mobile lateral wall (the backstop; Figs. 5 and 6). Model deformation was achieved by moving the backstop in compression using a screw-jack controlled by a stepper motor (push-from-behind type model design). Models were deformed at rates between 3 and 10 cm h^{-1} (corresponding to about a few cm yr^{-1} in nature) until a total amount of shortening of 15–20 cm is reached (Table 2). Different models performed differ on the presence or not of basal viscous layer and on the angle of the tapered sandpack (ranging between 3.6 and 6.8°, Table 2). The architecture of resulting thrust wedges varies as a function of these parameters.

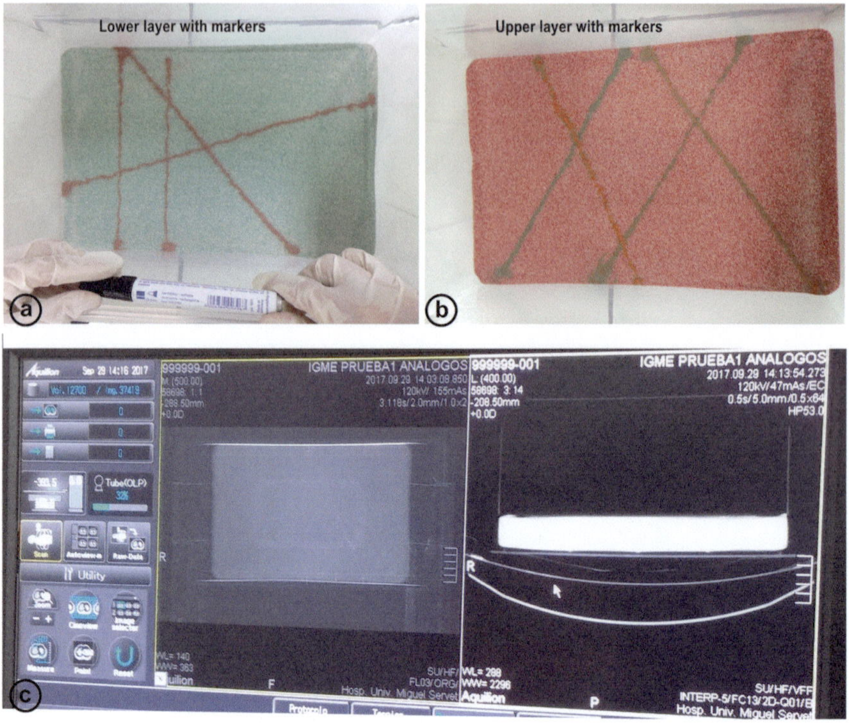

Fig. 3 Topview of Experiment 1: **a** Checking box performed in the University of Zaragoza where sand layers with linear markers with different proportions of red lead were let ready to scan at the HUMS CT scanner (plastic box dimensions: (20 × 30 × 15 cm). The picture shows the lower (green) layer of experiment 1. The linear markers of red-lead mixtures were placed by hand on top of the standard sand. **b** This layer was later covered by a second red layer with additional markers (see also Fig. 7) and covered again with a final green sand layer without markers. **c** Scanning parameters applied to the checking box

2.1 Materials Tested

Pure red lead powder (lead tetraoxide; Pb_3O_4, density ~8300 kg m^{-3}) was obtained from a Spanish manufacturer (Nitroparis S.L. [http://nitroparis.com/]), although red lead powder can be easily found by many providers elsewhere). It was mixed in different proportions with quartz sand ("Silice", product ID: L-7080S] by the company SCR-Sibelco N.V. https://es.sibelco.com/materials/silica/) regularly used in the Analogue Modelling Laboratory of the University of Zaragoza (Román-Berdiel et al., 2019). Special safety working conditions (protection equipment including EN 166 eyewear protection, EN 420 protective gloves and FFP3 particle filter mask) were taken because of the harmfulness of the very fine powdered red lead. Mixtures of quartz sand and red lead at 2, 5, 10, and 25% displayed densities of $\rho =$ ~1832,

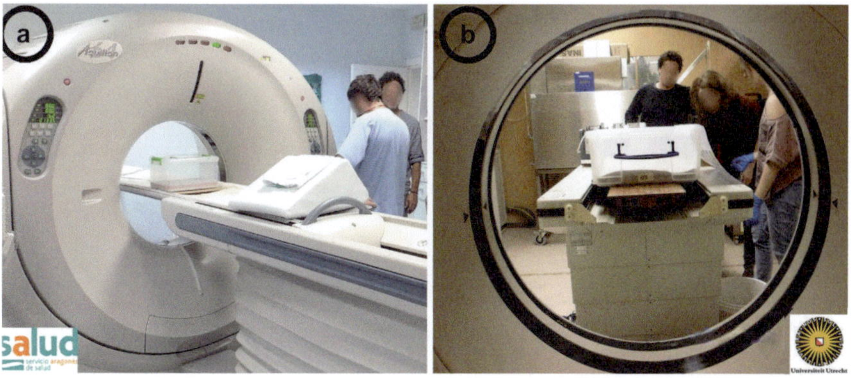

Fig. 4 Running experiments 1 and 2 in the CT scanners: **a** Final box under the CT scanner at HUMS Zaragoza–SALUD (Toshiba Aquilion 64). **b** UU TecLab previous CT scanner (Philips Tomoscan LX)

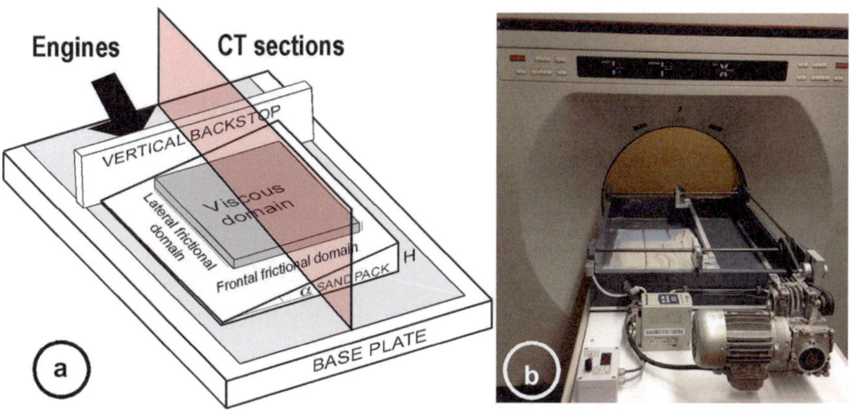

Fig. 5 **a** Concept and experiment setup (modified after Storti et al., 2007) adapted to the TecLab (UU). **b** Model M01 being scanned during its development (later stages)

2030, 2360, and 3350 kg m^{-3}, respectively, in contrast to the bulk density of pure quartz sand with 1610–1650 kg m^{-3} of density.

On the other hand, the mixture of iron powder (Beekman et al., 2019) and quartz sand (Willingshofer et al., 2018b) is taken with a weight ratio of 1:3. The bulk density of the sieved material is $\rho =$ ~2270 kg m^{-3}. The quartz sand is sold under the product name "PURESIL® MAM1ST300" by the company Sibelco (www.sibelco.eu) and is further described in Willingshofer et al. (2018b). The iron powder (density around 7800 kg m^{-3}) is sold under the product ID: 128.225.50 by the Netherlander company De Hazelaar (https://www.hazelaar.nl/home). In this case, safety conditions were not necessarily due to the harmlessness of this material.

Table 1 Experiments carried out (mostly at UU but some at UZ) specifying model variables and conditions as well as the screening techniques used in this technical note. We show examples (and provide files) from experiments MO2 and MX2

Model	Date	Sand			Silicone					Sand		Markers		Monitoring procedure			
		Maximum thickness	Angle	Dimension (cm)	Thickness (cm)	Dimension (cm)	Convergence rate (cm/h)	Total shortenning (cm)	Time (h)	Composition	Number of thickness (cm) of every layer	Composition	Number of beds	Surface scan (min)	Cenital photos (min)	CT scan (min)	Cross-sections
MO1	25 October 2017	2.8	3.6	45 × 50	No	No	5	20	4	Quartz	3:1, 1, 0.8	Pb (10%)	2 + Superficial	No	No	60	No
MO2	**26 October 2017**	**2.8**	**3.6**	**45 × 50**	**No**	**No**	**10**	**20**	**2**	**Quartz + Plastic (thin layers)**	**3: 1, 1, 0.8**	**Pb (10%) + Fe (25%, 2nd bed)**	**2**	**No**	**No**	**30**	**No**
MX1	27 October 2017	5.4	6.8	45 × 50	0'5	35 × 25	3	20	6.7	Quartz + Plastic (thin layers)	7: 0.1, 0.8, 0.8, 0.8, 0.8, 0.8, 0.8	Fe (25%)	6 + Superficial	50	Final	50	Yes
MZ1	28 October 2017	2.8	3.6	45 × 50	0'5	35 × 25	3	15	5	Quartz + Plastic (thin layers)	3: 0.5, 1, 0.8	No	Superficial	50	50	No	Yes
MY	30 October 2017	4.4	5.6	45 × 50	0'5	35 × 25	3	15	5	Quartz + Plastic (thin layers)	5: 0.5, 1, 1, 1, 0.4	Fe (50%)	On silicone + 4 + Superficial	25	25	50	No
MZ2	31 October 2017	2.8	3.6	45 × 50	0'5	35 × 25	3	15	5	Quartz + Plastic (thin layers)	3: 0.5, 1, 0.8	Fe (50%)	On silicone + 2 + Superficial	25	25	50	No

(continued)

Table 1 (continued)

Model	Date	Sand			Silicone					Sand		Markers		Monitoring procedure			
		Maximum thickness	Angle	Dimension (cm)	Thickness (cm)	Dimension (cm)	Convergence rate (cm/h)	Total shortenning (cm)	Time (h)	Composition	Number of thickness (cm) of every layer	Composition	Number of beds	Surface scan (min)	Cenital photos (min)	CT scan (min)	Cross-sections
M03	2 November 2017	2.8	3.6	45 × 50	No	No	3	15	5	Quartz + Plastic (thin layers)	3: 1, 1, 0.8	Fe (50%)	2 + Superficial	15	15	30	No
MX2	**3 November 2017**	**5.4**	**6.8**	**45 × 50**	**0'5**	**35 × 25**	**3**	**15**	**5**	**Quartz + Plastic (thin layers)**	**6: 0'5, 1, 1, 1, 0'4**	**Fe (50%)**	**On silicone + 5 + Superficial**	**25**	**25**	**50**	**No**

Table 2 CT scan setting used in the experiments

Variable	Utrecht	HUMS	Units
kVp	120	120	kV
Exposure time	7600	500	msg
Xray tube current	150	80	mA
Exposure	1140	40	mAs
Generator power		9	kW
Image spacing	50	0.8	mm
Equipment	Philips	Toshiba	
Model	Tomoscan LX	Aquilion 64	

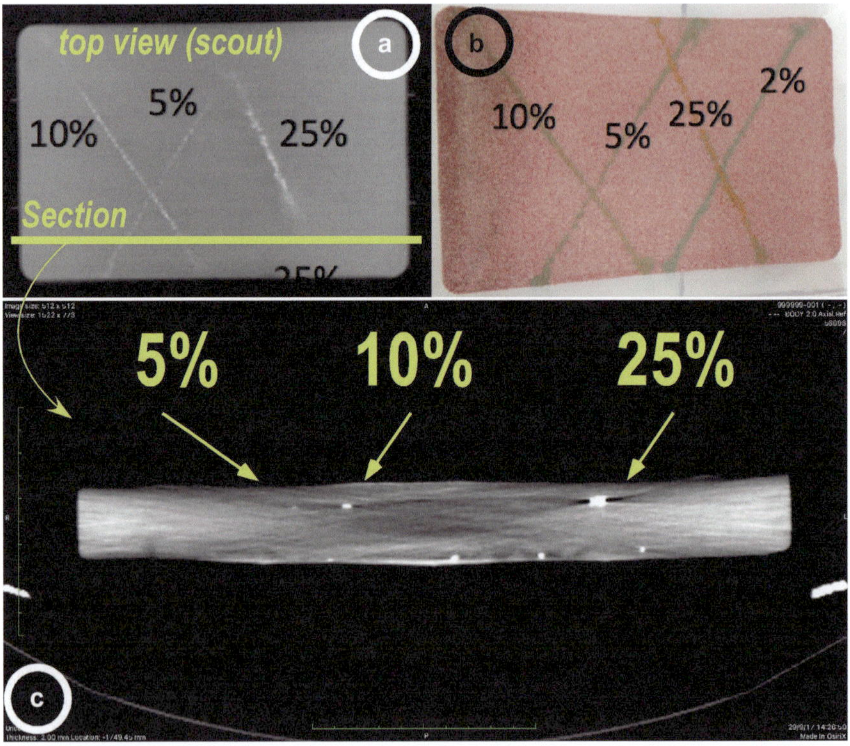

Fig. 6 **a** Picture of the 2nd layer from model M02 during its building. **b** Picture of the model in an advanced stage of development. In this experiment we used the two types of linear markers in order to see their radiologic contrast

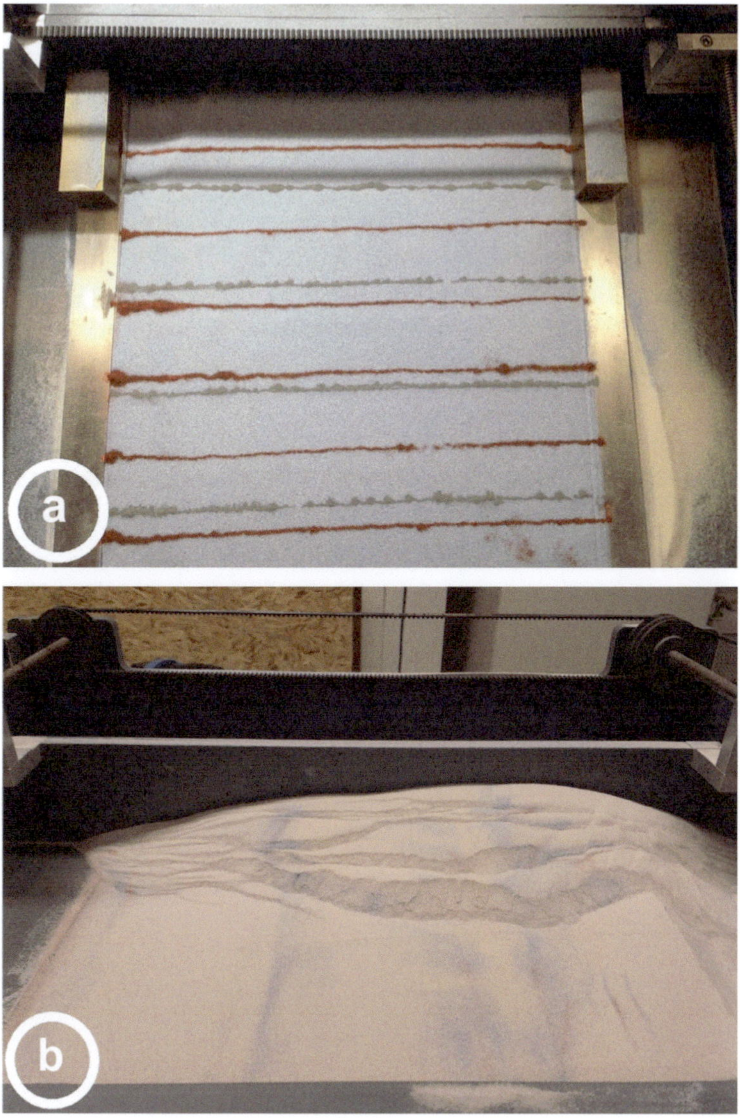

Fig. 7 Experiment 1 (red lead only): **a** Top view (scout) radiogram of experiment 1 where the upper layer is only visible. **b** Picture of the upper sand layer with markers (equivalent to the radiographic image shown in Fig, 6a). **c** Section example that shows the brightness of the different proportions. Note that the 2% mixture is not visible in the radiogram. Markers from the green (lower) and upper (pink) layer are all visible in the radiograms sections (**c**)

2.2 Measurement Procedures

In experiment 1, a Toshiba CT scanner model Aquilion 64 was used at the Miguel Servet University Hospital (HUMS- Radiology Service, Fig. 4a). A thorax protocol was used to scan the entire box due to its large size. Coronal and axial sections (SU/FF routines) were taken with peak kilo voltage (kVp) output of the X-ray generator of 120 kV and exposures of 47 mAs. Parallel sections were taken at 0.8 and 2 mm spacings and were measured during 0.5 s. each (Table 1, Fig. 3c). In experiment 2 and sand boxes series (Table 1), a Philips Medical Systems scanner model Tomoscan LX was used in the tectonic modelling laboratory (TecLab) of the Earth Sciences Department, Utrecht University (Fig. 4b). There, the kVp was the same, but the exposure time of the sections was longer (7.6 s), due to the larger size of the models. The spacing between sections was larger as well (every 50 mm) to avoid the excess of heat generation caused by X-ray emissions during the scanning process (Table 1). The optimal gantry temperature is around 25 °C but may range between 15–20 °C and 35–40 °C maximum, depending upon the models, but never above, otherwise is automatically stopped.

2.3 Analysis Method: Radiogram Analyses

Evaluation of the brightness of the linear markers (different proportions of both red lead and iron powder with quartz sands) was carried out by visual inspection in all cases directly on the digital radiograms (DICOM files) using the software Osirix viewer lite version (https://www.osirix-viewer.com/) where the image contrast and intensity can be adjusted to enhance the features of the model interior (Limberg, 2008).

3 Results

Experiments 1 and 2

In experiment 1, five mixtures were prepared: 2% of lead tetraoxide (LT) +98% of quartz sand (QS); 5% LT, 10% LT, 25% LT, and 50% LT. The radiograms from the HUMS CT scanner clearly show that <5% the markers are almost radio transparent, between 10 and 25%, they are clearly visible and sharp, but above 25% the mixtures absorb too much radiation and dazzle the radiograms preventing an accurate location of the markers (blurry points) (Fig. 7). In experiment 2 as well as in some sand boxes series (Table 1), linear markers of optimized red lead mixtures (25%) on top of sand layers were combined with linear markers of iron powder.

Rest of experiments: Linear markers with the optimal LT/sand (10%) and IP/sand (25 and 50%) ratios at different depths (Fig. 7 and Table 1), as well as thin layers of plastic sand, were drawn by hand inside the model. In this chapter, we only show few results from the model MO2 where both type of markers were tested (Figs. 6 and 8).

Both markers are clearly visible in the radiograms (Fig. 8), although a better precision is needed in the future to accurately drawn the grid in the models with the markers during the set up. In static models (Ramón et al., 2013), the red lead was dissolved in turpentine (50%) and acurately screen-painted on EVA foam plates. In the case of sand boxes, a robotor may help in this critical task to ensure an accurate amount of material (key point for an homogenous X-ray absorption), and to guarantee the perfect parallelism and the uniform spacing between markers.

Red lead lines are finer than the iron sand, which often dazzle the images. This may be caused by an excess of iron in the mixture. With the results that we have obtained so far, optimized lead tetraoxide mixtures (10%) require a lower amount of powder. Thus they have a lower impact in the density of the markers. Induced nucleation of deformation in the marker position during the model development must be prevented, approaching the density of the marker as much as possible to the sand layer one. In our experiments we have seen such effect, but this possible bias must be considered in future studies. Additional trials with other iron proportions have to be done in the near future to find the optimized proportions.

4 Conclusions and Future Work

Lead tetraoxide and iron powder mixtures with quartz sand can be used as linear markers to control deformation patterns in the interior of physical analogue models since they can be clearly identified in the radiograms (DICOM files) and can be tracked in different time-slices (scans) during the model development. The implementation of this methodology will allow a more accurate and a full 4D monitoring and quantification of strain using sophisticated particle image velocimetry techniques (volume correlation).

The optimal marker/sand ratio depends on the marker; lead tetraoxide (red lead) has better definition between 10 and 25% in weight; higher ratios dazzle the images. On the other hand, the iron powder/sand mixtures require a larger proportion 25–50% (having a potential impact in the density of the marker). Initially, the lead tetraoxide has a better performance. However, in view of the harmlessness of the iron powder against red lead, the first one would be preferable in future studies if optimized proportions are found. Alternatively, conventional rust-proof primer paints made of red lead, are much less harmful and can be still found for commercial use because it is dissolved in turpentine (and then the exposure to the dust is negligible). They can be a much safer alternative to the use of powdered minimum and could be used together with conventional modeling sands to produce safer markers for the health and the environment.

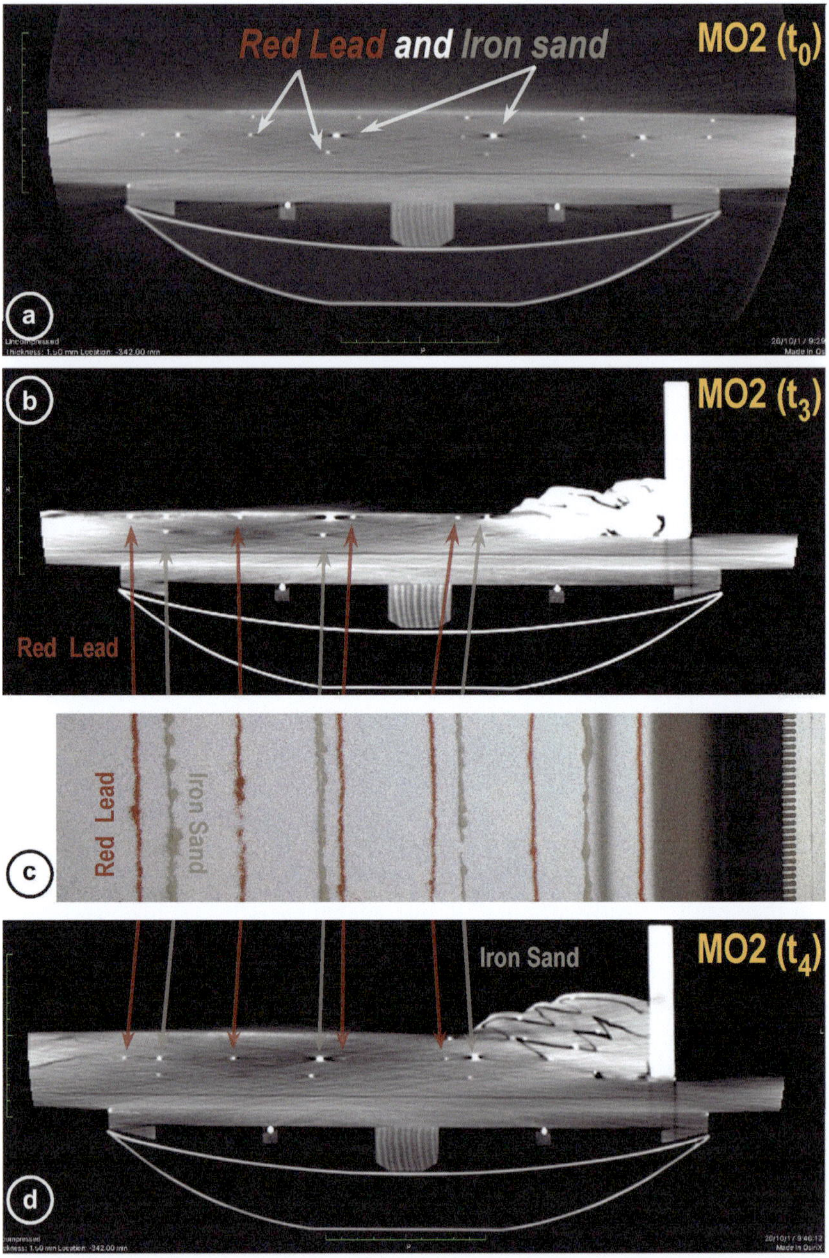

Fig. 8 Comparison of red lead (garnet color) and iron sand (grey-brown) markers scanned at different times during the model development. **a** Initial stage (t0). **b** and **d** Intermediate (t2 and t4) stages. **b** A top-view image of layer 2 during the model building used as example to make-easy the location of the markers

Future actions should also include the development of an automatic (robotized) system allowing for the accurate placement of the linear markers on top of the analogue model layers. A very critical point in the CT scanning of these models is the high-resolution acquisition of the so-called scout (top view) image of the model at every stage of the model monitoring. This zenithal image allows relating the markers along different sections of the model and thus the achieving of a genuine 3D reconstruction where the linear markers and stratigraphic horizons can be unequivocally tracked.

From a mechanical point of view, iron-sand markers (and similar metal [e.g., lead] powder-sand mixtures) are too small to act as a significant strength contrast in a model. The high density, however, should be taken into account when scaling the experiments. We did not run rheological experiments on the red lead and quartz sand mixtures because of the harmful character of this material.

Acknowledgements This research was supported by the project "Truly 4D control of analogue models under the CT scan using high X-Ray absorption materials" financed by the Transnational Access (TNA) Grant of the Earth Plate Observing System Project (EPOS)-WP16- Multi-scale laboratories (EPOS-TNA-MSL-2017-006). Additional financial support came from the projects CGL2014-54118-C2-2-R (DR3AM), UKRIA4D (PID2019-104693GB-I00/CTA) and KINETRANS- (PID2024-157800NB-C21) funded by MCIN/AEI/https://doi.org/10.13039/501100011033 from the Spanish Ministry of Science and Innovation and "ERDF A way of making Europe." The authors also thank the support given by the Geotransfer (E32 17R) and GeoApp (E0117R) Research Groups of the Aragonian Government. Soumyajit Mukherjee invited us to submit this chapter and provided a comprehensive and critical review that has helped improving the original manuscript. Mukherjee (2026) summarized this chapter.

Abbreviations

CSIC	Consejo Superior de Investigaciones Científicas
CT	Computed Tomography
DICOM	Digital Imaging and Communication In Medicine
EVA	Ethylene–vinyl acetate
FFP3	Filtering Face Piece (3rd level of protection in the European standard EN 149)
HUMS	Miguel Servet University Hospital (Zaragoza)
IGME	Instituto Geológico y Minero de España
IP	Iron Powder
IUCA	Instituto Universitario de Ciencias Ambientales
kVp	Kilovoltage peak
LT	Lead Tetraoxide
SALUD	Aragonian public health system
TecLab	Tectonic Modelling Laboratory at UU
UU	University of Utrecht

References

Adam, J., Klinkmüller, M., Schreurs, G., & Wieneke, B. (2013). Quantitative 3D strain analysis in analogue experiments simulating tectonic deformation: Integration of X-ray computed tomography and digital volume correlation techniques. *Journal of Structural Geology, 55*, 127–149.

Apotria, T. G. (1995). Thrust sheet rotation and out-of-plane strains associated with oblique ramps: An example from the Wyoming salient USA. *Journal of Structural Geology, 17*(5), 647–662.

Beekman, F., Willingshofer, E., Sokoutis, D., Pueyo, E. L., Casas, A. M., Peiro, A., Román-Berdiel, T., Soto, R., Schönebeck, J. M., Warsitzka, M., & Rosenau, M. (2019). Ring shear test data of iron powder–quartz sand mixture: A new marker material for analog modelling in a CT scanner at Utrecht University (The Netherlands) (EPOS TNA call 2017). V. 1. GFZ Data Services. https://doi.org/10.5880/fidgeo.2019.019

Colletta, B., Letouzey, J., Pinedo, R., Ballard, J. F., & Balé, P. (1991). Computerized X-ray tomography analysis of sandbox models: Examples of thin-skinned thrust systems. *Geology, 19*(11), 1063–1067.

Fedorik, J., Zwaan, F., Schreurs, G., Toscani, G., Bonini, L., & Seno, S. (2019). The interaction between strike-slip dominated fault zones and thrust belt structures: Insights from 4D analogue models. *Journal of Structural Geology, 122*, 89–105.

Harris, L. B., Yakymchuk, C., & Godin, L. (2012). Implications of centrifuge simulations of channel flow for opening out or destruction of folds. *Tectonophysics, 526–529*, 67–87.

Limberg, T. (2008). *Osirix as a resource* (pp. 1–23). Biomedical Visualization, University of Illinois.

Mukherjee, S. (2026). Introduction to "Structural Geology & Tectonics". In S. Mukherjee (Ed.), *Structural Geology & Tectonics*. Springer. ISBN: 978-981-95-4743-2.

Panien, M., Schreurs, G., & Pfiffner, A. (2005). Sandbox experiments on basin inversion: Testing the influence of basin orientation and basin fill. *Journal of Structural Geology, 27*(3), 433–445.

Pedrera, A., García-Senz, J., Pueyo, E. L., López-Mir, B., Silva-Casal, R., & Díaz-Alvarado, J. (2023). Inhomogeneous rift inversion and the evolution of the Pyrenees. *Earth-Science Reviews*, 104555.

Pueyo, E. L., Oliva-Urcia, B., Sánchez-Moreno, E. M., Arenas, C., Silva-Casal, R., Calvín, P., Santolaria, P., García-Lasanta, C., Oliván, C., Compaired, F., Casas, A. M., & Pocoví, A. (2021). The geometry and kinematics of the Southwestern termination of the Pyrenees; A field guide to the Sto. Domingo anticline. In S. Mukherjee (Ed.), *Structural Geology & Tectonics Field Guidebook-1* (pp. 49–102 (54 pp.)). Springer. ISBN: 978-3-030-60142-3

Pueyo, E. L., Pocoví, A., Millán, H., Sussman, A. J., & Weil, A. B. (2004). *Map-view models for correcting and calculating shortening estimates in rotated thrust fronts using paleomagnetic data* (Vol. 383, pp. 57–71). Geological Society of America Special Paper.

Pueyo, E. L., Santolaria, P., Calvín, P., Casas, A. M., Sánchez, E., Oliva-Urcia, B., Silva-Casal, R., Scholger, R., Rubio, F., Rey-Moral, C., Rodríguez-Pintó, A., Ayala, C., Egli, R., & Pocoví, A. (2025, in press). Vertical-axis rotations and salt tectonics in the southwestern Pyrenean termination. In T. Lawton, J. Geissman, A. Iriondo, et al. (Eds.) *Geophysical and geological studies of the proterozoic to recent tectonics, sedimentary basins, and sediment provenance of mexico and regions beyond: a volume to honor the legacy of Roberto S. Molina Garza*. Geological Society of America Special Paper.

Pueyo, E. L., Sussman, A. J., Oliva-Urcia, B., & Cifelli, F. (2016). Palaeomagnetism in fold and thrust belts: Use with caution. *Geological Society* (Vol. 425, No. 1, pp. 259–276). Special Publications.

Ramón, M. J. (2013). *Flexural unfolding of complex geometries in fold and thrust belts using paleomagnetic vectors*. Unpublished Ph.D., University of Zaragoza, Spain. http://zaguan.unizar.es/record/11750

Ramón, M. J., Briz, J. L., Pueyo, E. L., & Fernández, O. (2016b). Horizon restoration by best fitting of finite elements and rotation constraints: Sensitivity to the mesh geometry and pin-element location. *Mathematical Geosciences, 48*, 419–437.

Ramón, M. J., Pueyo, E. L., Briz, J. L., Pocoví, A., & Ciria, J. C. (2012). Flexural unfolding of horizons using paleomagnetic vectors. *Journal of Structural Geology, 35*, 28–39.
Ramón, M. J., Pueyo, E. L., Caumon, G., & Briz, J. L. (2016a). Parametric unfolding of flexural folds using palaeomagnetic vectors. In *Geological society* (Vol. 425 No. 1, pp. 247–258). Special *Publications.*
Ramón, M. J., Pueyo, E. L., Rodríguez-Pintó, A., Ros, L. H., Pocoví, A., Briz, J. L., & Ciria, J. C. (2013). A computed tomography approach to understanding 3D deformation patterns in complex flexural folds. *Tectonophysics, 593*, 57–72.
Rodríguez-Pintó, A., Pueyo, E. L., Calvín, P., Sánchez, E., Ramajo, J., Casas, A. M.,... & Román, T. (2016). Rotational kinematics of a curved fold: The Balzes anticline (Southern Pyrenees). *Tectonophysics, 677*, 171–189.
Román-Berdiel, T., Casas, A. M., Pueyo, E. L., Peiro Chamarro, A., Soto, R., Pohlenz, A., Warsitzka, M., & Rosenau, M. (2019). Ring shear test data of quartz sand and colored quartz sand used for analogue modelling in the Laboratorio de modelización analógica, Universidad de Zaragoza, Spain (EPOS TNA call 2017). *GFZ Data Services*. https://doi.org/10.5880/fidgeo.2019.025
Schmid, T. C., Schreurs, G., & Adam, J. (2022). Characteristics of continental rifting in rotational systems: New findings from spatiotemporal high resolution quantified crustal scale analogue models. *Tectonophysics, 822*, Article 229174.
Schreurs, G., Hänni, R., Panien, M., & Vock, P. (2003). The analysis of analogue model experiments by helical X-ray computed tomography. In F. Mees, R. Swennen, M. Van Geet, & P. Jacobs (Eds.), *Applications of X-ray computed tomography in the geosciences* (Vol. 215, pp. 213–223). Geological Society. Special Publications.
Soto, R., Casas-Sainz, A. M., & Pueyo, E. L. (2006). Along-strike variation of orogenic wedges associated with vertical axis rotations. *Journal of Geophysical Research: Solid Earth, 111*(B10).
Storti, F., Soto, R., Rossetti, F., & Casas-Sainz, A. M. (2007). Evolution of experimental thrust wedges accreted from along-strike tapered, silicone-floored multilayers. *Journal of the Geological Society, 164*(1), 73–85.
Strayer, L. M., & Suppe, J. (2002). Out-of-plane motion of a thrust sheet during along-strike propagation of a thrust ramp: A distinct-element approach. *Journal of Structural Geology, 24*(4), 637–650.
Sussman, A. J., Pueyo, E. L., Chase, C. G., Mitra, G., & Weil, A. B. (2012). The impact of vertical-axis rotations on shortening estimates. *Lithosphere, 4*(5), 383–394.
Warsitzka, M., Ge, Z., Schönebeck, J.-M., Pohlenz, A., & Kukowski, N. (2019). *Ring-shear test data of foam glass beads used for analogue experiments in the Helmholtz Laboratory for Tectonic Modelling (HelTec) at the GFZ German Research Centre for Geosciences in Potsdam and the Institute of Geosci-ences.* Friedrich Schiller University Jena. https://doi.org/10.5880/GFZ.4.1.2019.002
Willingshofer, E., Sokoutis, D., Beekman, F., Schönebeck, J.-M., Warsitzka, M., & Rosenau, M. (2018b). Ring shear test data of feldspar sand and quartz sand used in the Tectonic Laboratory (TecLab) at Utrecht University for experimental Earth Science applications. V. 1. GFZ Data Services. https://doi.org/10.5880/fidgeo.2018.072
Willingshofer, E., Sokoutis, D., Kleinhans, M., Beekman, F., Schönebeck, J-M., Warsitzka, M-. Rosenau, M. (2018a). Ring-shear test data of plastic sand, a new rock analogue material used for experimental Earth Science applications at Utrecht University, The Netherlands. V. 1. GFZ Data Services. https://doi.org/10.5880/fidgeo.2018.022
Yakymchuk, C., Harris, L. B., & Godin, L. (2012). Centrifuge modelling of deformation of a multi-layered sequence over a ductile substrate: 1. Style and 4D geometry of active cover folds during layer-parallel shortening. *International Journal of Earth Sciences, 101*, 463–482.
Yang, M., Zhang, W., Wu, X., Wei, D., Zhao, Y., Zhao, G., Han, X., & Zhang, S. (2016). Consecutive short-scan CT for geological structure analog models with large size on in-situ stage. *PLoS One, 11*(8), Article e0161358.

Zwaan, F., & Schreurs, G. (2023). A novel method for analogue modelling of lithospheric-scale rifting, monitored through X-Ray CT-scanning, at the University of Bern Tectonic Modelling Laboratory (Switzerland). *GFZ Data Services*. https://doi.org/10.5880/fidgeo.2023.005

Zwaan, F., Schreurs, G., & Adam, J. (2018a). Effects of sedimentation on rift segment evolution and rift interaction in orthogonal and oblique extensional settings: Insights from analogue models analysed with 4D X-ray computed tomography and digital volume correlation techniques. *Global and Planetary Change, 171*, 110–133.

Zwaan, F., Schreurs, G., & Rosenau, M. (2020). Rift propagation in rotational versus orthogonal extension: Insights from 4D analogue models. *Journal of Structural Geology, 135*, Article 103946.

Zwaan, F., Schreurs, G., Ritter, M., Santimano, T., & Rosenau, M. (2018b). Rheology of PDMS-corundum sand mixtures from the Tectonic Modelling Lab of the University of Bern (CH). V. 1. GFZ Data Services. https://doi.org/10.5880/fidgeo.2018.023

Structural Form of Stratovolcanoes on Planetary Bodies: A Theoretical Perspectives and Model Validation

Sambit Sahoo, Saishree Subhadarshini, Ashish Kumar Jena, and Bhaskar Kundu

Abstract Landforms associated to stratovolcanoes on Earth produced by the consecutive eruption episodes provide a history of the erupted material and magmatic compositions. Due to the natural shape formed by the piling up of the erupted material, it provides an opportunity to explore the flow conditions and physical parameters associated to the eruptive episodes of the particular volcano. In addition, with the availability of topography of volcanoes across different planetary bodies and the estimated physical parameters, it has been a great prospect to explore the landform evolution processes for the terrestrial bodies. Although previous studies have used the similarity of the lava flow to groundwater flow and validation of Darcy's law along with Dupuit's approximation in exploring the seamounts and ideal symmetric volcanoes like Mount Fuji, the outcomes are limited in a broader aspect to the sub-aerial volcanoes across different tectonic settings. In this study, we have further extended the approximation to provide a theoretical solution for modelling the 2D geometric shape of volcanoes in the terrestrial bodies and determining the role of different physical parameters (i.e. fluid flow, permeability, viscosity and gravitational acceleration). We have also tried to validate the theoretical model and controlling parameters for the geometrical shape of volcanoes on Earth, Mars and different planetary bodies.

Keywords Stratovolcano · Edifice topography · Gravitational acceleration · Viscosity · Lava flow · Dupuit's approximation

1 Introduction

Based on the distinct landforms associated by volcanism, different morphological features can be formed from erupted magma, for e.g. cinder cones, spatter cones, volcanic plateaus and lava domes. Based on the edifice structure formed by the magmatic compositions of successive eruptions, volcanoes can be mainly classified

S. Sahoo · S. Subhadarshini · A. K. Jena · B. Kundu (✉)
Department of Earth and Atmospheric Sciences, National Institute of Technology, Rourkela, Odisha, India
e-mail: kundub@nitrkl.ac.in

S. Mukherjee (ed.), *Structural Geology and Tectonics*,
https://doi.org/10.1007/978-981-95-4743-2_8

into two distinct varieties—stratovolcanoes and shield volcanoes. Stratovolcanoes have steep edifice structures that are formed due to repeated eruption and piling up of viscous lava and pyroclastic material in the alternative layers. Shield volcanoes with comparatively less steep edifices are formed as a result of repeated eruption of low viscous basaltic or andesitic lava which cannot easily pile up. It has been suggested that the size, temperature and composition of the material ejected can also determine the shape of the volcano (Milne, 1878, 1879; Griffiths, 2000). However, the volcano landforms can be referred to the result of the balance between factors such as age, growth stage, composition, eruption rate, vent position and migration, degree of erosion, lava/tephra ratio, deformation and ultimately on other intrinsic factors such as magma flux and tectonic setting (Becker, 1885; Gregg, 2017).

The volcanic landforms are also observed on other planetary bodies such as Mars, Venus and Moon (Fig. 1). But their structure varies distinctly from those on Earth due to atmospheric conditions such as temperature and pressure and geochemical composition of the associated fluid (Plescia, 2004). Volcanoes on the Mars and Moon show distinct differences in topography than on Earth as Lunar volcanoes have a relatively low elevation than those on Earth whereas Mars volcanoes are relatively larger due to the flow of low viscosity basaltic magma on the lunar surface (Francis et al., 1993). Moreover, in the case of lunar volcanism lower gravity and lack of atmosphere add up to the lower viscosity, as there is less cohesion and low external pressure on the magma (Spudis, 2015). In contrast, the magma on Mars is typically more viscous due to its higher silica content which provides a resistance to the flow (Carr, 1976; Greeley & Spudis, 1981).

The geometric forms of volcanoes can be both symmetric and asymmetric. The symmetric forms are exceedingly difficult to preserve in nature due to different geomorphologic processes (e.g. erosion and wind directions during the eruption) and tectonic forces (e.g. chamber collapse and crater destruction), acting regularly on the structures (Milne, 1879; Eaton & Murata, 1960). Lacey et al. (1981) attempted to model the composite central volcanoes and displayed the concave profiles, that were probably formed as the result of the repeated deposition of pyroclastic flows, ash fall deposits and lava flows, along with the effects of erosion (Fig. 2). Later, Angevine et al. (1984) observed that a majority of the composite volcanoes along with seamounts reflect the same concave upward shape with the largest slopes near the summit. However, his model identified the convex-up geometry of aseismic ridges with the largest slopes on the lower flanks. The complicacy of these two basic shapes is a result of several factors acting over the structure altogether. As per the studies carried out by Grosse et al. (2009), the evolutionary history of a volcano also relies on factors such as magma flux, edifice strength, structure and tectonics. Specifically, two balancing factors control its form, magma pressure versus lithostatic pressure and conduit resistance versus edifice resistance.

The previous studies have suggested that flow of magma in the subsurface has similarities to that of groundwater as Magma flows radially outwards and follows the path of least resistance. It also obeys Darcy's law for fluid flow, assuming the slope to be very small, i.e. considering Dupuit approximation to be valid. The radial flow of magma (Q) varies with the area of cross-section (A) and hence the radial

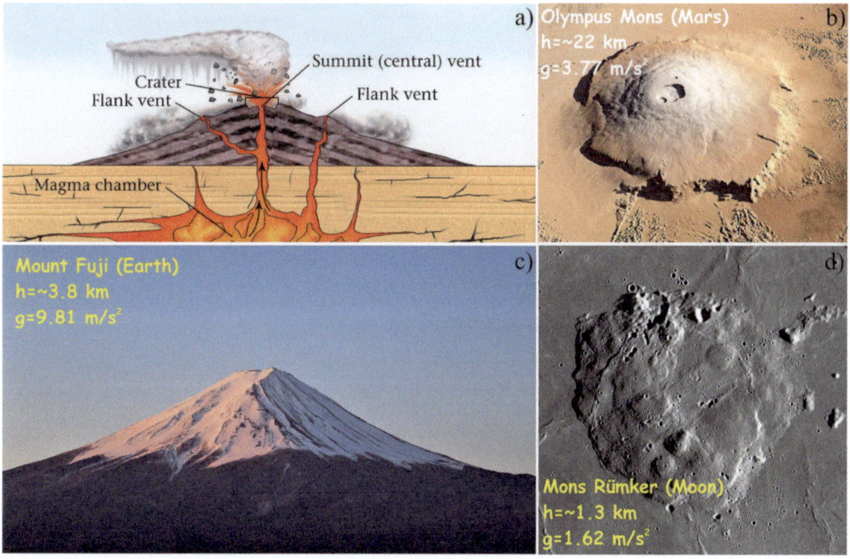

Fig. 1 **a** Conceptual diagram of a stratovolcano on Earth and high-resolution images from volcanoes on Earth and different planetary bodies. **b** Mount Fuji volcano (Earth) and **c** Olympus Mons (Mars) and **d** Mons Rumker (Moon)

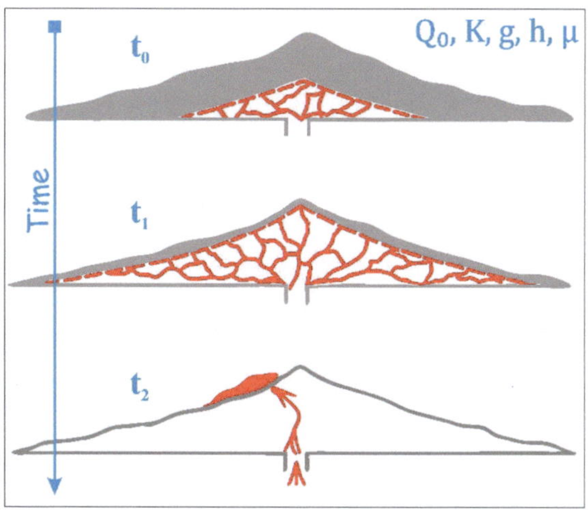

Fig. 2 Schematic diagram shows the process by which a volcanic edifice is extended by a surface flow. At t_0, magma seeps through the structures in pursuit of the easiest route to the surface. At time t_1, the magma rises to the surface at its lowest point of resistance. The edifice is extended at time t_2 by the surface flow. The flow depends on the mentioned physical parameters, i.e. Q_0, K, g, h, μ (modified from Lacey et al., 1981)

distance (r), permeability (k), magma density (p), the viscosity of magma (μ) and acceleration due to gravity (g) (Fig. 3). The porous flow model can be further used to determine the topographic geometry of different volcanoes theoretically, when the above-mentioned parameters are accurately known, both on Earth and planetary bodies. Though the model is an incredible step to predict the theoretical form of the

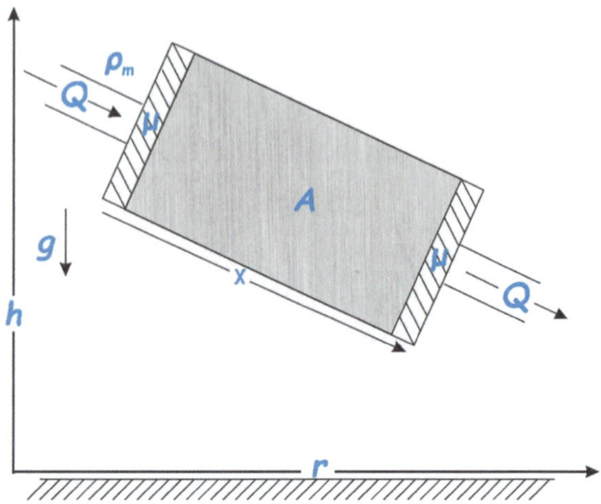

Fig. 3 Radial flow of magma follows the flow of ground water (Q), that varies with the area of cross-section (A), flow distance (x), height (h), gravity (g), permeability (k), density (ρ_m), viscosity (μ) and acceleration due to gravity (g)

volcanoes, it highly depends on the accuracy of the physical parameters considered, which is difficult job for a geoscientist to gather. Moreover, the geomorphologic factors such as erosion and denudation, that may largely affect its slope and structure are not considered in this model.

In this study, we provide a theoretical solution for modelling the 2D geometric shape of volcanoes in the terrestrial bodies and determining the role of different physical parameters (i.e. fluid flow, permeability, viscosity and gravity). Validation of the theoretical model has been also performed for the geometrical shape of volcanoes on Earth, Mars and other physical parameters to different planetary bodies.

2 Fluid Flow in Porous Material and Geometrical Form of Stratovolcanoes: Theory

In order to simplify the axisymmetric volcanic edifice equations from Lacey et al. (1981), we derived the solution for the 2D structure of a volcano. For the overall derivation, we consider that the magma is carried to the volcano's summit along the path that presents the least amount of resistance. Further, like the phreatic surface in an unconfined aquifer, the volcano's surface has a constant hydraulic head. For an easier calculation, we have approximated that the volcanic edifice has a small slope, i.e. $\partial h / \partial r \ll 1$, where the height is represented by 'h' and radial distance as 'r'. For the initial approximation the height is taken as very small, or initial height or $h_r = 1$. Now, using the viscous flow theory, some idealized models for the channel's geometrical arrangement in a porous material allow us to consider the Darcy's law for the fluid flow to derive the flow dynamics for the volcano. For the derivation, the common parameters used in Darcy's law along with Dupuit's approximation are utilized that involves the following parameters as Permeability (K), Gravity (g),

Viscosity (μ), *Density*(ρ) in deriving the theory of the idea of flow in an open, porous aquifer (Fig. 3).

Consider a 1D flow of lava with a hydraulic gradient against the gravity, modelled through a volcanic pathway or chamber which is assumed as the aquifer having radius (r) and a constant cross-sectional area (A) (Fig. 3). The volumetric flow rate (Q_r) through the chamber defined as the product of Darcy velocity (U_r) and cross-sectional area:

$$Q_r = AU_r \tag{1}$$

where cross sectional area of the chamber is defined as, $A = 2\pi rh$ and Darcy's velocity (U_r) as: $U_r = -\frac{K\rho_m g}{\mu}\frac{\partial h}{\partial r}$ (Lacey et al., 1981).

Replacing the values of A and U_r in Eq. (1):

$$Q_r = \frac{-2\pi rhgK\rho_m}{\mu}\frac{\partial h}{\partial r} \tag{2}$$

K is the permeability of chamber, ρ_m is density of magma, h is thickness of chamber and r is radius of chamber (Lacey et al., 1981).

Any net outflow from the magma pathway or chamber (assumed as cylindrical structure) must be balanced by a decrease in the height of the phreatic surface in the chamber in accordance with the law of conservation of fluids. Hence, mass balance is represented as:

$$2\pi r\frac{\partial h}{\partial t} = -\frac{\partial Q_r}{\partial r} \tag{3}$$

Now using the derived form of Q_r from Eq. (2) (Lacey et al., 1981):

$$\frac{\partial h}{\partial t} = \frac{K\rho_m g}{r\mu}\frac{\partial}{\partial r}\left(rh\frac{\partial h}{\partial r}\right) \tag{4}$$

Based on the physical parameters for a single volcano, we have assumed a constant B (cubic meters multiplied by meter square per second), which is a product of multiple constant parameters:

$$B = \frac{Kg\rho_m}{\mu} \tag{5}$$

$$\left.\begin{aligned} \frac{\partial h}{\partial t} &= \frac{B}{r}\left[\frac{\partial}{\partial r}(rh)\cdot\frac{\partial h}{\partial r} + rh\frac{\partial^2 h}{\partial r^2}\right] \\ \frac{\partial h}{\partial t} &= \frac{B}{r}\left[\left(h + r\frac{\partial h}{dr}\right)\frac{\partial h}{\partial r} + rh\frac{\partial^2 h}{\partial r^2}\right] \\ \frac{\partial h}{\partial t} &= B\left[\frac{h}{r}\frac{\partial h}{\partial r} + \left(\frac{\partial h}{\partial r}\right)^2 + rh\frac{\partial^2 h}{\partial r^2}\right] \end{aligned}\right\} \tag{6}$$

Now introducing similarity variables to solve Eq. (4), where Q_0 is the value of Q_r at $r = 0$ is the magma supply rate (Lacey et al., 1981):

$$\left.\begin{aligned} f &= \left(\frac{K\rho_m g}{\mu Q_0}\right)^{\frac{1}{2}} \cdot h \\ \varepsilon &= \left(\frac{\mu}{K\rho_m g Q_0}\right)^{\frac{1}{4}} \frac{r}{t^{\frac{1}{2}}} \end{aligned}\right\} \tag{7}$$

Considering the time derivative of Eq. (5)

$$\left.\begin{aligned} \frac{\partial h}{\partial t} &= \left(\frac{Q_0}{B}\right)^{\frac{1}{2}} \frac{\partial f}{\partial t} \\ &\mathit{and} \\ \frac{\partial h}{\partial r} &= \left(\frac{Q_0}{B}\right)^{\frac{1}{2}} \frac{\partial f}{\partial r} \end{aligned}\right\} \tag{8}$$

Taking second order derivative of Eq. (8)

$$\frac{\partial^2 h}{\partial r^2} = \left(\frac{Q_0}{B}\right)^{\frac{1}{2}} \frac{\partial^2 f}{\partial r^2} \tag{9}$$

deriving from Eq. (9)

$$h = \left(\frac{Q_0}{B}\right)^{\frac{1}{2}} f \tag{10}$$

$$r = (BQ_0)^{\frac{1}{4}} \xi \cdot t^{\frac{1}{2}} \tag{11}$$

dividing Eqs. (10) and (11),

$$\frac{h}{r} = \frac{\left(\frac{Q_0}{B}\right)^{\frac{1}{2}} f}{(BQ_0)^{\frac{1}{4}} \xi t^{1/2}} \tag{12}$$

Using the derived form of Eq. (12) in Eqs. (4) and (8):

$$\left.\begin{aligned} \left(\frac{Q_0}{B}\right)^{\frac{1}{2}} \frac{\partial f}{\partial t} &= B\left[\frac{h}{r}\left(\frac{Q_0}{B}\right)^{\frac{1}{2}} \frac{\partial f}{\partial r} + \left(\frac{Q_0}{B}\right)\left(\frac{\partial f}{\partial r}\right)^2 + h\left(\frac{Q_0}{B}\right)^{\frac{1}{2}} \frac{\partial^2 f}{\partial r^2}\right] \\ \frac{\partial f}{\partial t} &= B\left[\frac{h}{r}\frac{\partial f}{\partial r} + \left(\frac{Q_0}{B}\right)^{\frac{1}{2}}\left(\frac{\partial f}{\partial r}\right)^2 + h\frac{\partial^2 f}{\partial r^2}\right] \end{aligned}\right\} \tag{13}$$

Taking partial derivative of Eq. (7):

$$\left.\begin{aligned} \frac{\partial \xi}{\partial t} &= -\frac{1}{2}\left(\frac{1}{BQ_0}\right)^{\frac{1}{4}} \frac{r}{t^{3/2}} \\ \frac{\partial \xi}{\partial r} &= \left(\frac{1}{BQ_0}\right)^{\frac{1}{4}} \frac{1}{t^{1/2}} \\ \frac{\partial^2 \xi}{\partial r^2} &= 0 \end{aligned}\right\} \quad (14)$$

Expanding the initial form of derivatives from Eq. (13) (Lacey et al., 1981):

$$\frac{\partial f}{\partial t} = \frac{\partial f}{\partial \xi} \times \frac{\partial \xi}{\partial t} = f' \frac{\partial \xi}{\partial t} \quad \left(\text{where } f' = \frac{\partial f}{\partial \xi}\right) \quad (15)$$

Replacing the values from Eq. (14):

$$\left.\begin{aligned} \frac{\partial f}{\partial t} &= -\frac{1}{2}\left(\frac{1}{BQ_0}\right)^{\frac{1}{4}} \frac{r}{t^{3/2}} f' \\ \frac{\partial f}{\partial t} &= -\frac{1}{2}\frac{\xi}{t} f' \end{aligned}\right\} \quad (16)$$

Similarly expanding the initial form of derivatives from Eq. (13):

$$\frac{\partial f}{\partial r} = \frac{\partial f}{\partial \xi} \times \frac{\partial \xi}{\partial r} = f' \frac{\partial \xi}{\partial r} \quad (17)$$

Taking the second-order derivative,

$$\frac{\partial^2 f}{\partial r^2} = \frac{\partial}{\partial r}\left[f' \frac{\partial \xi}{\partial r}\right] = \frac{\partial}{\partial r}(f')\left(\frac{\partial \xi}{\partial r}\right)^2 + f' \frac{\partial^2 \xi}{\partial r^2} = f''\left(\frac{\partial \xi}{\partial r}\right)^2 + f'\left(\frac{\partial^2 \xi}{\partial r^2}\right) \quad (18)$$

Replacing the values of $\frac{\partial f}{\partial t}$ and $\frac{\partial^2 f}{\partial r^2}$ from Eqs. (16) and (18) in Eq. (13):

$$-\frac{1}{2}\frac{\varepsilon f'}{t} = B\left[\frac{h}{r} \cdot \left(f' \cdot \frac{\partial \xi}{\partial r}\right) + \left(\frac{Q_0}{B}\right)^{\frac{1}{2}} \left(\frac{f' \partial \xi}{\partial r}\right)^2 + h\left\{f''\left(\frac{\partial \xi}{\partial r}\right)^2 + f' \partial^2 \xi / \partial r^2\right\}\right]$$

Further taking the values from Eqs. (12) and (14) (modified after Lacey et al., 1981):

$$\left.\begin{aligned} -\frac{1}{2}\frac{\varepsilon f'}{t} &= B\left[\frac{f}{\xi}\frac{Q_0^{\frac{3}{4}}}{B^{\frac{3}{4}}}\frac{1}{t^{\frac{1}{2}}}\frac{f' B^{\frac{3}{4}}}{Q_0^{\frac{1}{4}}}\frac{1}{t^{\frac{1}{2}}} + B\frac{Q_0^{\frac{1}{2}}}{B^{\frac{1}{2}}} B^{-\frac{1}{2}} Q_0^{-\frac{1}{2}} (f')^2 \frac{1}{t} + \frac{Q_0^{\frac{1}{2}}}{B^{\frac{1}{2}}} f'' \frac{B^{\frac{1}{2}}}{Q_0^{\frac{1}{2}}}\frac{1}{t}\right] \\ -\frac{1}{2}\frac{\xi f'}{t} &= \left[\frac{ff'}{\xi t} + \frac{(f')^2}{t} + \frac{f''}{t}\right] \\ f'' + (f')^2 + \frac{1}{\xi}ff' + \frac{1}{2}\xi f' &= 0 \end{aligned}\right\} \quad (19)$$

Applying boundary conditions on the magma supply, we reconsider, $r \to 0$ and $Q_r \to Q_0$

where

$$Q_0 = -\frac{2\pi K \rho_m g r h}{\mu} \frac{\partial h}{\partial r} \tag{20}$$

Now replacing the values from Eqs. (8) and (14) in Eq. (17) (Lacey et al., 1981):

$$\begin{aligned} Q_0 &= -2\pi B r h \left(\frac{Q_0}{B}\right)^{\frac{1}{2}} f' \left(\frac{1}{BQ_0}\right)^{\frac{1}{4}} \frac{1}{t^{1/2}} \\ Q_0 &= -2\pi f f' \xi Q_0 \\ ff'\varepsilon &= -\frac{1}{2\pi} \end{aligned} \tag{21}$$

When $r \to 0$ it leads to$\varepsilon \to \varepsilon_0$, now, we can reconsider Eq. (6) as:

$$r_0 = \left(\frac{K \rho_m g Q_0}{\mu}\right)^{\frac{1}{4}} \xi_0 t^{1/2} \tag{22}$$

where the non-dimensional radius of the volcano is denoted byε_0. According to Eq. (22) the radial distance of volcanoes increases proportionally with the square root of time. Now taking the derivative of Eq. (22) with respect to time:

$$\frac{\partial r_0}{\partial t} = U_0 = \frac{1}{2}\left(\frac{K \rho_m Q_0 g}{\mu}\right)^{\frac{1}{4}} \frac{\xi_0}{t^{1/2}} \tag{23}$$

We already have the Darcy velocity described as:

$$U_0 = -\frac{K g \rho_m}{\mu} \frac{\partial h}{\partial r}$$

Taking the values $\frac{\partial h}{\partial r}$ from Eqs. (8) and (14) (Lacey et al., 1981):

$$U_0 = -B\left[\left(\frac{Q_0}{B}\right)^{\frac{1}{2}} f' \left(\frac{1}{BQ_0}\right)^{\frac{1}{4}} \frac{1}{t^{1/2}}\right] = \left(\frac{K \rho_m Q_0 g}{\mu}\right)^{\frac{1}{4}} \frac{f'}{t^{1/2}} \tag{24}$$

Considering these values of U_0 from Eqs. (23) and (24) (Lacey et al., 1981):

$$\begin{aligned} \frac{1}{2}\left(\frac{K \rho_m Q_0 g}{\mu}\right)^{\frac{1}{4}} \frac{\xi_0}{t^{1/2}} &= \left(\frac{K \rho_m Q_0 g}{\mu}\right)^{\frac{1}{4}} \frac{f'}{t^{1/2}} \\ \frac{1}{2}\xi_0 &= f' \end{aligned} \tag{25}$$

But when $r \to 0$ it leads to $h \to 0$, therefore $f = 0$ at $\xi = \xi_0$, which leads to $\varepsilon_0 = 1.16$

Replacing this value ε_0 in Eq. (22):

$$r_0 = 1.16\left(\frac{K\rho_m g Q_0}{\mu}\right)^{\frac{1}{4}} t^{1/2} \tag{26}$$

Again from Eqs. (8) and (14) (Lacey et al., 1981):

$$\frac{\partial h}{\partial r} = \left(\frac{Q_0}{B}\right)^{\frac{1}{2}} f' \left(\frac{1}{BQ_0}\right)^{\frac{1}{4}} \frac{1}{t^{1/2}} \tag{27}$$

Replacing Eq. (26) in Eq. (27) (Lacey et al., 1981):

$$\left(\frac{\partial h}{\partial r}\right)_0 = -0.58\left(\frac{\mu}{Kg\rho_m}\right)^{3/4} \frac{Q_0^{1/4}}{t^{1/2}} \tag{28}$$

Taking the negative product of Eq. (26) and multiplying with Eq. (28) (Lacey et al., 1981):

$$h_r = 0.673\left(\frac{\mu Q_0}{Kg\rho_m}\right)^{1/2} \tag{29}$$

For the modelling purpose, we have used Eqs. (26), (28) and (29) frequently to theoretically estimate the two-dimensional profile of the volcanoes. Equation (28) helps to determine the radial distance of the volcano, whereas Eq. (29) helps to determine the height of the volcano at distance 'r'. The flank slope of the volcano can be represented from the height and the radial distance of the volcano. We have also observed from the synthetic profile curves that the radial distance is dependent on time but the height does not depend on time.

3 Data Sets Used for the Theoretical Model Validation on Actual Volcanoes

3.1 The 2D Volcano Topographic Profile

To visualize the relationship between height and distance, we created a synthetic dataset using MATLAB version R2018a. We have made a custom script that takes different physical parameters (e.g. flow rate, permeability, gravitational acceleration, viscosity, time and density) as input and produces a synthetic dataset with distance and elevation. The axisymmetric form of volcanoes represents the various physical processes that have dominated during their genesis and evolution. But, in some cases non-uniform ensemble of subsurface dikes promoting to the irregular contribution of subsurface magma intrusion to the growth and evolution of a volcano creates deviations in the ideal axisymmetric form. Although this was considered in the recent studies by Cosburn and Roy (2020), but the actual fits of the theoretical curves to the magmatic evolutionary curves were not well fit and found with rather deviations due to the intrusions. For this work, we have only considered a porous flow model to determine the flow of magma through the edifice and have tried to emphasize the essential role of gravity and viscosity in designing its overall topography. The model assumes an equipotential volcanic surface for magma to flow through a uniform porous medium. Using the synthetic data sets, we can predict that under suitable conditions the volcanoes would all have the same shape but will vary in the topography depending on the various involving parameters in the fluid movement.

3.2 *Physical Parameters and Topographic Profiles of Volcanoes*

Table 1 compiles the associated physical parameters for magma and magnitudes of magma flow belonging to particular volcanoes from several literatures. The heights of the stratovolcanoes on Earth is taken from the public archive (Internet ref-1), and the heights of the volcanoes in the solar system are taken from various literatures compiled in Table 1. Topographic profile of the six analysed axisymmetric volcanoes on Earth is taken using Global Multi-Resolution Topography (GMRT) Map tool—GMRT v4.1.1 (internet ref-2). This synthesis is a multi-resolution Digital Elevation Model (DEM) that provides the required infrastructure for accessing the DEM of various features such as grids, images, points and profiles. Regions under United States are acquired at a 10 m resolution from the USGS National Elevation Dataset (NED) (internet ref-3) whereas, for acquiring the Global data, a 30 m resolution from the National Aeronautics and Space Administration (NASA)'s Advanced Space borne Thermal Emission and Reflection Radiometer Global Digital Elevation Model (ASTER) are considered. The topographic profiles of the three volcanoes in Mars planet are taken from Internet ref-4. The Mars orbiter laser altimeter (MOLA) is an instrument on the Mars Global Surveyor (MGS) spacecraft. MGS carried the Laser altimeter on the board, which determines the elevation of the surface features and various landforms by mapping process on the Mars.

The swaths of the different volcanoes and accompanying regions on the Earth were taken by using the Global Mapper software (Version 17.0, 2015). The average elevation along the perpendicular method was considered by varying swath widths for a smooth profile. The swaths are analysed in different directions, but we have considered the direction with best axisymmetric profile of the topography where the theoretical model curve can fit. In case of the volcanoes of Earth 1–5 km wide swaths are taken, whereas for the Mars the swaths are 20 km wide.

Table 1 Parameters considered for modelling the different volcanoes on Earth and Mars

Volcano (earth/mars)	Gravity (m/s^2)	Height (m)	Age (ka/Ma)	Flow rate (m^3/s)	Viscosity (Pa s)	Permeability (m^2)	References
Ayelu, Ethiopia	9.81	2145	71×10^2 ka	0.00025	100,000–1,000,000	10^{-9}	Smithsonian Institution, Global Volcanism Program; Barberi et al. (1974)
Etna, Italy	9.81	3357	500 ka	0.05	1000–11,000	10^{-9}	Smithsonian Institution, Global Volcanism Program; Branca et al. (2008)
Fuji, Japan	9.81	3776	100 ka	0.19	100–4000	10^{-9}	Jet Propulsion Laboratory, NASA; Aoki et al. (2019), Sato (2005)
Merapi, Indonesia	9.81	2910	170 ka	0.05	500–10,000	10^{-9}	Smithsonian Institution, Global Volcanism Program; Gertisser et al. (2012), Kushnir et al. (2016), Siswowidjoyo et al. (1995)
Villarrica, Chile	9.81	2847	95 ka	0.1	100–50,000	10^{-9}	Smithsonian Institution, Global Volcanism Program; Lara et al. (2004)
St. Helens, USA	9.81	2950	23×10^3 ka	0.0005	10,000–1,000,000	10^{-9}	(United States Geological Services; Evarts et al. (1987)

(continued)

Table 1 (continued)

Volcano (earth/mars)	Gravity (m/s^2)	Height (m)	Age (ka/Ma)	Flow rate (m^3/s)	Viscosity (Pa s)	Permeability (m^2)	References
Olympus Mons	3.72	21,900	350–150 Ma	0.09	1000–11,000	10^{-10}	Carr (2007), Plescia (2004)
Ascraeus Mons	3.72	14,900	1416–1320 Ma	0.01	1000–11,000	10^{-10}	Plescia (2004)
Elysium Mons	3.72	12,600	150 Ma	0.07	1000–11,000	10^{-10}	Plescia (2004)

4 Model Validation from Stratovolcanoes on Planetary Bodies

4.1 Volcanic Landforms on Earth

The prior studies have simply compared the theoretically generated curve to topographic profiles for various stratovolcanoes on Earth (Table 2). However, in order to evaluate our theoretical model more accurately, we have chosen six distinct stratovolcanoes from various locations around the world. The diverse stratovolcanoes were picked up based on the different tectonic environment at tectonic boundaries as well as intracontinental regions (Fig. 4). For the variations in the controlling parameters volcanoes at intracontinental region are picked for acidic behaviour of magma and at the tectonic boundary for the basaltic behaviour. The volcanoes on the Earth that we have considered for the study are Mount Etna (Italy), Merapi (Indonesia), Villarrica (Chile), St. Helens (USA), Mount Fuji (Japan) and Ayelu (Ethiopia) (Fig. 4, Table 1). Some of the landforms show perfect axisymmetric stratovolcano characteristics where others do not have smooth profiles. Figure 4 inset shows the locations of these volcanoes (top left) and the swaths taken of the six representative stratovolcanoes. For generating a smooth topographic profile for Etna the swath taken is averaged for 2 km, 1 km for Merapi, 4 kms for Villarrica, 3 km for St. Helens, 1 km for Fuji and for Ayelu the swath is averaged at 0.5 km (Fig. 4). The swaths are taken in such a way that they can produce the smooth and ideal axisymmetric profiles to validate the model.

By using the theoretically derived solution in MATLAB, we have generated the model curves of or these six axisymmetric volcanoes on Earth as well as on Moon and Mars. The controlling physical parameters used to obtain the flank slope have been given in Table 1. The theoretically modelled profiles were observed in a good fit with the actual topographic profiles (Fig. 5). These theoretical generated curves show

Table 2 Studies involving the geometrical form of volcanoes

Name of the volcano	Geometry type	Theory	References
Stratovolcanoes on Earth	Non-linear	Hydraulic resitance to magma flow	Lacey et al. (1981)
Aseismic ridges, Seamounts	Non-linear	Porous flow through Dupuit's approximation	Angevine et al. (1984)
Composite volcanoes on Earth	Conical	Primary profile and steady state profiles	Davidson and Silva (2000)
Stratovolcanoes on Earth	Non-linear	Extension of Porous flow through Dupuit's approximation	Arjona and Díaz (2015)
88 stratovolcanoes on Earth	Axisymmetric	Linear and non-linear arms of edifice using elliptical base	Cosburn and Roy (2020)

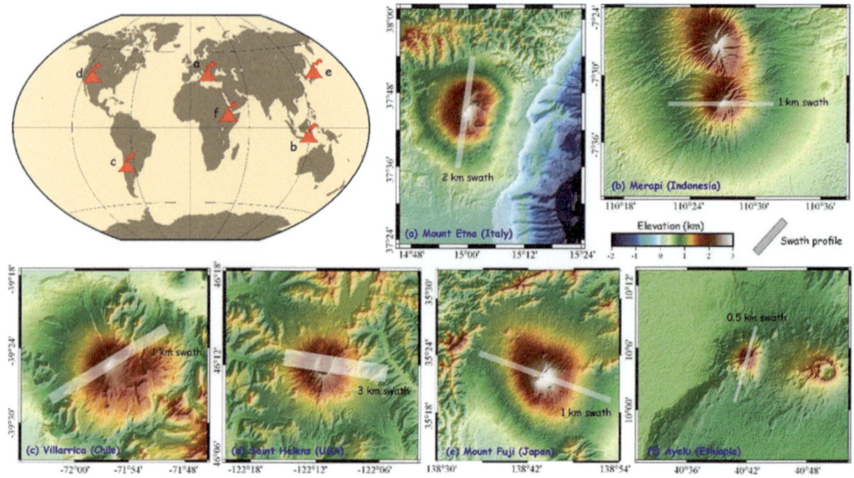

Fig. 4 Global figure on the top left corner represents the locations of the considered stratovolcanoes of the world. Swaths are taken using Global Mapper from GMRT topography datasets of six different axis-symmetric volcanoes of Earth these are **a** Mount Etna. **b** Merapi. **c** Villarrica. **d** Saint Helens. **e** Mount Fuji. **f** Ayelu. The colour bar represents the contour elevation of the map and the grey bands represent the swaths taken across the individual volcanoes

a good agreement towards the smooth flank slopes and also satisfy that the concave shape theory for the sub-aerial volcanoes as referred by Angevine et al. (1984). But, we have also observed some irregularities in the actual flank slopes, such as the Mount Etna and Mount Villarrica, where the theoretically modelled curves are slightly mismatched. An explanation to that can be the theoretical modelled slopes are smooth whereas under natural conditions the volcanoes do not have a smooth slope. The natural irregular shapes of the volcanic flanks may suggest the evolution of the flanks due to various natural conditions and processes such as weathering and endogenic processes going on within the volcanic edifice. Although the processes cannot be neglected, we have tried to fit the theoretical curves with the natural slope so that we can compare the parameters associated with the volcanic process. This can give an idea about the temporal evolution of the parameters related to the volcanoes.

Different parameters like the viscosity of the lava, age of the volcano and tectonic environment can be predicted from the fitting of the theoretically modelled curve with the actual flank slope of the volcanoes. As from the theoretical solution, we have found that the radial distance of the volcanoes is also dependent on the age of the volcano, so from this model we can assume that the stratovolcano having a large radial distance has a quite long eruption history which are also consistent with the previous assumptions of the piling up process (Cosburn & Roy, 2020).

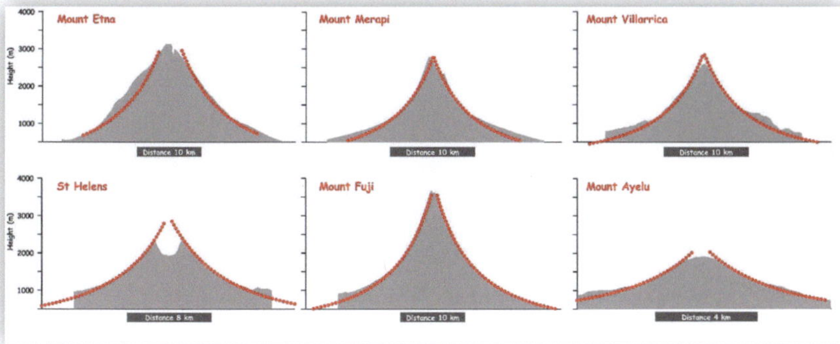

Fig. 5 Theoretically modelled profile (red dashed lines) in comparison with the Topographic profile (grey fillings) plotted from the swaths of axis-symmetric Volcanoes of Earth. Note that the theoretical curves fit well with actual topographic profile. The slight mismatch in curve represents geomorphological and endogenic factors. Scale for the radial distance is marked below each profile

4.2 Volcanic Landforms on Mars

The MOLA data sets provide the detailed elevation map for the Mars. Figure 6 presents MOLA-derived important features of the planet. We can observe that the elevated volcanic features are mostly distributed at the Eastern part of the Mars from the central meridian line, which can be referred to as the most tectonically active region in the past. Due to the absence of active tectonics on Mars, we have not considered the volcanoes on associated tectonic environment method as like on Earth. We considered three different prominent volcanoes—Elysium Mons, Olympus Mons and Ascraeus Mons having enormous height of ~20 km for Mars (Fig. 7). We have followed a similar methodology to model the topography for the selective volcanoes on Mars as we have provided for the Earth. Using the appropriate value of the parameters mentioned in Table 1, theoretical curves were generated for all the three volcanoes, which were further compared with their real ones.

The swaths taken for the three volcanoes are in form of ~10 km widths so as to observe the required smooth profiles. From the results it is observed that the theoretically generated curves fit well to their original topographic ones except for the Olympus Mons (Fig. 7). This may be due to the inefficiency in values of some parameters such as time which were taken in the range of 150–350 Ma (Table 1). Moreover, the excess height of the volcanoes may be a result of the very low gravitational acceleration acting on the surface of Mars or may be due to excess viscous nature of the magma. This also supports the fact that the Mars having a calculated low gravitational acceleration value (3.72 which is around 1/3rd of the Earth) and an estimated high viscosity range (Table 1).

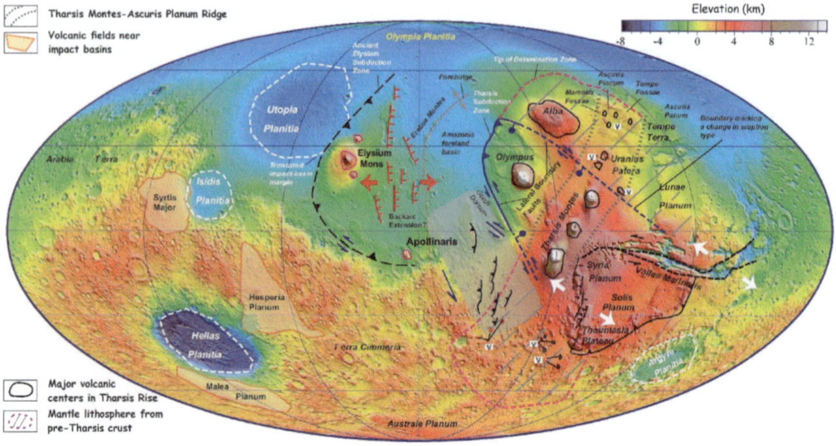

Fig. 6 Tectonic map of Mars showing the different features on the Martian surface (taken from Yin, 2012). The colour bar represents the contour elevation of the map. The various elevational (volcanoes and mounds) and depressional (craters and valleys) features are marked prominently

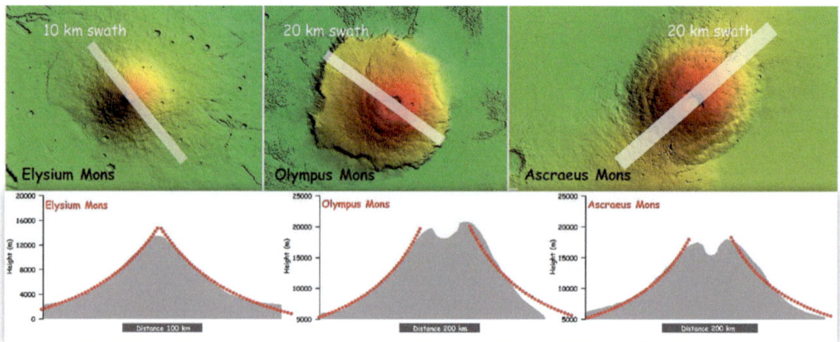

Fig. 7 Comparison of the theoretically modelled profile with the Topographic profile of Volcanoes on Mars. The upper panel represents different swath profiles whereas the lower portion shows the observed fit. Note that the theoretical curves (red dashed lines) fit well with the grey filled actual topographic profile but there is a. slight mismatch in case of the Olympus Mons. The scale for the radial distance is marked below each profile

5 Relation Among Volcanic Heights, Magma Viscosity and Gravity

5.1 Relation Between Volcanic Height and Viscosity

Although several physical, chemical and natural factors affect a volcano's height, viscosity as a chemical parameter can have significant impact on the lava flow off the volcano. As we have already discussed about the major role of the flow of the lava from the volcano in shaping the landform (Sect. 4.1), the viscosity becomes a major factor on determining the shape of the volcano. The viscosities of volcanoes on Earth varies widely according to the tectonic environment and other natural factors. The pre-eruptive viscosities of the basaltic to rhyolitic magmas range from 10^1 to 10^8 Pa s. (Giordano et al., 2008). Viscosities for andesitic magmas range from 10^2 to 10^7 Pa s, which is lower or higher than the 10^5 to 10^6 Pa s. for phenocryst-poor rhyolitic magmas (Giordano et al., 2008). Despite the fact that the majority of predicted viscosities of erupted magmas are $<10^6$ Pa s, the composition of the magma affects the viscosities. Incidentally, a relationship between a range of magma viscosity with the heights of stratovolcanoes worldwide has remained a due.

We have illustrated the relationship between the volcano height and the magma viscosity using Eq. (30):

$$h_r = 0.673\left(\frac{\mu Q_0}{Kg\rho_m}\right)^{1/2} \tag{30}$$

Here, we observe that the height of the volcanoes is directly proportional to the viscosity of the magma flow. We have varied only two parameters here (i.e. Viscosity and Height) and considered the other parameters as constant. Simplified them into logarithmic form and we find:

$$\begin{aligned} Log_{10}h_r &= Log_{10}0.673 + \frac{1}{2}Log_{10}\left(\frac{\mu Q_0}{Kg\rho_m}\right) \\ Log_{10}h_r &= +0.5(Log_{10}g) + [c] \end{aligned} \tag{31}$$

where

$$[c] = \left[Log_{10}0.673 + 0.5(Log_{10}Q_0 + Log_{10}\mu - Log_{10}K - Log_{10}\rho_m)\right] \tag{32}$$

Using Eq. (31), we have plotted (Fig. 9) the theoretical predicted slope for different possible values of the controlling parameters for a volcanic landform. The natural database for height and fluid flow parameters are collected from different databases for various stratovolcanoes around the globe (Table 1). As the volcanoes with same height have a range of viscosities (10^5–10^6 for heights ranging from 2000 to 2200 m, 10^3–10^5 for heights ranging from 2800 to 3000 m, 10^2–10^3 for heights ranging from

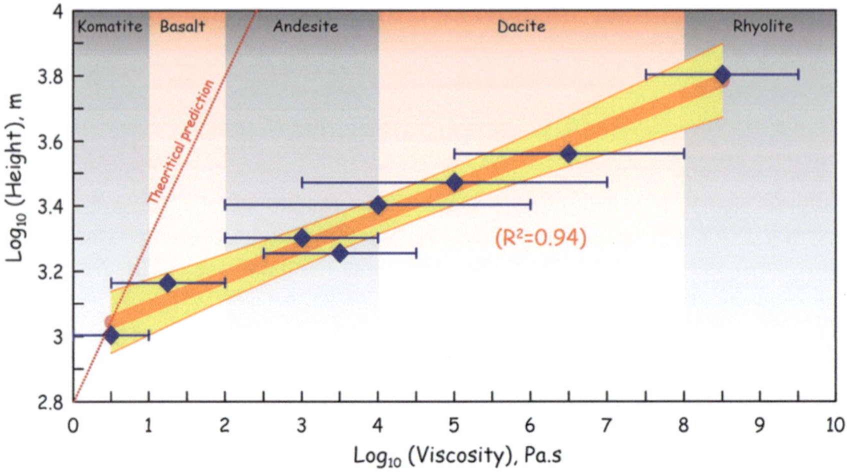

Fig. 8 Correlation between Log (Height) and Log (Viscosity) of different volcanoes on Earth. The dotted red line represents the theoretically predicted slope and the thick orange line represents the average slope of all the points between the 90% confidence interval. The error bars are represented in blue

3000 to 4000 m), it was so difficult to plot them in a linear relation. So, we have taken a particular viscosity range (as mentioned earlier) for the volcanoes with similar height. We have plotted the graph between log (height) and log (viscosity) for the actual observational slope of different volcanoes around the globe (Fig. 8).

The R^2 value for Fig. 8 establishes the accuracy of the relationship, where the percentage of the variation of the dependent variable (the range of viscosity in the eruptive history) that can be predicted from the independent variable(s) (present height). The goodness of fit is termed as the R^2 value, or coefficient of determination. R^2 values range from 0 to 1, with a value of 0 indicating that the regression prediction does not fit the data and a value of 1 indicating that it does. From Fig. 8, we get the R^2 value of the relationship between the log (height) and log (Viscosity) to be 0.94. The average slope of the graph is plotted with 90% interval. It is observed that the calculated slope from the real height and viscosity slightly deviates (R^2 value of 0.94) from the theoretically predicted slope, as the viscosity varies in a wide range. We can also argue that keeping other parameters constant while establishing relationship in between height of volcanoes and viscosity can be the reason for this deviation. But the negative slope with good correlation with the theoretically predicted curve and supports the assumption of the height and viscosity dependency. We can also note the positive correlation between viscosity and height of a volcano convey that they are directly proportional, and we can observe an increase in height with the increase in viscosity of fluid associated with the volcanic eruption.

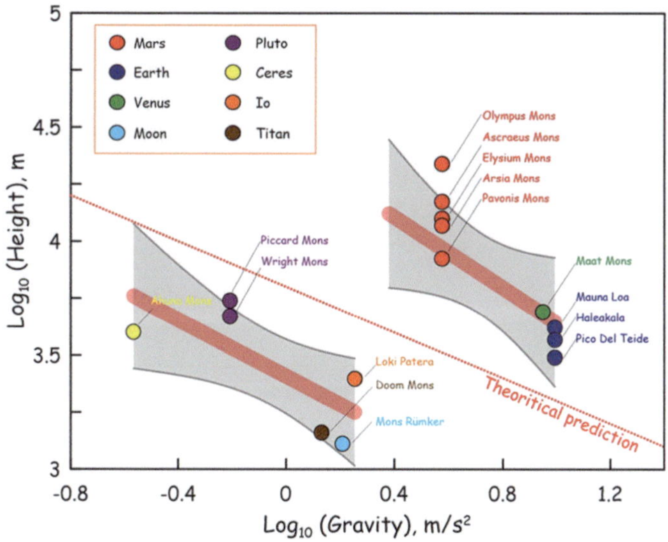

Fig. 9 Correlation between Log (Height) and Log (Gravity) of tallest volcanoes of terrestrial and other planetary bodies. The slope for the different planetary bodies have been segregated in to two different slopes between the 90% percent confidence interval where the terrestrial bodies are represented above the theoretical prediction (red dotted line) and other planetary bodies are represented below the prediction line

5.2 Relationship Between Volcanic Height and Gravity

From the theoretical definition, we can infer that the height of the volcanoes varies inversely with the square root of gravity (Eq. 30) taking flow rate and permeability as the constant parameters. This can be tested on the volcanic landforms present on planet Mars, which has a lower gravity (~3.77 m s^{-2}) and currently, has the tallest volcanoes of the solar system with heights ~20 km whereas height of volcanic landforms on earth are maximum ~10 km e.g. the Hawaiian shield. Thus to determine the actual relationship between gravity and volcanic height of planetary bodies through real observations we plotted the height versus gravity graph for different planetary bodies with a simplified equation (Eq. 31) from the previous results (Fig. 9). Using the graphical representation (Fig. 9), we have tried to illustrate the relationship between the volcano height and the gravitational acceleration by using the values from Table 3 for the parameters in Eq. (30).

We have varied only two parameters gravity and height and considered the other parameters as constant. Simplifying them into logarithmic scales:

$$Log_{10}h_r = Log_{10}0.673 + \frac{1}{2}Log_{10}\left(\frac{\mu Q_0}{K g \rho_m}\right)$$
$$Log_{10}h_r = -0.5(Log_{10}g) + [c] \qquad (33)$$

Table 3 Height and gravity data of tallest volcanoes of the Universe

Name of planet	Name of volcano	Gravitational acceleration (m/s^2)	Height (km)	Type	References
Terrestrial bodies					
Venus	Maat Mons	8.87	4.9	Volcanic	Robinson et al. (1995)
Earth	Mauna kea and Mauna loa	9.81	10.2 (4.2 km of this is above sea level)	Volcanic	Smithsonian Institution, Global Volcanism Program
	Haleakalā		9.1 (Rises 3.1 km above sea level)	Volcanic	Smithsonian Institution, Global Volcanism Program
	Pico del teide		7.5 (Rises 3.7 km above sea level)	Volcanic	Smithsonian Institution, Global Volcanism Program
Moon	Mons rümker	1.62	1.3	Volcanic	Wöhler et al. (2007)
Mars	Olympus Mons	3.72	21.9	Volcanic	Plescia (2004), Carr (2007)
	Ascraeus Mons		14.9	Volcanic	Plescia (2004)
	Elysium Mons		12.6	Volcanic	Plescia (2004)
	Arsia Mons		11.7	Volcanic	Plescia (2004)
	Pavonis Mons		8.4	Volcanic	Plescia (2004)
Other planetary bodies					
Ceres	Ahuna Mons	0.284	4	Cryovolcanic	Ruesch et al. (2016)
Io	Unnamed	1.796	2.5	Volcanic	Moore et al. (1984)
Titan	Doom Mons	1.353	1.45	Cryovolcanic	Lopes et al. (2013)
Pluto	Piccard Mons	0.42	5.5	Cryovolcanic	Schenk et al. (2018)
	Wright Mons		4.7	Cryovolcanic	Schenk et al. (2018)

where

$$[c] = \left[Log_{10}0.673 + 0.5(Log_{10}Q_0 + Log_{10}\mu - Log_{10}K - Log_{10}\rho_m)\right]$$

With 90% confidence intervals for the theoretical slope, the slope represented above the theoretical slope represents the terrestrial bodies whereas those below it represents extra-terrestrial bodies (Fig. 9). This can be corresponded to the higher g value in the terrestrial planets due to higher density than the planetary bodies and satellites those are present far away from the Sun in the Solar system. The observed profiles are very similar to the theoretical profile showing a negative slope. The slope value is −0.44, which is slightly greater than theoretically predicted slope −0.5. From this, we can infer that physical parameters such as gravity has a relationship with the height of the volcanoes on different planetary bodies. In other words, this relation can be used to remotely determine the gravity of the planet or height of a volcanic feature in the solar system.

6 Conclusions

- The theoretical modelled curves show a good fit with the topographical profiles of different volcanoes derived from satellite observations on Earth's and Mars's surface.
- The modelled values follow the trend of viscosity correlation of global volcanoes with volcano height but the modelled slope value is steeper than the observed values for volcanoes at different tectonic environment.
- Gravitational acceleration has a better fit to volcanic height correlation in comparison to viscosity. This can be explained by the consistency of gravity as a physical parameter all over the planetary body as compared to chemical parameters such as viscosity.

7 Limitations

- Effect of different geomorphologic factors such as erosion and wind direction during eruption could not be considered and the swath was taken in such a way that the axisymmetric profile of the volcano can be observed prominently by avoiding erosional factors.
- Endogenic processes such as caldera collapse and flank eruption of the magma chamber could not be theoretically considered.
- The theory is largely applicable to stratovolcanoes. Other volcanoes such as shield, and lava domes do not fit in to this observation and modelling.

Acknowledgements The presented work in the article has been performed within the framework of Sambit Sahoo's Ph.D. thesis at NIT Rourkela. SS is supported by CSIR-UGC fellowship. We thank Soumyajit Mukherjee (IIT Bombay) for inviting, handling and reviewing this article. Mukherjee (2026) summarized this work.

Authorship Contribution Statement B.K. has provided the research ideas. S.S have written the original draft and performed the theoretical computation and correlation analysis. Figures are generated by S.S. A.K.J. and S.S. performed theoretical computation and correlation analysis. B.K. reviewed and edited the draft and interpreted the results. All authors have taken part in finalizing the manuscript.

Data Sets and Codes Availability Statement Topographic map presented in Fig. 1 has been generated using the General Bathymetric Chart of the oceans by rectangular area search with GeoTIFF output (https://download.gebco.net/). The Figs. are generated by Grapher graphical application (version 8.7.844 URL: https://www.goldensoftware.com/products/grapher), Surfer graphical application (version 8.7.844 URL: https://www.goldensoftware.com/products/surfer), Generic Mapping Tools (Wessel et al., 2019) (version 5.2.1; URL: https://www.generic-mapping-tools.org/) and Corel Draw graphical application (version 18; URL: https://www.coreldraw.com/en).

Declaration of Competing Interest The authors have declared that the present work is true and not been influenced by any means and they truly have no known conflict of interests or competing financial interests or personal relationships.

References

Aoki, Y., Tsunematsu, K., & Yoshimoto, M. (2019). Recent progress of geophysical and geological studies of Mt. Fuji Volcano, Japan. *Earth-Science Reviews, 194*, 264–282.

Angevine, C. L., Turcotte, D. L., & Ockendon, J. R. (1984). Geometrical form of aseismic ridges, volcanoes, and seamounts. *Journal of Geophysical Research: Solid Earth, 89*(B13), 11287–11292.

Arjona, A., & Díaz, J. I. (2015). Geometrical evolution of volcanoes: A theoretical approach. *Revista de la Real Academia de Ciencias Exactas Físicas y Naturales. Serie a. Matemáticas, 109*, 511–534.

Barberi, F., et al. (1974) Evolution of Eolian arc volcanism (southern Tyrrhenian Sea). *Earth and Planetary Science Letters, 21*(3), 269–276.

Becker, G. F. (1885). The geometrical form of volcanic cones and the elastic limit of lava. *American Journal of Science, 3*(178), 283–293.

Branca, S., Coltelli, M., De Beni, E., & Wijbrans, J. (2008). Geological evolution of Mount Etna volcano (Italy) from earliest products until the first central volcanism (between 500 and 100 ka ago) inferred from geochronological and stratigraphic data. *International Journal of Earth Sciences, 97*, 135–152.

Carr, M. H. (1976). The volcanoes of Mars. *Scientific American, 234*(1), 32–43. http://www.jstor.org/stable/24950258

Carr, M. H. (2007, September). Water on early Mars. In *Ciba Foundation Symposium 202—Evolution of Hydrothermal Ecosystems on Earth (and Mars?) Evolution of Hydrothermal Ecosystems on Earth (and Mars?): Ciba Foundation Symposium* (Vol. 202, pp. 249–272). Wiley.

Cosburn, K., & Roy, M. (2020). Analysing the topographic form of stratovolcanoes. *Journal of Volcanology and Geothermal Research, 407*, Article 107123.

Davidson, J., & De Silva, S. (2000). Composite Volcanoes. *Encyclopedia of Volcanoes, 1*, 663–681.

Eaton, J. P., & Murata, K. J. (1960). How volcanoes grow. *Science, 132*(3432), 925–938. https://doi.org/10.1126/science.132.3432.925

Evarts, R. C., Ashley, R. P., & Smith, J. G. (1987). Geology of the Mount St. Helens area: Record of discontinuous volcanic and plutonic activity in the Cascade Arc of southern Washington. *Journal of Geophysical Research: Solid Earth, 92*(B10), 10155–10169.

Francis, P., Oppenheimer, C., & Stevenson, D. (1993). Endogenous growth of persistently active volcanoes. *Nature, 366*(6455), 554–557. https://doi.org/10.1038/366554a0

Gertisser, R., Charbonnier, S. J., Keller, J., & Quidelleur, X. (2012). The geological evolution of Merapi volcano, central Java, Indonesia. *Bulletin of Volcanology, 74*, 1213–1233.

Giordano, D., Russell, J. K., & Dingwell, D. B. (2008). Viscosity of magmatic liquids: A model. *Earth and Planetary Science Letters, 271*(1–4), 123–134. https://doi.org/10.1016/j.epsl.2008.03.038

Greeley, R., & Spudis, P. D. (1981). Volcanism on Mars. *Reviews of Geophysics, 19*(1), 13–41. https://doi.org/10.1029/RG019i001p00013

Gregg, T. K. (2017). Patterns and processes: Subaerial lava flow morphologies: A review. *Journal of Volcanology and Geothermal Research, 342*, 3–12.

Griffiths, R. W. (2000). The dynamics of lava flows. *Annual Review of Fluid Mechanics, 32*(1), 477–518. https://doi.org/10.1146/annurev.fluid.32.1.477

Grosse, P., van Wyk de Vries, B., Petrinovic, I. A., Euillades, P. A., & Alvarado, G. E. (2009). Morphometry and evolution of arc volcanoes. *Geology, 37*(7), 651–654.

Internet ref-1: Retrieved June 5, 2024, form https://volcano.si.edu/

Internet ref-2: Retrieved June 5, 2024, form https://www.gmrt.org/GMRTMapTool/

Internet ref-3: Retrieved June 5, 2024, form https://gdg.sc.egov.usda.gov/Catalog/ProductDescription/NED.html

Internet ref-4: Retrieved June 5, 2024, form https://attic.gsfc.nasa.gov/mola/

Kushnir, A. R., Martel, C., Bourdier, J. L., Heap, M. J., Reuschlé, T., Erdmann, S., … & Cholik, N. (2016). Probing permeability and microstructure: Unravelling the role of a low-permeability dome on the explosivity of Merapi (Indonesia). *Journal of Volcanology and Geothermal Research, 316*, 56–71.

Lacey, A., Ockendon, J. R., & Turcotte, D. L. (1981). On the geometrical form of volcanoes. *Earth and Planetary Science Letters, 54*(1), 139–143.

Lara, L. E., Naranjo, J. A., & Moreno, H. (2004). Lanín volcano (39.5 S), Southern Andes: geology and morphostructural evolution. *Revista geológica de Chile, 31*(2), 241–257.

Lopes, R. M., Kirk, R. L., Mitchell, K. L., LeGall, A., Barnes, J. W., Hayes, A., … & Malaska, M. J. (2013). Cryovolcanism on Titan: New results from Cassini RADAR and VIMS. *Journal of Geophysical Research: Planets, 118*(3), 416–435.

Milne, J. (1878). I.—On the form of Volcanos. *Geological Magazine, 5*(8), 337–345.

Milne, J. (1879). VI.—Further Notes upon the Form of Volcanos1. *Geological Magazine, 6*(11), 506–514.

Moore, J. M., Albin, E. F., & Greeley, R. (1984). Topographic evidence for shield volcanism on Io. In *Bulletin of the American astronomical society* (Vol. 16, p. 655).

Mukherjee, S. (2026). Introduction to "Structural Geology & Tectonics". In S. Mukherjee (Ed.), *Structural Geology & Tectonics*. Springer. ISBN: 978-981-95-4743-2.

Plescia, J. B. (2004). Morphometric properties of Martian volcanoes. *Journal of Geophysical Research: Planets, 109*(E3).

Robinson, C. A, Thornhill, G. D, & Parfitt, E. A. (1995). Large-scale volcanic activity at Maat Mons: Can this explain fluctuations in atmospheric chemistry observed by Pioneer Venus? *Journal of Geophysical Research: Planets, 100*(E6) 11755–11763. https://doi.org/10.1029/95JE00147

Ruesch, O., Platz, T., Schenk, P., McFadden, L. A., Castillo-Rogez, J. C., Quick, L. C., … & Russell, C. T. (2016). Cryovolcanism on ceres. *Science, 353*(6303), aaf4286.

Sato, H. (2005). Viscosity measurement of subliquidus magmas: 1707 basalt of Fuji volcano. *Journal of Mineralogical and Petrological Sciences, 100*(4), 133–142.

Schenk, P. M., Beyer, R. A., McKinnon, W. B., Moore, J. M., Spencer, J. R., White, O. L., … & Olkin, C. (2018). Basins, fractures and volcanoes: Global cartography and topography of Pluto from New Horizons. *Icarus, 314*, 400–433.

Siswowidjoyo, S., Suryo, I., & Yokoyama, I. (1995). Magma eruption rates of Merapi volcano, Central Java, Indonesia during one century (1890–1992). *Bulletin of Volcanology, 57*, 111–116.

Spudis, P. D. (2015). *The encyclopedia of volcanoes volcanism on the moon* (pp. 689–700). Elsevier. https://doi.org/10.1016/B978-0-12-385938-9.00039-0

Taylor, P. T., Wood, C. A., & O'Hearn, T. J. (1980). Morphological investigations of submarine volcanism: Henderson Seamount. *Geology, 8*(8), 390–395.

Wessel, P., Luis, J. F., Uieda, L. A., Scharroo, R., Wobbe, F., Smith, W. H., & Tian, D. (2019). The generic mapping tools version 6. Abstract plain language summary key points. *Geochemistry Geophysics Geosystems, 20*(11), 5556–5564. https://doi.org/10.1029/2019GC008515

Wöhler, C., Lena, R., & Phillips, J. (2007). Formation of lunar mare domes along crustal fractures: Rheologic conditions dimensions of feeder dikes and the role of magma evolution. *Icarus, 189*(2), 279–307. https://doi.org/10.1016/j.icarus.2007.01.011

Yin, A. (2012). An episodic slab-rollback model for the origin of the Tharsis rise on Mars: Implications for initiation of local plate subduction and final unification of a kinematically linked global plate-tectonic network on Earth. *Lithosphere, 4*(6), 553–593.

Deformation Models Induced by Major Regional Scale Structures Beneath the Northeast India and the Contemporary Seismotectonics

Debasis D. Mohanty and Ruchika Bajpai Mohanty

Abstract The northeast (NE) India represents a complex geodynamics with an active seismicity, known as one of the major seismic potential zones of the world. The seismic deformation pattern beneath this region is disparate and is mostly controlled by complex tectonics and large geological structures surrounded by this region. It is well proved that the long extension of major geologic structures (mountain belts, folds and faults) contributes largely towards the deformation patterns beneath a particular region. This chapter presents deformation models for the lithospheric-scale deformation mechanism controlled by major geological structures present in and around the Shillong plateau, Assam Foredeep and the northeast Himalayan arc region. The shear wave splitting analysis at the closest proximity of these deep-rooted major faults prove that the lithospheric deformation patterns are strongly influenced and controlled by their geological alignments. Collisional tectonics resulted from the rapid and continuous convergence of the Indian plate beneath the Eurasian plate and vertical coherent deformation due to fault movements are the responsible mechanisms for lithospheric scale-deformations associated with the major faults in the Himalayan arc and southern Shillong regions. Contrary to this, the Absolute Plate Motion related strain due to the asthenosphere drag at the base of Indian lithosphere acts as a major source of anisotropy beneath the Assam Foredeep and the central Shillong region, where no influences from the shallow or crustal-scale structures could be attributed towards deformation mechanisms.

Keywords Seismic anisotropy · Deformation · Shear wave splitting · Shillong Plateau · Assam Foredeep · Northeast Himalaya

D. D. Mohanty (✉)
Department of Geosciences and Technology, CSIR—North East Institute of Science and Technology, Jorhat, Assam, India
e-mail: devlinkan06@yahoo.com; debasis@neist.res.in

Academy of Scientific and Innovative Research (AcSIR), New Delhi, India

R. B. Mohanty
Assam Agriculture University, Jorhat, Assam, India
e-mail: ruchikabajpai27@rediffmail.com

S. Mukherjee (ed.), *Structural Geology and Tectonics*,
https://doi.org/10.1007/978-981-95-4743-2_9

1 Introduction

Large-scale geological structures play an essential role in lithospheric-scale deformation patterns, which are exclusively dominant in the active tectonic regions (e.g. Biswas et al., 2022). Dynamic tectonics affect the configuration of geological settings from crustal to mantle scale, and the enormous tectonic force associated therein is capable of grand changes in the Earth's structures. As a consequence of lithospheric-scale tectonics, the maximum compressional strain (SH_{max}) tends to align the major structures locally/regionally. These alignments in regional scale stress or strain patterns control the deformation pattern on a broader scale. Situated in the north-eastern corner of the Indian subcontinent, NE India signifies a most complex tectonic terrain with significant seismicity (Phukan & Mohanty, 2023), which leads this part of the nation to be placed as Zone VI of the seismic zoning map (Agrawal et al., 2023; Biswas et al., 2024); BIS, 2023; Gogoi et al., 2022; Goswami et al., 2022a, 2022b). The large-scale geological structures, which tend to originate from collisional tectonics/subduction mechanism/pop-up geodynamics/strike-slip tectonics throughout the NE region are the Main Boundary Thrust (MBT), Main Central Thrust (MCT), Kopili Fault (KF), Brahmaputra Fault (BF), Oldham Fault (OF) and Dauki Fault (DF) along with a number of regional or local scale structures. These large-scale structures are causative sources for several (moderate) earthquakes in the NE region.

Seismic anisotropy plays an important role in modelling the deformation patterns of the Earth (Wüstefeld et al., 2008). The directional dependencies of shear wave velocity inside the Earth with a particular direction in wave propagation and polarization, is termed as the seismic anisotropy, where the shear wave splits into two components with a particular time lag in between them. Shear wave splitting (SWS), as a consequence of seismic anisotropy is characterized by two parameters, viz., fast polarization direction (FPD, Φ) and delay time lag (δt) between the two components of shear wave. These two parameters are frequently used to determine the lithospheric deformation patterns in various geologic provinces with a lateral resolution of <50 km (Savage, 1999). The core-refracted phases such as SKS, SKKS and PKS (henceforth 'XKS') are frequently used to derive the lithospheric-scale anisotropy due to their near vertical ray path propagation throughout the Earth. The A-type Lattice Preferred Orientation (LPO) of abundant olivine minerals are responsible for azimuthal anisotropic behaviour in the mantle due to a large horizontal flow of minerals (Mainprice & Nicolas, 1989). Though LPO of minerals are the major causes of seismic anisotropy in the lithospheric mantle, the vertical coherent deformations subparallel to the direction of the strike of mountain ranges or fault zones associated with the large orogenic sutures or regional structures are capable of producing large deformation patterns in the lithosphere (Silver et al., 2001).

This article provides the deformation models associated with the major seismic structures in the northeast Indian region with an emphasis on the Shillong, Assam Foredeep and Himalayan arc regions (Fig. 1). These models explain the influences of the major structures in controlling the deformation patterns in lithosphere, from crustal to mantle scale, responsible for contemporary tectonics of NE India. Several seismologic research in the form of the seismic anisotropy have been made in the recent past decades from the NE India (Mohanty, 2022, 2023, 2025); Mohanty & Mondal, 2020, 2021); Mohanty & Singh, 2021; Mohanty et al., 2024), 2016; Mondal & Mohanty, 2021; Singh et al., 2006; Roy et al., 2014). The present study critically summarizes the existing deformation models and allows for a constructive comparison between them on the basis of the seismic influences associated with the major geological structures to understand the concurrent tectonics of the NE Indian region.

2 Modelling the Seismic Anisotropy

Seismic anisotropy relies on the directional dependencies of seismic velocity in a geological medium and depends upon the constituents of a particular medium. The approach of the seismic anisotropy modelling is best suited for the shear wave analysis as the wave splits into two constituent components at the onset of an anisotropic medium. Contrary to the isotropic medium, where there is no splitting in the shear wave, the time lag between the two components of the shear wave in the anisotropic medium represents the magnitude of heterogeneity or anisotropy in the medium and termed as delay time (δt), which suggests the thickness of the anisotropy medium. The direction of fast axis, associated with the faster component, is responsible to constrain the deformation alignment as the faster component prefers to travel in this direction. This particular direction associated with the fast wave is termed as the *Fast Polarization Direction* (FPD, Φ). Both Φ and δt are known as anisotropic parameters and are utilized to understand the deformation patterns of the lithosphere (Savage, 1999; Silver & Chan, 1991).

2.1 Techniques for Shear Wave Splitting (SWS) Analysis

This present study encompasses all the SWS measurements from the NE Indian region particularly in the surroundings of major geological structures present at Shillong and Assam Foredeep region. The contemporary robust techniques used for determination of inverse anisotropy parameters (ϕ and δt) are the Transverse Energy Minimization method (SC) and Cross correlation Method (RC). The Cross Correlation Method (Fukao, 1984) rotates and shifts the times of two horizontal components of the seismogram to reverse the effect of splitting. The parameters are set by observing the greatest correlation between these two splitting parameters (Φ and δt) between two components. The core-refracted phases such as SKS, SKKS and PKS, which have similar waveforms and near normal incidence are more favoured for this method to infer the splitting occurred from mantle. The Transverse Energy Minimization Method (Silver & Chan, 1991) applies an inverse splitting operator to

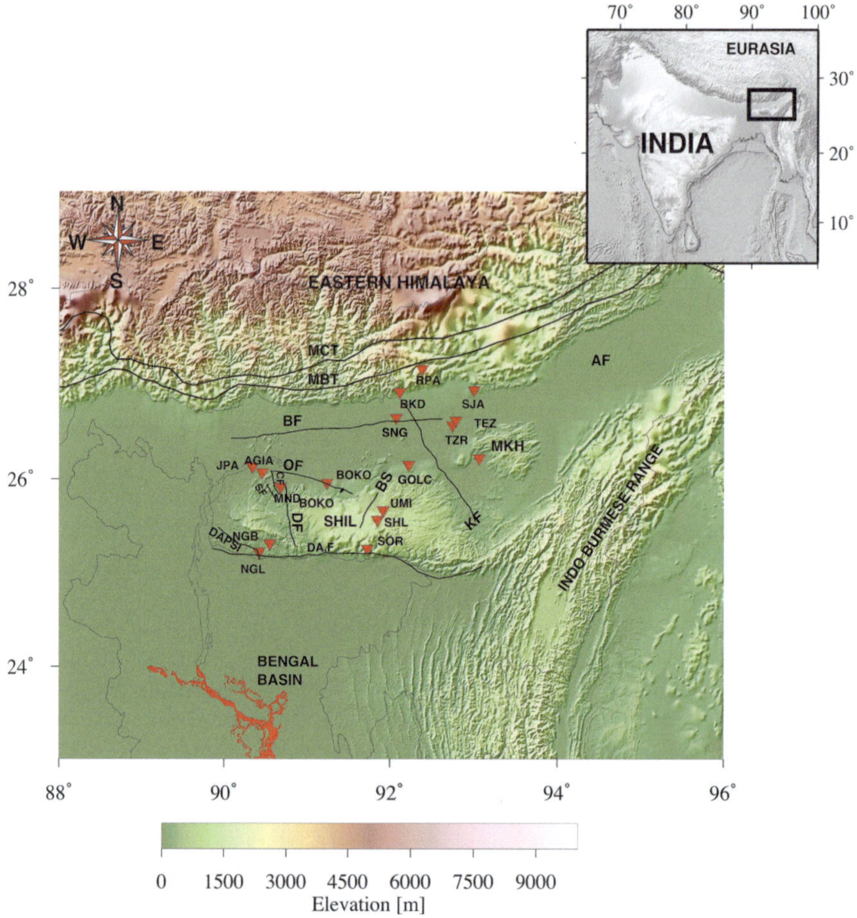

Fig. 1 Simplified tectonic and topographic map of northeast India. Location of broadband seismic stations (BBS) are shown by red inverted triangles along with their name codes, whose data are used in the present shear wave splitting measurements. Inset map (right top corner) of India shows the location of study area in Indian subcontinent, demarcated by black rectangle. Major tectonic divisions are labelled in bold black letters. MBT: Main Boundary Thrust, MCT: Main Central thrust, SHIL: Shillong Plateau, KF: Kopili Fault, DF: Dudhnoi Fault, DA.F: Dauki Fault, AF: Assam Foredeep, DAPSI: Dapsi Fault, SF: Samim Fault, CF: Chedrang Fault, OF: Oldham Fault, BF: Brahmaputra Fault

reverse the effects of splitting and assumes the original wave before splitting. This method based on the assumption that there is no existence of initial energy in the transverse component. A grid search for various pairs of Φ and δt are initiated and the pair that best removes the effects of splitting by minimizing the energy on the transverse component is chosen. The broadband Seismometer (BBS) sensors within the frequency range of 0.08–50 Hz (details in Fig. 1) are used to estimate the SWS of shear-waves in the NE India region. All the splitting parameters estimated from the

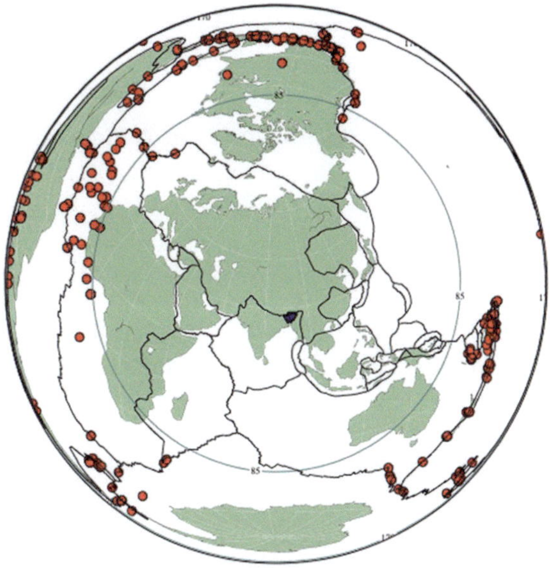

Fig. 2 Epicentral distribution of teleseismic earthquake events used for XKS splitting measurements. Only those events are mentioned which are utilized after primary scrutiny and processing steps to produce splitting measurements (compiled from all the available literatures in the study region). Red circles represent the events and the inverted solid black triangle represent the location of broadband seismic stations in the study area

close approximation of major seismic/geological structures are compiled and represented in Fig. 3. The teleseismic events used to estimate the lithospheric deformation patterns for the whole NE Indian region are represented in Fig. 2.

The SWS measurements obtained from NE India are compiled in this chapter for a comprehensive understanding and knowledge of structure induced deformation mechanisms (e.g. Misra & Mukherjee, 2017). The measurements from available studies (Mohanty & Mondal, 2020; Mohanty & Singh, 2021; Mondal & Mohanty, 2021; Roy et al., 2014; Singh et al., 2006) are characterized as best, fair and null results, based on different criteria mentioned in the corresponding studies, which depend on the ratio and certain relations between anisotropic parameters measured from the two methods, e.g. SC and RC, as mentioned above. All the available studies have chosen the SC measurements for interpretation of splitting parameters over the RC measurements because of their robustness in the highly complex geologic terrains (Wüstefeld et al., 2008). Further, to unify anisotropic measurements, the significant splitting results for a particular station location are stacked to produce a composite or resultant pair of splitting parameters (Φ and δt), based on the minimization of normalized error surface corresponding to the 95% confidence region for the entire splitting measurements (Wolfe & Silver, 1998).

3 Deformation Models Associated with Major Seismic Structures

3.1 Deformation Along the Himalayan Arc

The Himalayan arc, MCT and MBT occur in the northeast India along ~ E-W trend (Angellier and Baruah, 2009) in this corner of the subcontinent as a consequence of the active continent–continent collision between the Indian and Eurasian plate. The splitting measurements from this part of the area (Fig. 3, Table 2), suggest a dominance of the fast polarization directions in E-W or ENE-WSW directions (Mohanty & Mondal, 2020; Singh et al., 2006). The analysis is performed with three BBSs (RPA, SJA and BKD) present at the onset of these collisional arcs. The N-S compression due to the under-thrusting of Indian plate beneath the Eurasian mass is responsible for this large strain-induced maximum shear along E-W direction (Angelier & Baruah, 2009; Kumar et al., 2023). This maximum shear is the major reason for the current alignment of the MBT and MCT in the NE India. The variational collisional stress modifies the E-W trend of Himalayan arcs to ENE-WSW directions at the northern corner, north of 27° N latitude (Angelier & Baruah, 2009). This trend of alignment of major structures, like MCT and MBT, as a result of collisional dynamics is associated with the LPO of strained minerals (predominantly olivine minerals) of the uppermost mantle (Christensen & Crosson, 1968; Silver, 1996) which tend to deform by dislocation creep under a high temperature and pressure condition as evidenced from the laboratory experiments (Zhang & Karato, 1995). The upper mantle is predominantly composed of the minerals like olivine, which are prone to crystalline dislocations due to massive strains produced by huge mantle flows, causing preferred mineral orientations and therefore anisotropy (Nicolas & Christensen, 1987). The LPO derived anisotropy in the upper mantle is the major contributor towards the lithospheric deformation. Hence, the observed fast polarization directions (Φ) at the Himalayan arcs (MBT, MCT) are manifestation of the collisional shear induced deformations derived from the upper mantle anisotropy in this northeast corner of the subcontinent.

The anisotropic inversion models produced at the vicinity of Himalayan arcs (MBT and MCT) in the northeast Indian corner are capable of drawing the inference that the huge collisional tectonics, due to the indentation of Indian mantle front beneath the Eurasian plate, is responsible for an ENE-WSW deformation in the lithosphere. The inversion of splitting parameters strongly produces the FPDs (Fig. 3), aligned parallel to the major structures in this corner derived from the strong compressional strain due to the large collision in this region. The composite results (Fig. 4) at these stations reflect same patterns of FPDs as the individual splitting measurements.

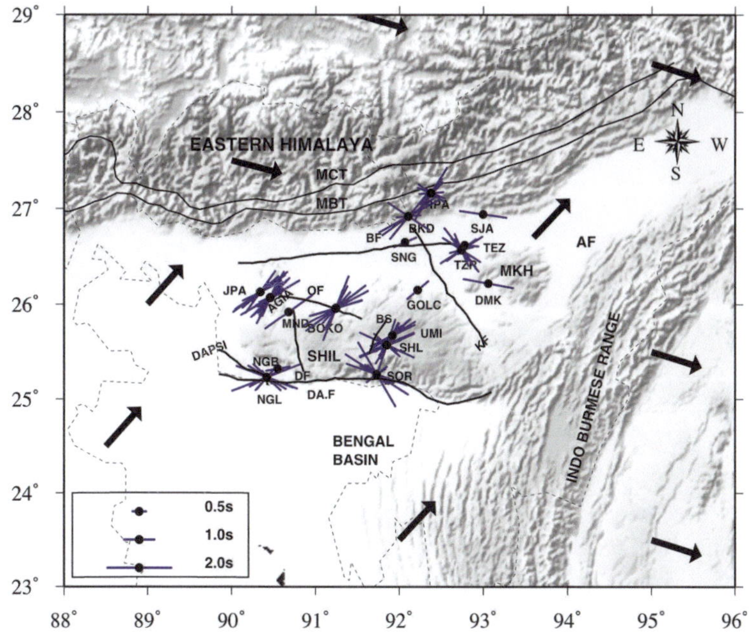

Fig. 3 Representation of individual XKS splitting measurements at broadband seismic stations (solid black circles) installed in the study region. The direction and length of solid blue bars represent the fast polarization directions (FPD, ϕ) and delay times (δt) respectively at those particular stations. Results from SC method have shown only and compiled from the available studies (Mohanty & Mondal, 2020; Mohanty & Singh, 2021; Mondal & Mohanty, 2021; Roy et al., 2014; Singh et al., 2006). The black thick arrows represent the APM directions in a no net rotation frame [*NUVEL-1A*, De Mets et al., 1994]. All the other abbreviations used in this figure are same as Fig. 1

3.2 *Deformation Models for the Oldham and Brahmaputra Fault*

Away from the active collision zone towards south, two major geological faults define their existence in the northeast Indian region. A comparatively E-W elongated Brahmaputra fault (BF), regarded as a river fault due to the existing flow of the major Brahmaputra River in this part of the northeast region. This fault acts as a differentiation mark between the active collisional zone towards its north and separates the Shillong plateau situated at its southern proximity. Further south of it, in the northern periphery of the Shillong mass, a major blind reverse fault known as the Oldham Fault (OF) is dipping steeply away from the Himalaya with an extension of ~110 km and depth varying in between 9 and 45 km (Bilham & England, 2001). This NW–SE elongated major reverse fault is the centre of debate for quite a long time and thought to be the cause of the great 1897 Shillong earthquake, responsible for the pop-up tectonics of the Shillong mass (Bilham & England, 2001). Considered as a hidden fault, without any surface exposure, the Oldham fault's contribution

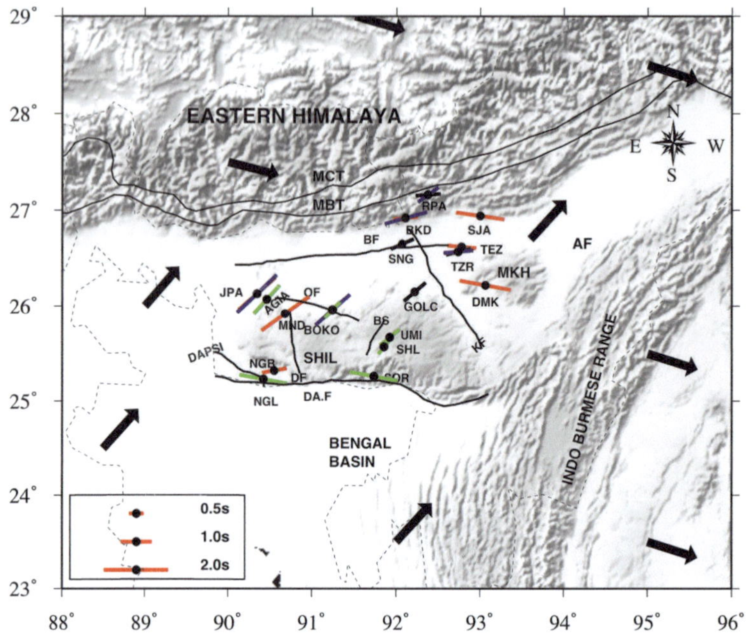

Fig. 4 Analysis and comparison of composite XKS splitting measurements from different studies available in the Shillong plateau, Assam Foredeep and Himalayan arc regions. The direction and length of solid-coloured bars represent the fast polarization directions (FPD, ϕ) and delay times (δt) respectively at the particular seismic stations (black solid circles). Red and blue solid bars represent composite splitting results from Singh et al. (2006) and Mohanty and Mondal (2020), respectively. Black solid bars are the measurements from Roy et al. (2014). The green solid bars represent the composite splitting parameters from Mohanty and Singh (2021), Mondal and Mohanty (2021). All the abbreviations are similar as used in Fig. 1. APM directions are represented by thick black arrows in a no net reference frame [*NUVEL-1A*, DeMets et al., 1994]. Only the composite results of particular stations are plotted in this figure

towards the 1897 big earthquake was challenged and consequently proved that the Brahmaputra fault (north of Oldham fault) was the major cause of this big earthquake (Kayal et al., 2006; Rajendran et al., 2004). Further, the recent numerical modelling has confirmed the Oldham fault as a rupture caused by the 1897 earthquake or any other previous seismic events (Islam et al., 2011). Though the fault is blind, this hidden blind reverse fault has been attributed towards the current tectonics of the Shillong plateau. Apart from these two major structures, two small-scale shallow depth faults named as Chedrang Fault (CF) and Samim Fault (SF) with the NW–SE trend in their surface exposures are also existing nearby the OF in the northern Shillong mass.

SWS analysis at the close proximity of BF, OF, SF and CF are performed at four BBSs locations (Figs. 1, 3 and Table 2) in the Shillong plateau named as Boko (BOKO), Agia (AGIA), Jogighopa (JPA) and Manikganj (MND) to understand the deformation patterns and the influence of these structures on the inversion models.

These results (Mohanty & Mondal, 2020; Mohanty & Singh, 2021; Singh et al., 2006) are the prime findings till date to understand the deformation patterns in the northern Shillong region. The significant delay times (an average δt of 1.2 s and higher) at all these stations in this part of the plateau, represent the more anisotropic nature of the lithosphere. The FPDs at all these stations are dominantly aligned in NE-SW direction and subparallel to the Absolute Plate Motion (APM) direction of the Indian plate in a no net rotation frame (NNR) (DeMets et al., 1994). It is quite interesting to note that, not a single deformation model achieved from the SWS analysis of data from the corresponding stations, show any pattern of deformation aligning to the geological extent/strike of the structure or faults. At the onset of the OF, the stations AGIA and BOKO have produced significant delay times with FPDs aligning in APM directions. There is no correlation of the FPDs with the NW–SE alignment of the OF. These results also support the long-range debate of existence of OF (Kayal et al., 2006) and cannot ascertain the real existence of this fault on the basis of splitting results only. The same consequences can be attributed from the nearby stations, MND and JPA, representing significant delay times with a NE-SW alignment in FPDs, parallel to APM of Indian plate. All the splitting measurements at the closest proximity of OF, BF and the minor faults like SF and CF, strongly discard the effect of these structures on the lithospheric deformation. These results, further prove the shallow origin of these faults for which they could not contribute their effects up to lithospheric scale. The APM of Indian plate is proved to be the major source of deformation in this region. The asthenospheric drag at the base of Indian lithosphere, responsible for APM of Indian plate in NE direction, is controlling the deformation pattern in this region of Shillong. A recent study (Mohanty & Singh, 2021; Mohanty et al., 2021) has proved the source localization of anisotropy at a depth of ~100 km, which corresponds to the asthenospheric source beneath Shillong. The crust beneath Shillong plateau is very thin (~32 km, Singh et al., 2017) with a comparatively shallow upper mantle and thinner lithosphere of ~100 km, which establishes the Lithosphere-Asthenosphere Boundary (LAB) at ~100 km (Devi et al., 2011). This further strengthens the understanding that the major anisotropic source at LAB is contributing towards the deformation. This source of anisotropy for this region is actually the consequence of the asthenosphere drag at the base of the Indian lithosphere responsible for the APM strain. Hence, the deformation model at the northern Shillong mass is controlled by the APM of Indian plate, where there is no influence from the regional structures or faults from its surroundings.

3.3 Deformation Caused by the Dauki Fault

At the southern extreme of the Shillong plateau, the north-dipping Dauki fault (DF) plays an important role in the pop-up geodynamics of this plateau mass. DF was responsible for the great 1897 Shillong earthquake (Bilham & England, 2001). This fault draws a major distinction of ~11 m in between the topographies of Shillong Plateau (in north) and Bengal Basin (in south) with a clear identification of southward downgrading in basin mass from the plateau. Moreover, this major and large E-W trending disconformity plays a vital role in accommodating a substantial part of shortening in between the Himalaya and Indian mass and contributes towards a contraction of Indian plate at a rate of about 4 ± 2 mm y^{-1} (Bilham & England, 2001). Traversing towards west, a conjugate of the DF called the Dapsi fault, is a north-dipping reverse fault (Fig. 1). This conjugate fault is often considered to be the extension of the DF.

SWS of seismic data (Table 2, Fig. 3) at the closest proximity of the DF is analysed at three BBSs throughout its geologic extension, e.g. Sora (SOR) and Nangalbibra (NGL and NGB) (Mohanty & Singh, 2021; Mohanty & Mondal, 2020; Singh et al., 2006). At the southern proximity of eastern Shillong, the station SOR has produced a comparatively higher anisotropic character as compared to other parts of the region with a delay time of 1.7 s (Mohanty & Singh, 2021), with the FPDs aligned parallel to the extension of DF in E-W directions (composite $\Phi = 100°$, Fig. 4). Traversing through the geologic extension from east to west, the two BBSs at NGL and NGB are situated just exactly at the conjunction of Dauki and Dapsi Faults at the western fringe of Shillong. The inversion of anisotropic parameters represents a significant delay time and FPD as 1.5 s and 100°, respectively (composite values, Mohanty & Singh, 2021). Similarly, the splitting measurements at NGB represent E-W aligned FPDs and a comparatively significant anisotropy ($\Phi = 80°$and $\delta t = 0.8$ s as composite values, Singh et al., 2006, Fig. 4). Hence, an E-W deformation along the whole extension of Dauki and Dapsi Faults strongly related to the anisotropic fabric at the initiation of this faulting along this southern proximity of Shillong, estimated as Pliocene (Najman et al., 2016). A strong attribution towards this deformation behaviour may be added from the compressional tectonics from the Himalayan zone, which is suggested to be accommodated substantially by the Shillong mass away from the active collision zone by producing a large-scale E-W deformation. This strong compression could initiate a substantial shape preferred orientation (SPO) of minerals present alongside the geological extension of major structures, resulting in a deformation pattern parallel to the extension of the structures. A possible contribution to the anisotropic fabric along the geological extension of Dauki Fault is also attributed from the overthrusting of the Indo-Burmese fold-belt at the eastern Shillong along which a lateral motion between Indo-Burmese Wedge and Shillong plateau causes a higher range of anisotropy (Mohanty & Singh, 2021). Considering the anisotropic measurements and the trend of splitting parameters, a clear understanding could be attributed that the coherent lithospheric deformation parallel to the geological extension of the Dauki fault is dominant in this southern proximity of the plateau mass.

3.4 Deformation at the Central Shillong and Assam Foredeep and Influence from the Kopili Fault

The Kopili Fault (KF), situated in the north-eastern region of India is treated as an active major transpressional feature responsible for generating lower crustal dextral strike-slip earthquakes. This 300 km long and 50 km wide lineament (Fig. 1) traverses through the Bhutan Himalaya in the northern corner to the western Manipur in the south extent, is supposed to be developed from the collisional tectonics between India and Eurasia. This fault region has witnessed two large earthquakes—the 1869 earthquake (M ~ 7.8) and the 1943 earthquake (M ~ 7.3), and has been considered as an active seismic zone due to an active tectonic depression filled up by the alluvium of the Kopili river and its tributaries, as a consequence of active Indo-Eurasian collision. This active fault/lineament separates the Shillong plateau from the Assam Foredeep Region (AF). The existing research on SWS analysis at AF and central Shillong region clearly gives a better picture of deformation patterns in and around the Kopili fault region and enables to understand the possible influence of this geologic structure on the lithospheric deformation beneath this active tectonic region.

Seismic anisotropy estimation through splitting analysis of core-refracted XKS phases have been performed at the central Shillong and Assam Foredeep regions at the BBSs, e.g. at Tezpur (TZR, TEZ), SNG, GOLC, DMK, UMI and SHL (Mondal & Mohanty, 2021; Mohanty and Mondal, 2020; Roy et al., 2014; Singh et al., 2006). In the eastern Shillong plateau and west to the KF, the seismic stations, e.g. UMI, SHL, SNG and GOLC are close-spaced and decodes the deformation patterns beneath the eastern Shillong (Figs. 1, 3 and 4) and influence from the KF. Little away from the KF, the stations UMI and SHL, signify the deformation patterns beneath eastern Himalaya. Both these stations are situated at a very close proximity of a shallow shear zone—the Barapani Shear (BS). The FPDs at SHL and UMI trend ~ NE (Fig. 3) and parallel to the direction of the extension of the BS. The composite splitting parameters after stacking of the individual measurements (Wolfe & Silver, 1991) represent significant delay times of 0.6 and 0.8 s with FPDs of 40° and 52° for SHL and UMI, respectively (Mohanty & Singh, 2021) (Fig. 4). The fast directions at both these stations clearly suggest NE-SW oriented deformation patterns in-line the APM of the Indian plate. The significant delay times further stress that the deformation mechanism in this region is governed by APM-related strain due to the basal drag at the base of Indian lithosphere. The influence form BF may be less attributed towards the deformation mechanism in this part, as the effect from this shallow origin small-scale structure may not be able to produce a significant anisotropy (delay time of 0.6 and 0.8 s). Hence, the effect of BS along with major contributions from APM-related strain cause the lithospheric deformation in this eastern Shillong region. These measurements certainly deny any contribution from KF. Further closer to the KF, the splitting measurements at two stations SNG and GOLC show composite FPDs (Fig. 4) along the same NE-SW trend with comparatively small delay time at SNG. The FPDs clearly resemble the deformation mechanism due to APM-related strain, same as SHL and UMI. All these splitting measurements (UMI, SHL, SNG and GOLC) suggest no contribution from KF towards the lithospheric deformation, whereas the APM of the Indian plate due to asthenospheric drag at the base of Indian lithosphere plays a vital role in controlling the deformation pattern in this region. The comparatively smaller delay time at SNG may be due to the influence from collisional strain cancelling the effect from the APM strain of the Indian plate, resulting in reduction in delay time magnitude.

In the eastern flank of KF situated in the Mikir Hill plateau, three close-spaced BBSs, e.g. Tezpur (TEZ, TZR) and Dokumukh (DMK) contribute majorly towards understanding the deformation patterns in this region. The two stations, TEZ and TZR are situated very closely in the Uguri Hills part of Mikir Plateau of the Tezpur city. The individual splitting measurements at these stations clearly indicate the FPDs in two directions—NE-SW and E-W, suggesting the influence from lithospheric strain due to the ongoing India-Eurasia collision and the APM strain due to asthenosphere drag in NE-SW direction. The composite splitting results show significant delay times (0.93 and 0.90 s) and FPDs (85° and 95°, in E-W directions) for the stations TZR and TEZ, respectively. The E-W pattern in fast directions of splitting parameters suggest the dominance of lithospheric strain due to N-S ongoing collisional tectonics between the Indian and the Eurasian plates, controlling the deformation patterns in this Assam Foredeep (AF) region. The composite measurement (Fig. 4) from the nearby station Dokmukh (DMK) is in-line and same as the measurements from TEZ and TZR. The significant delay time (1.6 s) represents a higher anisotropic region with a FPD parallel to ~ E-W direction (100°), suggesting the same mechanism of deformation due to collisional dynamics in this Assam Foredeep region. All the splitting measurements from the AF are neither in-line the strike of the Kolpili fault nor parallel to its geological extension. This signifies that, the fault has no influence on the deformation patterns. Further, observing the seismic patterns in just proximity of the KF, it suggests a distribution of clusters of micro-earthquakes majorly in the crustal-depth scales, suggests the lineament or fault as a crustal-scale structure (Kayal, 1996), without a deep root beyond the crust. There is no evidence of coherent deformation of the lithosphere parallel to the geological extension of KF is observed so far in this region. Hence, Kopili Fault has no contribution towards the lithospheric-scale deformation beneath the NE Indian region.

4 Conclusions

We model lithospheric deformation beneath the Shillong, Assam Foredeep and the Himalayan arc region of the NE India. We specifically study the influences from the major geological structures in controlling the deformation patterns. While the deep-rooted major structures, e.g. MCT, MBT and DF contributed majorly towards the lithospheric-scale deformations beneath the northeast India, the shallow origin and small-scale faults or lineaments produce no influence on the deformation mechanism. These small-scale features may be dominant in controlling the shallow/crustal level deformation patterns, but show no influence towards the lithospheric deformation phenomena beneath this region. The major structures control the deformation patterns as an effect of shape preferred orientation (SPO) of strained minerals along their geological extensions, which can be best described as the vertical coherent deformation of lithosphere along these structures. Majorly, the lithospheric strains due to huge collisional tectonics beneath the Himalayan arc are responsible for LPO of strained minerals, which in term represents E-W extensional deformation parallel to the arcs. In the southern Shillong region, the thrust faulting initiated beneath the DF and is controlling the shape preferred orientation of strained minerals for an E-W extensional deformation pattern.

Acknowledgements Authors are especially thankful to Prof. Soumyajit Mukherjee, IIT Bombay, for the invitation and review of this chapter. DDM is thankful to Prof. V. M. Tiwari, Director, CSIR-NEIST, Assam, for the permission towards this publication and data utilization. DDM acknowledges the research support from Science and Engineering Research Board, India in the form of Mathematical Research Impact Centric Support Grant (MATRICS), vide project grant no. MTR/2022/000343. Mukherjee (2026) summarized this work.

References

Agrawal, N., Gupta, L., Dixit, J., & Dash, S. K. (2023). Seismic risk assessment for the north eastern region of India by integrating seismic hazard and social vulnerability. *Sustainable and Resilient Infrastructure, 8*, 102–132.

Angelier, J., & Baruah, S. (2009). Seismotectonics in Northeast India: A stress analysis of focal mechanism solutions of earthquakes and its kinematic implications. *Geophysical Journal International., 178*, 303–326. https://doi.org/10.1111/j.1365-246X.2009.04107.x

Bilham, R., & England, P. (2001). Plateau "'pop-up"' in the great 1897 Assam earthquake. *Nature, 410*(6830), 806–809. https://doi.org/10.1038/35071057

BIS. (2023). *Seismic zoning map of India*, Bureau Indian Standard Publication.

Biswas, M., Mukherjee, S., Gogoi, M. P., & Barik, N. (2024). Assessment of active tectonics of the Tripura-Mizoram fold belt (Surma Basin), North East India. *Environment, Earth and Ecology*.

Biswas, T., Bose, N., Dutta, D., & Mukherjee, S. (2022). Arc-parallel shears in collisional orogens: Global review and paleostress analyses from the NW Lesser Himalayan Sequence (Garhwal region, Uttarakhand, India). *Marine and Petroleum Geology, 138*, Article 105530.

Christensen, N. I., & Crosson, R. S. (1968). Seismic anisotropy in the upper mantle. *Tectonophysics, 6*(2), 93–107. https://doi.org/10.1016/0040-1951(68)90013-9

DeMets, C., Gordon, R. G., Argus, D. F., & Stein, S. (1994). Effect of recent revisions to the geomagnetic reversal time scale on estimates of current plate motions. *Geophysical Research Letters, 21*(20), 2191–2194. https://doi.org/10.1029/94GL02118

Devi, E. U., Kumar, P., & Kumar, M. R. (2011). Imaging the Indian lithosphere beneath the Eastern Himalayan region. *Geophysical Journal International, 187*, 631–664. https://doi.org/10.1111/j.1365-246X.2011.05185

Fukao, Y. (1984). Evidence from core-reflected shear waves for anisotropy in the earth's mantle. *Nature, 309*(5970), 695–698. https://doi.org/10.1038/309695a0

Gogoi, M. P., Gogoi, B., & Mukherjee, S. (2022). Tectonic instability of the petroliferous upper Assam valley (NE India): A geomorphic approach. *Journal of Earth System Science, 131*, 18.

Goswami, T. K., Kalita, P., Mukherjee, S., Mahanta, B. N., Sarmah, R. K., Laishram, R., Saikia, H., Gogoi, M., Macharari, R., & Oza, B. (2022b). Basement cross-strike Bomdila fault beneath

Arunachal Himalaya: Deformation along curved thrust traces, seismicity and implications in hydrocarbon prospect of the Gondwana sediments. *Geological Journal, 57*, 4974–4999.

Goswami, T., Gogoi, M., Mahanta, B. N., Mukherjee, S., Saikia, H., Shaikh, M., Kalita, P., Baral, U., & Sarmah, R. (2022a). Brittle tectonics in the western Arunachal frontal fold belt, India: Change in stress regime from pre-collisional extension to collisional compression. *Geological Journal, 57*, 5019–5038.

Islam, M. S., Shinjyo, R., & Kayal, J. R. (2011). Pop-up tectonics of the Shillong plateau in northeastern India: Insight from numerical simulations. *Gondwana Research, 20*, 395–404.

Kayal, J. R. (1996). Earthquake source processes in north east India: A review. *Journal of Himalayan Geology, 17*, 53–69.

Kayal, J., Arefiev, S. S., Barua, S., Hazarika, D., Gogoi, N., Chowdhury, A., & Kailata, S. N. (2006). Shillong plateau earthquakes in northeast India region: Complex tectonic model. *Current Science, 91*, 109–114.

Kumar, P., Malik, J. N., & Gahalaut, G. (2023). Arc-parallel shear and orogenic deformation along the oblique Himalayan convergent plate margin: Implications from topographic and gradient anomaly profiling in the Himalaya. *Pure and Applied Geophysics., 180*, 2169–2189.

Mainprice, D., & Nicolas, A. (1989). Development of shape and lattice preferred orientations: Application to the seismic anisotropy of the lower crust. *Journal of Structural Geology., 11*, 175–189. https://doi.org/10.1016/0191-8141(89)90042-4

Misra, A. A., & Mukherjee, S. (2017). Dyke–brittle shear relationships in the Western Deccan Strike-slip Zone around Mumbai (Maharashtra, India). *Geological Society* (Vol. 445, pp. 269–295). Special Publications, London.

Mohanty, D. D. (2022). *Theory and practices involved in depth and source localization of anisotropy: Recent developments in subsurface investigation. Recent developments in using seismic waves as a probe for subsurface investigations.* CRC Press. https://doi.org/10.1201/9781003177692

Mohanty, D. D. (2025). Lithospheric deformation beneath the Sikkim Himalaya and tectonic implications: anisotropic contributions from crust and mantle. *Journal of seismology.* https://doi.org/10.1007/s10950-025-10329-5

Mohanty, D. D., & Mandal, P. (2021). Evidence of heterogeneous and unstable anisotropic settings beneath the Northeast Indian lithosphere from characterization of null splitting measurements. *Acta Geophysica.* https://doi.org/10.1007/s11600-021-00676-5

Mohanty, D. D., & Mondal, P. (2020). Disparate behaviour of deformation patterns beneath northeast Indian lithosphere inferred from shear wave splitting analysis. *Physics of the Earth and Planetary Interiors., 298*, Article 106315. https://doi.org/10.1016/j.pepi.2019.106315

Mohanty, D. D., & Singh, A. (2021). Shear wave birefringence and current configuration of active tectonics of Shillong plateau: An appraisal of Indian plate motion and regional structures. *International Journal of Earth Sciences.* https://doi.org/10.1007/s00531-021-02114-9

Mohanty, D. D., Biswal, S., & Yoshizawa, K. (2024). Decoupled deformation between crust and mantle beneath Indo-Burmese Wedge: a new seismotectonic model. *Earth & Planetary Science Letters (EPSL), 648.* https://doi.org/10.1016/j.epsl.2024.119089

Mohanty, D. D., Biswal, S., Phukan, M., & Eluyemi, A. (2021). Possible depth determination and source localization of seismic anisotropy beneath Shillong plateau and Himalaya foredeep region, Invited Paper. *Geological Journal.* https://doi.org/10.1002/gj.4334

Mohanty, D. D., Singh, A., O'Driscoll, L. J., Kumar, M. R., Srinagesh, D., & Humphreys, E. D. (2016). *P* wave velocity structure below India and Tibet incorporating anisotropic delay time effects. *Geochemistry, Geophysics, Geosystems, 17*, 725–738. https://doi.org/10.1002/2015GC006064

Mohanty, D.D. (2023). Source localized modelling to estimate the depth of anisotropy beneath eastern Himalaya using spatial coherency of shear-wave splitting parameters. *Physics of the Earth and Planetary Interiors, 336*, 106987. https://doi.org/10.1016/j.pepi.2023.106987

Mondal, P., & Mohanty, D. D. (2021). Mantle deformation and seismic anisotropy beneath Northeast India inferred from SKKS birefringence. *Studia Geophysica Et Geodaetica., 65*, 36–52.

Mukherjee, S. (2026). Introduction to "Structural Geology & Tectonics". In S. Mukherjee (Ed.), *Structural Geology & Tectonics*. Springer. ISBN: 978-981-95-4743-2.

Najman, Y., Bracciali, L., Parrish, R. R., Chisty, E., & Copley, A. (2016). Evolving strain partitioning in the eastern Himalaya: The growth of theShillong plateau. *Earth and Planetary Science Letters, 433*, 1–9. https://doi.org/10.1016/j.epsl.2015.10.017

Phukan, A., & Mohanty, D. D. (2023). Seismic hazard estimation and medium-term earthquake precursor analysis of North East India: An assessment on large earthquake scenario. *All Earth, 35*(1), 46–64. https://doi.org/10.1080/27669645.2023.2191917

Rajendran, C. P., Rajendran, K., Daurah, B. P., & Earnest, A. (2004). Interpreting the style of faulting and paleoseismicity associated with the Shillong, northeast India, earthquake: Implications for regional tectonism. *Tectonics, 23*, TC4009.

Roy, S. K., Kumar, M. R., & Srinagesh, D. (2014). Upper and lower mantle anisotropy inferred from comprehensive SKS and SKKS splitting measurements from India. *Earth and Planetary Science Letters., 392*, 192–206. https://doi.org/10.1016/j.epsl.2014.02.012

Savage, M. K. (1999). Seismic anisotropy and mantle deformation: What have we learned from shear wave splitting? *Reviews of Geophysics, 37*, 65–106. https://doi.org/10.1029/98RG02075

Silver, P. G. (1996). Seismic anisotropy beneath the continents: Probing the depths of geology. *Annual Review of Earth and Planetary Sciences, 24*, 385–432. https://doi.org/10.1146/annurev.earth.24.1.385

Silver, P. G., & Chan, W. W. (1991). Shear wave splitting and subcontinental mantle deformation. *Journal of Geophysical Research, 96*, 16429–16454. https://doi.org/10.1029/91jb00899

Silver, P. G., Gao, S., & Liu, K. H. (2001). Mantle deformation beneath southern Africa. *Geophysical Research Letters, 28*(13), 2493–2496. https://doi.org/10.1029/2000GL012696

Singh, A., Kumar, M. R., Mohanty, D. D., Singh, C., Biswas, R., & Srinagesh, D. (2017). Crustal structure beneath India and Tibet: New constraints from inversion of receiver functions. *Journal of Geophysical Research: Solid Earth., 122*, 7839–7859. https://doi.org/10.1002/2017JB013946

Singh, A., Kumar, M. R., Raju, P. S., & Ramesh, D. S. (2006). Shear wave anisotropy of the northeast Indian lithosphere. *Geophysical Research Letters, 33*. https://doi.org/10.1029/2006GL026106

Wolfe, C. J., & Silver, P. (1998). Seismic anisotropy of oceanic upper mantle: Shear wave splitting methodologies and observations. *Journal of Geophysical Research, 103*(B1), 749–771.

Wüstefeld, A., Bokelmann, G., Zaroli, C., & Barruol, G. (2008). SplitLab: A shear-wave splitting environment in Matlab. *Computer and Geosciences, 34*(5), 515–528. https://doi.org/10.1016/j.cageo.2007.08.002

Zhang, S., & Karato, S. (1995). Lattice preferred orientation of olivine aggregates deformed in simple shear. *Nature, 375*, 774–777. https://doi.org/10.1038/375774a0

Successive Studies of the 2011 Tohoku-Oki Earthquake Signatures from Global Navigation Satellite System Measurements

Kutubuddin Ansari

Abstract Modern geodesy can monitor the geodynamic motion of the Earth because of tectonic activities and modifications in the environment because of sea level variation and atmospheric changes. The Tohoku-Oki earthquake (Mw 9.1) occurred on March 11, 2011 in Japan and was the strongest devastating earthquake in the modern world history. This article discussed the 2011 Tohoku-Oki Japan earthquake signature captured by geodetic data on the Earth surface, the Pacific Ocean as well as in space.

1 Introduction

The global navigation satellite system (GNSS) is a constellation of orbiting satellites with a ground control station network of receivers and stations, which works together and calculate 3D position at the global level through a trilateration-adapted version. In recent times, the accuracy of GNSS positioning has increased because of technological progress in receiving signals, geophysical model refinement, and earth reference frame improvements (Ansari et al., 2021; Li et al., 2015; Mukul et al., 2018). The GNSS receiver position is determined with the basic positioning principle by evaluating the distances between four or more satellites with the pre-determined position. Precise point positioning technology is an absolute positioning method that caters to global and regional services based on observations from a single GNSS receiver. For this, a network of many GNSS stations has been established around the world publicly or privately, that provides temporal and spatial dislocations of these positioning on the surface of the earth and is used to understand the crustal deformation due to the geodynamic activity. The magnitude of the earthquake is one factor that affects the tectonic horizontal and vertical movements. A significant change in sea level is possible and destructive local tsunamis may occur near the earthquake

K. Ansari (✉)
Department of Geomatics Engineering, Faculty of Engineering, Karadeniz Technical University, Trabzon 61080, Türkiye
e-mail: kdansarix@gmail.com

S. Mukherjee (ed.), *Structural Geology and Tectonics*,
https://doi.org/10.1007/978-981-95-4743-2_10

epicenter. Most of the time thrust earthquakes are the cause of tsunami generation but in a few cases, large strike-slip earthquakes with < Mw 8.0 Mw may generate small tsunamis (He et al., 2022a)

Tsunami is one of the significant disasters, which occur due the earthquakes and can be well studied with the help of GNSS signals. The rise of sea level using tide gauges from geodetic observations has been estimated in several studies (Hannah & Bell, 2012; Raj et al., 2022). Absolute sea level changes cannot be measured directly by tide gauges but the sea surface height relative to the crustal reference point (GNSS station vertical position), which may move with local subsidence or tectonic activity can be measured accurately. Generally, tide gauge measurements are biased by regional and local procedures which can be linear or non-linear in a multi-decadal period (Ansari et al., 2022a; Parker, 2018).

Linear processes consist of strain accumulation because of inter-seismic events and glacial isostatic adjustment whereas non-linear process includes all the transient signals such as recorded time series of earthquake post-seismic relaxation (Panet et al., 2007). Therefore, to obtain the rise of precise absolute sea level, it is necessary to correct the tide gauge sea level measurements from the vertical land motions in the vertical direction (Denys et al., 2020). There are numerous kinds of models developed by analyzing the data of geodetic time series such as satellite altimetry and GNSS reflectometry (Ansari et al., 2020; Church & White, 2011; Hughes & Williams, 2010). Absolute sea level rise and sea level recorded by satellite altimetry also can be correlated for comparison purposes (He et al., 2022a). The tide gauge benchmark is situated on the land surface near the instruments while the satellite altimetry records the seal-level height above the benchmark or datum. In this way, the observation from the tide gauge measurements is the relative sea level measurements with respect to the benchmark elevation. GNSS reflectometry is also one of the valid options in remote sensing in the last decades (Larson et al., 2013; Tabibi et al., 2017). In this technique, GNSS stations are set up near to the coastal zones and the opportunity of reflected signals from the sea surface is utilized. The ability of exceptional spatio-temporal sampling over the Earth's surface is used and in parallel several reflection tracks measurements from different GNSS transmitters are taken. Nowadays multi-constellation GNSS signals are capable to measure highly precise sea level heights (Ansari, 2023; Wang et al., 2020).

The refraction from the troposphere and ionosphere significantly disturbs the high-precision GNSS positioning (Ansari et al., 2016; Jamjareegulgarn et al., 2020; Reddybattula et al., 2019; Seok et al., 2022; Sharma et al., 2018). GNSS signals are used as a source of information to investigate and regularly observe atmospheric activity. Such kind of data is collected from the GNSS receiver situated on the ground as well as in space. It provides important information about the atmospheric state and then by including several metrological models for weather prediction. For example, information about the troposphere is accessed by zenith total delays, and the information about the ionosphere is recognized by electron density per volume. The conjugation of ionospheric total electron content (TEC) anomalies prior to the earthquakes was detected in the stations and has been employed extensively for earthquakes in different earthquakes in the world (Priyadarshi et al., 2011; Oikonomou et al., 2020;

Sekertekin et al., 2020; Şentürk et al., 2020; Thomas et al., 2022; Ulukavak & Inyurt, 2020). Priyadarshi et al. (2011) derived total electron content (TEC) TEC from the GNSS in India between 2007 and 2010. They selected nine earthquakes that occurred within a diameter of 2000 km and studied TEC changes related to these earthquakes. Oikonomou et al. (2020) investigated pre-earthquake ionospheric and atmospheric disturbances of three large earthquakes occurred in Mexico, i.e., Mw 6.6 on 21–Jan–2016, Mw 8.2 on 08–Sept–2017, Mw 7.1 on 19–Sept–2017. They also examined the ionospheric TEC variations during the entire year of 2017 prior to 37 earthquakes. The analysis revealed that it is possible to distinguish pre-earthquake ionospheric disturbances related to the imminent seismic events in low-latitude regions.

Sekertekin et al. (2020) investigated the GNSS TEC behaviors of the potential earthquake before and after the 2013 Mw 7.7 Awaran, Pakistan earthquake. They identified TEC anomalies 3 and 5 days before the earthquake. Şentürk et al. (2020) studied ionospheric TEC before the Mw 7.3 Iran and Iraq border earthquake occurred on 12–Nov–2017. The analyses showed positive anomalies 8–9 days before the earthquake and some positive and negative anomalies 1–6 days before the earthquake. Ulukavak and Inyurt (2020) studied the ionospheric variations induced by six sequential earthquakes in Nepal in 2015 with a magnitude >6.0 and examined the TEC data from GNSS measurements. They performed a moving median analysis for 15 days to determine the pre- and post-earthquake abnormalities. Thomas et al. (2022) collected TEC data from three stations within the Mw 7.2 Haiti 2021 earthquake zone and used a statistical procedure for the differential and anomalous detection in daily TEC. The study displayed both enhancement and reduction in diurnal TEC before and after the earthquake within the investigating period. The dense network of ground Japan GNSS receivers was able to detect the changes in TEC around 30 min before the earthquakes along with the long last post-seismic drop of TEC (will discuss further in the current study). These kinds of studies support electric field models, which show a redistribution of electron density immediately before large earthquakes within the ionosphere. In this way, satellite geodesy techniques are used to examine the geodynamics motion of the Earth, variations in the environment because of changes in climate, and modeling of the atmosphere.

The aim of the current study is to summarize the earthquake signatures captured by GNSS measurements on the surface of the earth, and variations of sea level as well as in space by taking an example of the Tohoku-Oki Japan 2011 earthquake. The summary of this big earthquake and the availability of the GNSS network in Japan has been discussed in Sect. 2. The effect of the earthquake captured by GNSS on the surface of the earth and sea level has been discussed in Sects. 3 and 4, respectively. The study of the Tohoku-Oki earthquake signature in space has been written in Sect. 5. Finally, the summary of the current study has been given in Sect. 6.

2 Summary of Tohoku-Oki Earthquake and Availability of GNSS Network in Japan

The Tohoku-Oki earthquake (Mw 9.1) struck on 11–March–2011, along Japan's northeastern coastal region near Honshu. It was caused by shallow thrust faulting at the subduction zone where the Pacific Plate meets the North American Plate (Fig. 1). This event was preceded by a significant foreshock sequence, including a Mw 7.4 earthquake on 09–March–2011, approximately 40 km from the main shock epicenter, followed by three other seismic events exceeding Mw 6 on the same day. On the day of the earthquake, the Pacific Plate, moving westward at a velocity of approximately 8.3 cm/year relative to the North American Plate, began subducting beneath Japan at the Japan Trench megathrust (USGS). The seismic deformation impacted not only Japan but also neighboring regions such as China, Russia, the Korean Peninsula, Southeast Asia, and other nearby areas (Ansari & Park, 2022; Wang et al., 2011). This earthquake, along with the subsequent tsunami, resulted in severe casualties and extensive damage.

Japan's islands are situated at the intersection of four tectonic plates: the Philippine Sea Plate, the Eurasian Plate, the Pacific Plate, and the North American Plate, forming the northwest rim of the Pacific Ocean. The subduction of the Pacific Plate beneath the North American Plate at the earthquake zone occurs at an approximate rate of 80 mm/year relative to the Eurasian Plate, eventually diving beneath the North American Plate (Ozawa et al., 2011).

The Geospatial Information Authority of Japan (GSI), formerly known as the Geographical Survey Institute, provides elevation and land position data referenced to mean sea level. Currently, GSI has developed a GNSS-based control station network for precise positioning observations. This network, called the GNSS Earth Observation Network (GEONET), is designed to monitor tectonic activity across Japan (https://terras.gsi.go.jp/). GEONET includes over 1300 ground-based GNSS sites, most of which are installed on bedrock to measure high-precision surface displacements (Takamatsu et al., 2023). Several of these sites are located in coastal regions, enabling sea level studies. Each GNSS site is equipped with high-quality geodetic antennas and receivers capable of accessing signals from multiple GNSS constellations. Additionally, the GSI has established around 25 tide observation stations along Japan's coasts to continuously monitor sea levels (https://www.gsi.go.jp/kanshi/tide_furnish.html). These tide stations can also detect coastal vertical displacement, record tsunami waves, and support disaster management efforts. The tide levels are observed based on fixed altitudes, with the height system in Japan defined relative to mean sea level (Ansari & Bae, 2021). The tide gauges are calibrated to measure sea levels above a local reference datum (Fig. 1). These stations are strategically located to face open seas and near large rivers. For this study, 36 GNSS stations and 12 tide gauge stations situated in northeastern Japan were selected, ensuring sufficient spatial coverage near the earthquake's epicenter.

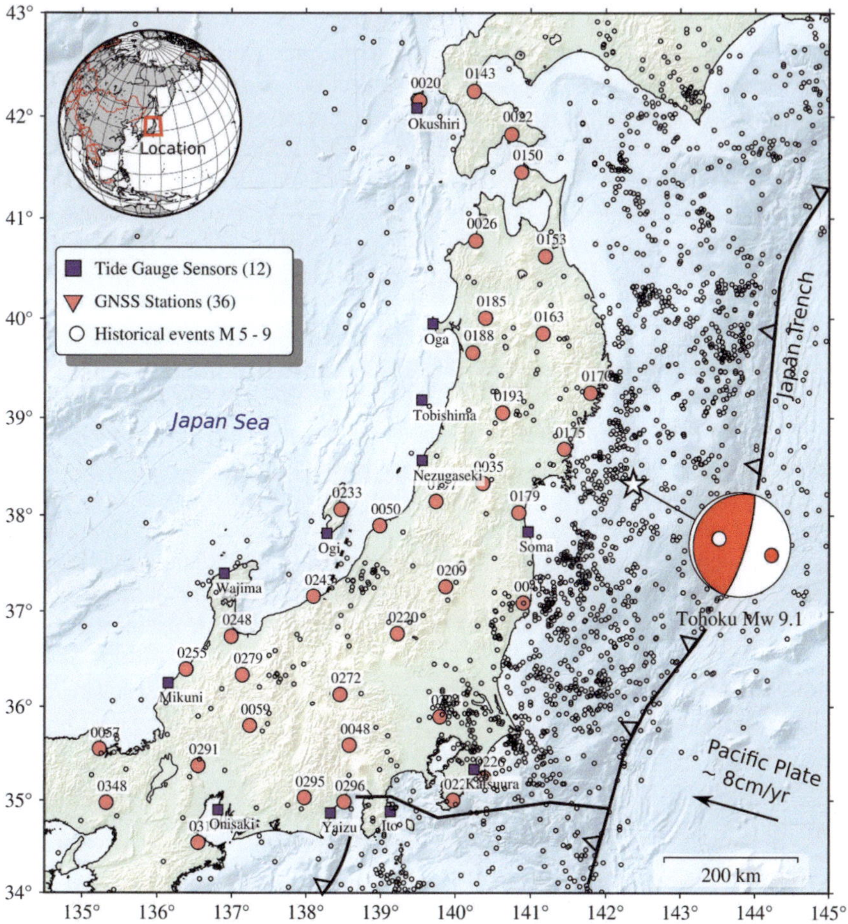

Fig. 1 Location of Tohoku-Oki earthquake (Mw 9.1) occurred on 11–March–2011 in the north-eastern coastal zone of Honshu, Japan due to the faulting of shallow thrust on the zone of subduction at the boundary between the North American plate and Pacific plate. The location of the tide gauge station in the coastal area of Northeast Japan has been shown with purple color while the GNSS sites are shown with red color. (After Ansari et al. 2024)

3 Study of Earthquake Signatures on Earth Surface

The GEONET observation data in RINEX format has been taken from GSI Japan websites and later it was processed with GAMIT-GLOBK post-processing software for the estimation of daily precise point positioning (Herring et al., 2018). Differential phase observation methodology was used to eliminate source errors and perfectly estimate the station coordinates, phase ambiguity and satellite orbit parameters for each day of observation. Daily precise position measurements of GNSS sites were calculated with respect to the International Terrestrial Reference Frame 2014 (ITRF14) along with one sigma uncertainty and GNSS time series was attained at each site. North and east offset of the obtained coordinate time series from 2002 to 2018 of 6 GNSS sites has been plotted in Figs. 2 and 3, respectively. There is a significant break in each time series, which corresponds to the signature of the Tohoku earthquake (Mw 9.1) on the Earth surface. The magnitude break in each time series is different depending on the location of the GNSS site. Earthquake main shock GNSS displacement offset was computed by previous studies which showed more than 4.30 m horizontal displacement with a subsidence of ~0.66 m in the north of Honshu, Japan. Roughly GNSS-measured horizontal displacements for the aftershock were ~44 cm (Simons et al., 2011). It was noticed that the slip was most intense in the range of depths around 10–20 km from the trench. This slip displacement was considerably decreasing in a small area toward the trench at the wedge toe (Minson et al., 2014).

Ansari and Park (2022) studied the Tohoku-Oki earthquake signature in South Korea by using the GNSS measurements. Analysis detail of Korean GNSS data showed the change of crustal deformation inside South Korean due to the Tohoku-Oki earthquake. They used the daily GNSS point positions for four days (07–10 March 2011) of the earthquake and estimated its average position. This average position was considered a pre-seismic position of GNSS sites. Similarly, they took the average of four days after (12–15 March 2011) the earthquake daily position of GNSS sites and considered this average position as a post-seismic position of GNSS sites. The difference between the post-seismic and pre-seismic positions of these stations is considered as co-seismic displacements of GNSS sites. Results showed that the maximum co-seismic displacement because of the Tohoku-Oki earthquake within South Korea reached up to 36.82 $\pm$ 0.21 mm in the east direction and 5.90 $\pm$

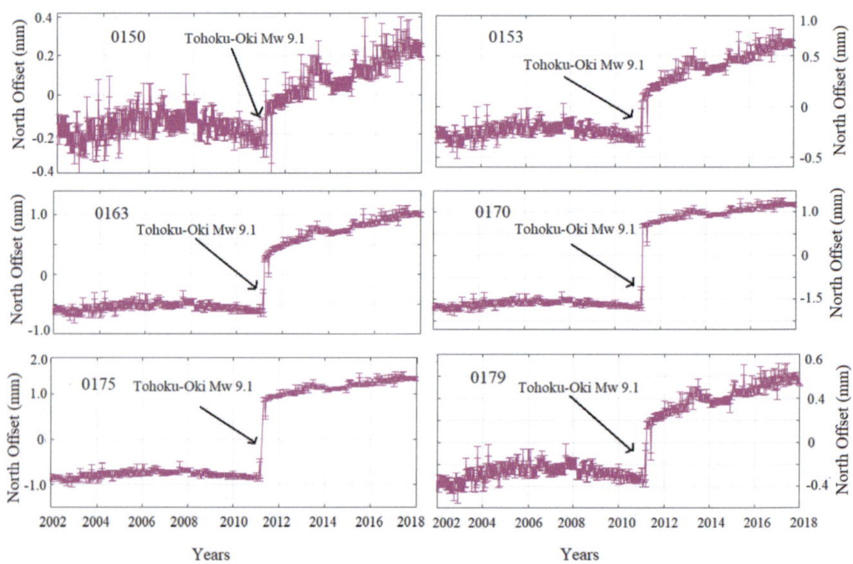

Fig. 2 North offset of the obtained coordinate time series from 2002 to 2018 of 6 GNSS sites

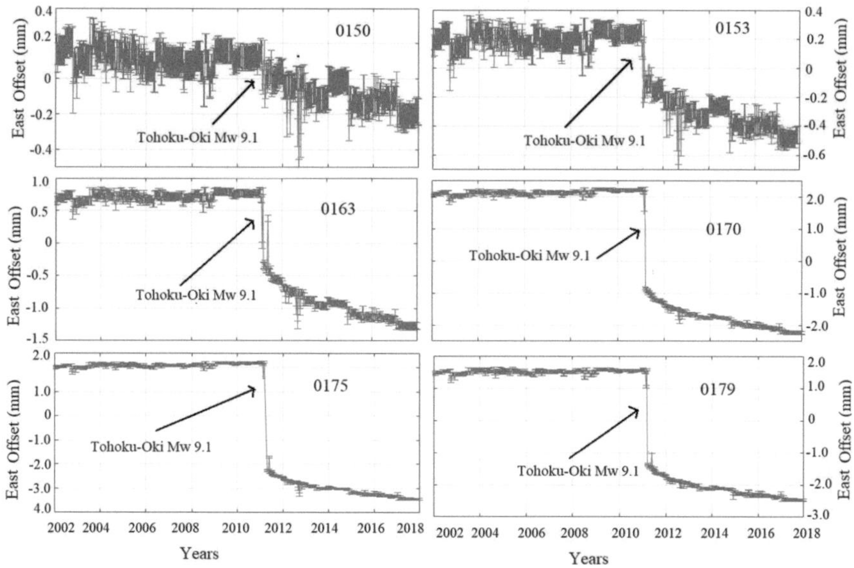

Fig. 3 East offset of the obtained coordinate time series from 2002 to 2018 of 6 GNSS sites

0.08 mm in the north direction. Kim et al. (2018) studied error ellipses before and after the earthquake in South Korean GNSS measurements and noticed the larger error after the earthquake compared to the error before the earthquake. They concluded that the high uncertainty in GNSS measurements because of the Tohoku-Oki earthquake is related to the stability of crustal deformation in Korea. Wang et al. (2011) used GNSS data in China to study far-field co-seismic displacement and to evaluate the Tohoku-Oki earthquake crustal deformation impact in eastern China. The obtained results revealed ~35 mm displacement in the south-southeast direction near the China, Russia and Korea triple border junction. Although there was decreasing co-seismic displacements magnitude in South Korea and China, there GNSS observation was successfully able to capture the earthquake signature on the Earth surface. Such an analysis was able to provide a good understanding of the directional change of crustal deformation because of the Tohoku-Oki earthquake in Japan as well as in the neighboring countries.

4 Study of Earthquake Signatures in Ocean

The seismic waveform inversion because of the Tohoku-Oki earthquake revealed that the earthquake rupture started from the location (38.1°N; 142.8°E;~26 km deep), followed by the rupture propagations in two ways. This propagation has the largest slip of ~25 m and is dominated by thrusting (Wang et al., 2011). Most of this rupture was located under the sea and was detected by telerecorded measurements. The

Pacific plate was moved around 83 mm yr^{-1} west direction relative to the North American plate at the location of the earthquake. USGS recorded that the Pacific plate started to move westward beneath to Japan at the Japan Trench megathrust on the day of the earthquake (11–March–2011), east of the epicenter. Simons et al. (2011) selected 12 sensors based on azimuth and spatial distribution in the Pacific Ocean east of the Japan Trench. The closest of them detected a maximum tsunami wave height of > 1.9 m. The location of the tide gauge station in the coastal area of Northeast Japan has been shown with purple color while the GNSS sites are shown with red color (Fig. 4). As we can see from the figure the SOMA station which is the nearest to the epicenter faces a huge break in sea level because of this earthquake. The same kind of huge break can be seen in the vertical offset of the nearest 0179 GNSS sites. In this way, relative displacements between these displacements will be the actual displacement of tide gauge GNSS sites located near the epicenter. Ozawa et al. (2011) studied the Tohoku-Oki earthquake effects at the coastal line and found that the movement of eastward co-seismic deplanements was ~ 5.3 m subsidence by up to 1.2 m. They noticed that the eastward displacements of the Pacific coastal area did not differ from the western coastal area significantly, whereas the eastward displacements on the Pacific coast were much larger than that of the western coast. Additionally, after the earthquake, the Pacific coast near the source region was uplifted.

Klos et al. (2019) used monthly data from 38 permanent service for mean sea level (PSMSL) tide gauge stations of which 30 stations were from Japan and 8 from the islands in the southern part of the Western North Pacific. They studied GNSS-derived time series with linear vertical land motion and affected non-linearity because of the Tohoku-Oki earthquake separately. It was noticed that prior to the Tohoku-Oki earthquake, GNSS-derived vertical land motion for the Japanese station generates velocities between 4.5 and 5.6 mm yr^{-1}. Southern Japan subsided while there was an uplift in northern Japan. The station located outside of Japan showed almost the same vertical land motion with variation from -1.3 to 0.8 mm yr^{-1}, probably having no influence of seismicity. Klos et al. (2019) used >5 years longer GNSS time series data and estimated the precise velocities. Velocities changed over time with a maximum difference between post and pre-seismic velocities ~15.3 mm yr^{-1} and a maximum co-seismic offset of 36 mm. The tide gauge derived absolute rise of sea level showed a maximum difference of ~3 mm yr^{-1} between optimized non-linear and standard linear vertical land motion GNSS time series estimates. There are five tide gauge stations in which two of them, Oga and Ogi, were mostly affected (Fig. 4). The misfit between the linear and non-linear estimates comes because of pre- and post-seismic changes of GNSS velocities and vertical offset due to the time of earthquake occurrence.

Ansari and Bae (2021) used wavelet analysis over six tide gauge sites and studied changes in sea level in the coastal area of Japan. They found that P109 (near Ogi; 37.815° N; 138.281° E; 44.70; Sado Island of Japan) has relatively negative tide gauge velocities while other sites have relatively tide gauge positive velocities, which indicates a kind of subsidence in Sado Island of Japan. In further study, Ansari et al. (2022b) collected GNSS data from the P109 site from 01 October to 31

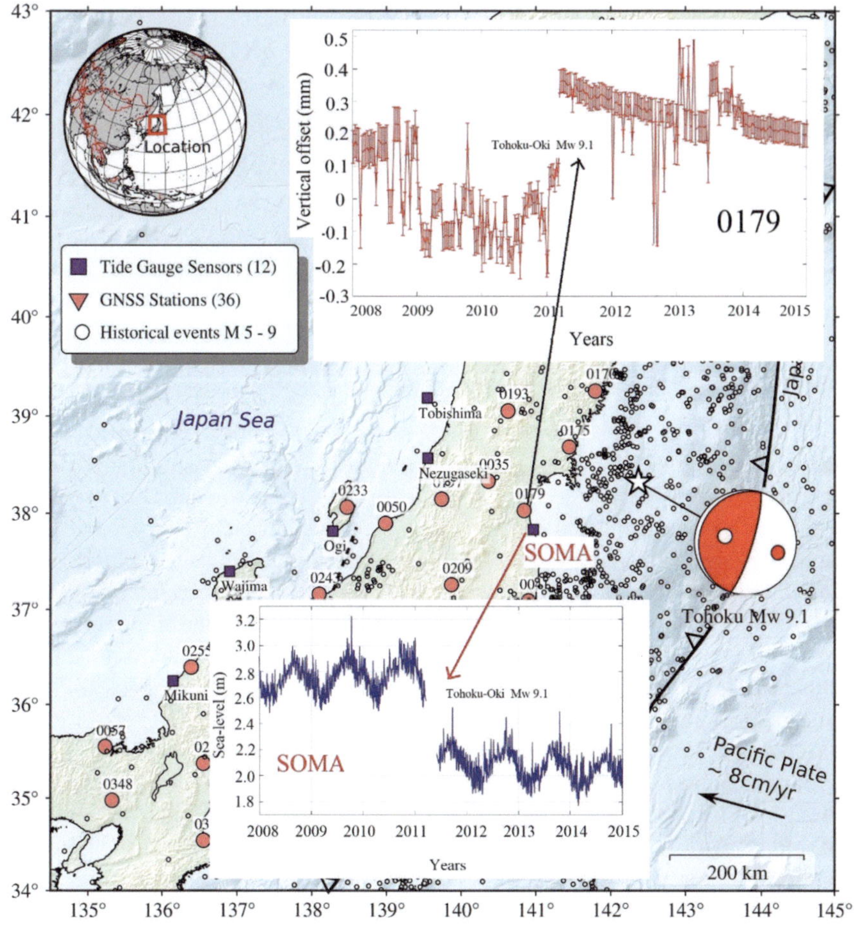

Fig. 4 Location of the tide gauge station in the coastal area of Northeast Japan has been shown with purple color while the GNSS sites are shown with red color. The SOMA station which is the nearest to the epicenter faces a huge break in sea level because of this earthquake. The same kind of huge break can be seen in the vertical offset of the nearest 0179 GNSS sites

December 2019 and studied sea level changes by using the Quasi-Zenith Satellite System (QZSS) reflectometry. Fujiwara et al. (2017) used a multibeam bathymetric survey by crossing the northern Japan Trench offshore Sanriku. They studied the seafloor displacements by taking the bathymetry difference between before and after the Tohoku-Oki earthquake. They noticed that there were <20 m trench-normal horizontal disablements and several meters of vertical displacements. After the earthquake, comparatively, the mid-slop terrace was depressed, the middle slope subdivided, and the landward lower slope was uplifted near the trench. Therefore, there was a very large submarine landslide, localized very large fault slip was unlikely. The plausible explanation for the devastating tsunami source in the northern Japan

Trench was possibly because of extending broader along the trench instead of smaller uplift of the lower slope near the trench (Fujiwara et al., 2017).

5 Study of Earthquake Signatures in Atmosphere

The progress in ground and space-based instrumental facilities provides a better understanding of the atmosphere and space environment because of the earthquake and tsunami. The Tohoku-Oki earthquake excited unprecedented ionospheric perturbations as well as a devastating tsunami. The ionospheric responses because of Tohoku-Oki earthquake have been widely studied by several authors in near and far fields by using GNSS TEC measurements (Astafyeva et al., 2011; Coïsson et al., 2015; Le et al., 2013; Occhipinti et al., 2013; Rakoto et al., 2018; Rolland et al., 2011). Astafyeva et al. (2011) used GNSS TEC data with 1 Hz frequency and studied perturbation of the ionosphere over the seismic fault area. These kinds of high-rate data were suitable to determine the first arrived co-seismic high-precise co-seismic ionospheric perturbations. They were able to detect perturbations over rupture ~464 s before the earthquake. He et al. (2022b) used a GNSS station located in Australia and detected the conjugation of ionospheric TEC anomalies around 41.5 min prior to the Tohoku-Oki earthquake. A gradual change in ionospheric TEC and geomagnetic declination signals in Japan were noticed ~40 min before the main shock followed by impulsive improvements of ~10 min after the main shock (Enomoto et al., 2020). This indicates TEC changes timing was very close in Japan and Australia. Chou et al. (2020) used Japanese GNSS data with 30 s sampling rates and studied clustering in the ionospheric disturbance. They studied the Hilbert spectra of GNSS TEC from 10 to 12 March 2011 (one day before to one day after the earthquake). Results showed a significant improvement in GNSS TEC on the day of the earthquake (Fig. 5). In Fig. 5, high-frequency signals drawn by a white rectangle with a frequency range from ~0.5 to 5.0 mHz at ~05:50–08:00 UT and ~0.6–3 mHz at ~12:00–19:00 UT (Chou et al., 2020). Here, the initial improvements from ~5:50 to 08:00 UT were related to the Tohoku earthquake/tsunami-induced acoustic and gravity waves.

Rolland et al. (2011) analyzed the ionospheric TEC disturbance after the Tohoku-Oki earthquake. They used high density and wide coverage of the GNSS network of Japan to image the ionosphere just after the main shock. They observed a first intense signal ~10 min after the seismic rupture, which consisted of two modes: one propagating beyond 3 km s^{-1} and the other at ~1 km s^{-1}. Further analysis of the TEC time series of the latter mode near the source showed the typical frequencies of acoustic resonance. Beyond 400 km from the source, both the tsunami-induced gravity wave and a third mode are imaged. They showed that the pattern of this slow (225 ± 10 m s^{-1}) and long period gravity wave (1.8 ± 0.2 mHz) are most prevalent in the North-West of the epicentral area. Le et al. (2013) investigated GNSS TEC's abnormal behavior before the Tohoku-Oki earthquake and noticed a significant enhancement in TEC on 08–March–2011 as well as an increment in solar activity in the same period. They used theoretical and empirical models and

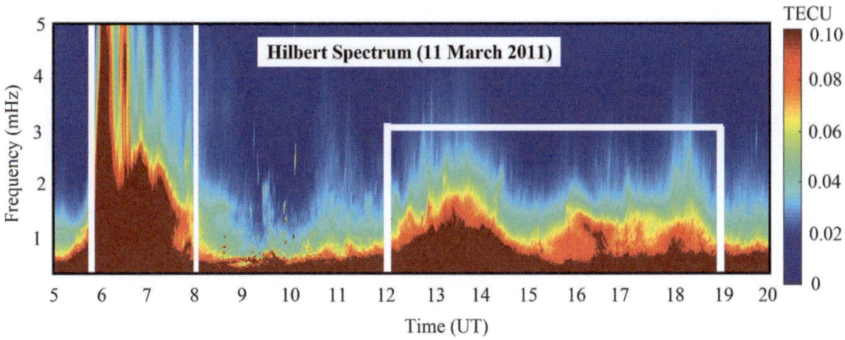

Fig. 5 Hilbert spectra of GNSS TEC from 10 to 12 March 2011 (one day before to one day after) the earthquake. The analysis results showed a significant improvement in GNSS TEC on the day of the earthquake (edited after Chou et al., 2020)

checked the TEC anomalies. The comparison between simulation and observation results concluded that only solar radiation enhancement was not enough to produce such enhancement in TEC values. There must be additional mechanisms such as geomagnetic activity and before the Tohoku-Oki earthquake ionospheric disturbance which was responsible for the TEC significance on 08–March–2011.

6 Conclusions

This chapter covers the study of how current geodetic techniques can detect earthquakes signature on surface, ocean and space. The study took an example of the 2011 Tohoku earthquake (Mw 9.1) that occurred in Japan. The following three points are important.

- A huge break in each time series is noticed, which corresponds to the signature of the Tohoku earthquake (Mw 9.1) on the Earth's surface. The magnitude break in each time series differs depending on the location of the GNSS site. It was noticed that the slip was most intense within 10–20 km depth from the trench. This slip displacement considerably decreases in a small area toward the trench at the wedge toe.
- The location of the tide gauge station in the coastal area of Northeast Japan nearest to the epicenter has a huge break in sea level because of this earthquake. These tide gauge observations can successfully be detected by GNSS-Reflectometry technique at a high-precise level. The same kind of huge break can be seen in the vertical offset of the nearest GNSS sites. Relative displacements between these displacements will be the actual displacement of tide gauge GNSS sites located near the epicenter. In this way, earthquake signatures can be noticed in the ocean by using current geodetic techniques.

- The ionospheric responses because of Tohoku-Oki earthquake were able to detect perturbations over a rupture ~464 s before the earthquake. The GNSS station located in Australia detected the conjugation of ionospheric TEC anomalies around 41.5 min prior to the Tohoku-Oki earthquake. A gradual change in ionospheric TEC and geomagnetic declination signals in Japan were noticed around 40 min before the main shock followed by impulsive improvements around 10 min after the main shock.

Acknowledgements Soumyajit Mukherjee (IIT Bombay) invited, handled and reviewed this article. Mukherjee (2026) summarized this work.

References

Ansari, K. (2023). Review on role of multi-constellation global navigation satellite system-reflectometry (GNSS-R) for real-time sea-level measurements. In *Structural geology and tectonics field guidebook* (Vol. 2, pp. 333–358). https://doi.org/10.1007/978-3-031-19576-1_13

Ansari, K., & Bae, T. S. (2021). Modelling and mitigation of real-time sea level measurement over the coastal area of Japan. *Marine Geophysical Research, 42*(4), 40. https://doi.org/10.1007/s11001-021-09460-y

Ansari, K., & Park, K. D. (2022). Geodetic analysis inside the South Korean Peninsula and Impact of the 2011 Tohoku-Oki (TO) earthquake. *Acta Geologica Sinica-English Edition, 96*(2), 631–647. https://doi.org/10.1111/1755-6724.14885

Ansari, K., Althuwaynee, O. F., & Corumluoglu, O. (2016). Monitoring and prediction of precipitable water vapor using GPS data in Turkey. *Journal of Applied Geodesy, 10*(4), 233–245. https://doi.org/10.1515/jag-2016-0037

Ansari, K., Bae, T. S., & Inyurt, S. (2020). Global positioning system interferometric reflectometry for accurate tide gauge measurement: Insights from South Beach, Oregon, United States. *Acta Astronautica, 173*, 356–362. https://doi.org/10.1016/j.actaastro.2020.04.060

Ansari, K., Bae, T. S., Seok, H. W., & Kim, M. S. (2021). Multiconstellation global navigation satellite systems signal analysis over the Asia-Pacific region. *International Journal of Satellite Communications and Networking, 39*(3), 280–293. https://doi.org/10.1002/sat.1389

Ansari, K., De Oliveira-Júnior, J. F., & Jamjareegulgarn, P. (2022a). Spatial changes of ocean circulation along the coast of gulf of thailand using tide gauge measurements. *IEEE Access, 10*, 130795–130805. https://doi.org/10.1109/ACCESS.2022.3229212

Ansari, K., Seok, H. W., & Jamjareegulgarn, P. (2022b). Quasi zenith satellite system-reflectometry for sea-level measurement and implication of machine learning methodology. *Scientific Reports, 12*(1), 21445. https://doi.org/10.1038/s41598-022-25994-6

Ansari, K., Walo, J., Simanjuntak, A. V., & Wezka, K. (2024). Evaluation of recent Tectonic movement in Northeast Japan by using long-term GNSS and Tide Gauge Measurements. *Journal of Structural Geology, 188*, 105258. https://doi.org/10.1016/j.jsg.2024.105258

Astafyeva, E., Lognonné, P., & Rolland, L. (2011). First ionospheric images of the seismic fault slip on the example of the Tohoku-oki earthquake. *Geophysical Research Letters, 38*(22). https://doi.org/10.1029/2011GL049623

Chou, M. Y., Cherniak, I., Lin, C. C., Pedatella, N. M. (2020). The persistent ionospheric responses over Japan after the impact of the 2011 Tohoku earthquake. *Space Weather, 18*(4), e2019SW002302. https://doi.org/10.1029/2019SW002302

Church, J. A., & White, N. J. (2011). Sea-level rise from the late 19th to the early 21st century. *Surveys in Geophysics, 32*, 585–602. https://doi.org/10.1007/s10712-011-9119-1

Coïsson, P., Lognonné, P., Walwer, D., & Rolland, L. M. (2015). First tsunami gravity wave detection in ionospheric radio occultation data. *Earth and Space Science, 2*(5), 125–133. https://doi.org/10.1002/2014EA000054

Denys, P. H., Beavan, R. J., Hannah, J., Pearson, C. F., Palmer, N., Denham, M., & Hreinsdottir, S. (2020). Sea level rise in New Zealand: The effect of vertical land motion on century-long tide gauge records in a tectonically active region. *Journal of Geophysical Research: Solid Earth, 125*(1), e2019JB018055. https://doi.org/10.1029/2019JB018055

Enomoto, Y., Heki, K., Yamabe, T., Sugiura, S., & Kondo, H. (2020). A possible causal mechanism of geomagnetic variations as observed immediately before and after the 2011 Tohoku-Oki earthquake. *Open Journal of Earthquake Research, 9*(02), 33. https://doi.org/10.4236/ojer.2020.92003

Fujiwara, T., dos Santos Ferreira, C., Bachmann, A. K., Strasser, M., Wefer, G., Sun, T., Kanamatsu, T., & Kodaira, S. (2017). Seafloor displacement after the 2011 Tohoku-Oki earthquake in the northern Japan trench examined by repeated bathymetric surveys. *Geophysical Research Letters, 44*(23), 11–833. https://doi.org/10.1002/2017GL075839

Hannah, J., & Bell, R. G. (2012). Regional sea level trends in New Zealand. *Journal of Geophysical Research: Oceans, 117*(C1). https://doi.org/10.1029/2011JC007591

He, X., Montillet, J. P., Li, Z., Kermarrec, G., Fernandes, R., & Zhou, F. (2022a). Recent advances in modelling geodetic time series and applications for earth science and environmental monitoring. *Remote Sensing, 14*(23), 6164. https://doi.org/10.3390/rs14236164

He, L., Wu, L., Heki, K., & Guo, C. (2022b). The conjugated ionospheric anomalies preceding the 2011 Tohoku-Oki earthquake. *Frontiers in Earth Science*, 810. https://doi.org/10.3389/feart.2022.850078

Herring, T. A., King, R. W., & McClusky, S. C. (2018). Introduction to GAMIT/GLOBK.

Hughes, C. W., & Williams, S. D. (2010). The color of sea level: Importance of spatial variations in spectral shape for assessing the significance of trends. *Journal of Geophysical Research: Oceans, 115*(C10). https://doi.org/10.1029/2010JC006102

Jamjareegulgarn, P., Ansari, K., & Ameer, A. (2020). Empirical orthogonal function modelling of total electron content over Nepal and comparison with global ionospheric models. *Acta Astronautica, 177*, 497–507. https://doi.org/10.1016/j.actaastro.2020.07.038

Kim, S., Ree, J. H., Yoon, H. S., Choi, B. K., & Park, P. H. (2018). Crustal deformation of South Korea after the Tohoku-Oki earthquake: Deformation heterogeneity and seismic activity. *Tectonics, 37*(8), 2389–2403. https://doi.org/10.1029/2018TC004967

Klos, A., Kusche, J., Fenoglio-Marc, L., Bos, M. S., & Bogusz, J. (2019). Introducing a vertical land motion model for improving estimates of sea level rates derived from tide gauge records affected by earthquakes. *GPS Solutions, 23*(4), 102. https://doi.org/10.1007/s10291-019-0896-1

Larson, K. M., Löfgren, J. S., & Haas, R. (2013). Coastal sea level measurements using a single geodetic GPS receiver. *Advances in Space Research, 51*(8), 1301–1310. https://doi.org/10.1016/j.asr.2012.04.017

Le, H., Liu, L., Liu, J. Y., Zhao, B., Chen, Y., & Wan, W. (2013). The ionospheric anomalies prior to the M9. 0 Tohoku-Oki earthquake. *Journal of Asian Earth Sciences, 62*, 476–484. https://doi.org/10.1016/j.jseaes.2012.10.034

Li, X., Zhang, X., Ren, X., Fritsche, M., Wickert, J., & Schuh, H. (2015). Precise positioning with current multi-constellation global navigation satellite systems: GPS, GLONASS, Galileo and BeiDou. *Scientific Reports, 5*(1), 1–14. https://doi.org/10.1038/srep08328

Minson, S. E., Simons, M., Beck, J. L., Ortega, F., Jiang, J., Owen, S. E., Moore, A. W., Inbal, A., & Sladen, A. (2014). Bayesian inversion for finite fault earthquake source models–II: The 2011 great Tohoku-Oki. *Japan Earthquake. Geophysical Journal International, 198*(2), 922–940. https://doi.org/10.1093/gji/ggu170

Mukherjee, S. (2026). Introduction to "Structural Geology & Tectonics". In S. Mukherjee (Ed.), *Structural Geology & Tectonics*. Springer. ISBN: 978-981-95-4743-2.

Mukul, M., Jade, S., Ansari, K., Matin, A., & Joshi, V. (2018). Structural insights from geodetic Global Positioning System measurements in the Darjiling-Sikkim Himalaya. *Journal of Structural Geology, 114*, 346–356. https://doi.org/10.1016/j.jsg.2018.03.007

Occhipinti, G., Rolland, L., Lognonné, P., & Watada, S. (2013). From Sumatra 2004 to Tohoku-Oki 2011: The systematic GPS detection of the ionospheric signature induced by tsunamigenic earthquakes. *Journal of Geophysical Research: Space Physics, 118*(6), 3626–3636. https://doi.org/10.1002/jgra.50322

Oikonomou, C., Haralambous, H., Pulinets, S., Khadka, A., Paudel, S. R., Barta, V., Muslim, B., Kourtidis, K., Karagioras, A., & İnyurt, S. (2020). Investigation of pre-earthquake ionospheric and atmospheric disturbances for three large earthquakes in Mexico. *Geosciences, 11*(1), 16. https://doi.org/10.3390/geosciences11010016

Ozawa, S., Nishimura, T., Suito, H., Kobayashi, T., Tobita, M., & Imakiire, T. (2011). Coseismic and postseismic slip of the 2011 magnitude-9 Tohoku-Oki earthquake. *Nature, 475*(7356), 373–376. https://doi.org/10.1038/nature10227

Panet, I., Mikhailov, V., Diament, M., Pollitz, F., King, G., De Viron, O., Holschneider, M., Biancale, R., & Lemoine, J. M. (2007). Coseismic and post-seismic signatures of the Sumatra 2004 December and 2005 March earthquakes in GRACE satellite gravity. *Geophysical Journal International, 171*(1), 177–190. https://doi.org/10.1111/j.1365-246X.2007.03525.x

Parker, A. (2018). Sea level oscillations in Japan and China since the start of the 20th century and consequences for coastal management-Part 2: China pearl river delta region. *Ocean & Coastal Management, 163*, 456–465. https://doi.org/10.1016/j.ocecoaman.2018.08.004

Priyadarshi, S., Kumar, S., & Singh, A. K. (2011). Changes in total electron content associated with earthquakes (M > 5) observed from GPS station, Varanasi, India. *Geomatics, Natural Hazards and Risk, 2*(2), 123–139. https://doi.org/10.1080/19475705.2011.563390

Raj, N., Gharineiat, Z., Ahmed, A. A. M., & Stepanyants, Y. (2022). Assessment and prediction of sea level trend in the South Pacific region. *Remote Sensing, 14*(4), 986. https://doi.org/10.3390/rs14040986

Rakoto, V., Lognonné, P., Rolland, L., & Coïsson, P. (2018). Tsunami wave height estimation from GPS-derived ionospheric data. *Journal of Geophysical Research: Space Physics, 123*(5), 4329–4348. https://doi.org/10.1002/2017JA024654

Reddybattula, K. D., Panda, S. K., Ansari, K., & Peddi, V. S. R. (2019). Analysis of ionospheric TEC from GPS, GIM and global ionosphere models during moderate, strong, and extreme geomagnetic storms over Indian region. *Acta Astronautica, 161*, 283–292. https://doi.org/10.1016/j.actaastro.2019.05.042

Rolland, L. M., Lognonné, P., Astafyeva, E., Kherani, E. A., Kobayashi, N., Mann, M., & Munekane, H. (2011). The resonant response of the ionosphere imaged after the 2011 off the Pacific coast of Tohoku Earthquake. *Earth, Planets and Space, 63*, 853–857. https://doi.org/10.5047/eps.2011.06.020

Sekertekin, A., Inyurt, S., & Yaprak, S. (2020). Pre-seismic ionospheric anomalies and spatio-temporal analyses of MODIS Land surface temperature and aerosols associated with Sep, 24 2013 Pakistan Earthquake. *Journal of Atmospheric and Solar-Terrestrial Physics, 200*, Article 105218. https://doi.org/10.1016/j.jastp.2020.105218

Şentürk, E., Inyurt, S., & Sertçelik, İ. (2020, October). Ionospheric anomalies associated with the M w 7.3 Iran–Iraq border earthquake and a moderate magnetic storm. In *Annales Geophysicae* (Vol. 38, No. 5, pp. 1031–1043). Copernicus GmbH. https://doi.org/10.5194/angeo-38-1031-2020

Seok, H. W., Ansari, K., Panachai, C., & Jamjareegulgarn, P. (2022). Individual performance of multi-GNSS signals in the determination of STEC over Thailand with the applicability of Klobuchar model. *Advances in Space Research, 69*(3), 1301–1318. https://doi.org/10.1016/j.asr.2021.11.025

Sharma, S. K., Ansari, K., & Panda, S. K. (2018). Analysis of ionospheric TEC variation over Manama, Bahrain, and comparison with IRI-2012 and IRI-2016 models. *Arabian Journal for Science and Engineering, 43*(7), 3823–3830. https://doi.org/10.1007/s13369-018-3128-z

Simons, M., Minson, S. E., Sladen, A., Ortega, F., Jiang, J., Owen, S. E., Meng, L., Ampuero, J. P., Wei, S., Chu, R., & Helmberger, D. V. (2011). The 2011 magnitude 9.0 Tohoku-Oki earthquake: Mosaicking the megathrust from seconds to centuries. *Science, 332*(6036), 1421–1425. https://doi.org/10.1126/science.1206731

Tabibi, S., Geremia-Nievinski, F., & van Dam, T. (2017). Statistical comparison and combination of GPS, GLONASS, and multi-GNSS multipath reflectometry applied to snow depth retrieval. *IEEE Transactions on Geoscience and Remote Sensing, 55*(7), 3773–3785. https://doi.org/10.1109/TGRS.2017.2679899

Takamatsu, N., Muramatsu, H., Abe, S., Hatanaka, Y., Furuya, T., Kakiage, Y., Ohashi, K., Kato, C., Ohno, K., & Kawamoto, S. (2023). New GEONET analysis strategy at GSI: Daily coordinates of over 1300 GNSS CORS in Japan throughout the last quarter century. *Earth, Planets and Space, 75*(1), 49. https://doi.org/10.1186/s40623-023-01787-7

Thomas, J. E., Ben, U. C., Ekanem, A. M., George, N. J., & Nathaniel, E. U. (2022). Seismo-ionospheric anomalies before M7. 2 Haiti earthquake of August 14, 2021, from GPS-TEC. *Acta Geophysica*, 1–7. https://doi.org/10.1007/s11600-022-00903-7

Ulukavak, M., & Inyurt, S. (2020). Seismo-ionospheric precursors of strong sequential earthquakes in Nepal region. *Acta Astronautica, 166*, 123–130. https://doi.org/10.1016/j.actaastro.2019.09.033

Wang, M., Li, Q., Wang, F., Zhang, R., Wang, Y., Shi, H., Zhang, P., & Shen, Z. (2011). Far-field coseismic displacements associated with the 2011 Tohoku-Oki earthquake in Japan observed by Global Positioning System. *Chinese Science Bulletin, 56*, 2419–2424. https://doi.org/10.1007/s11434-011-4588-7

Wang, X., Zhang, S., Wang, L., He, X., & Zhang, Q. (2020). Analysis and combination of multi-GNSS snow depth retrievals in multipath reflectometry. *GPS Solutions, 24*, 1–13. https://doi.org/10.1007/s10291-020-00990-3

Review of Crustal Deformation Studies in Western India by Using Global Navigation Satellite System Measurements

Kutubuddin Ansari

Abstract This chapter is a review of crustal deformation studies by using global navigation satellite system (GNSS) measurements in western India. GNSS velocity field in India-fixed reference frame (IREF) are reviewed and modeled by using well-known technique known as Singular Spectrum Analysis (SSA). The change in GNSS velocity field points out a considerable N-S contractional deformation near the location of the 2001 Bhuj earthquake. Moreover, the GNSS sites horizontal velocity trend at the northern part of western India is totally opposite to the velocity trend of southern part of western India, indicating an extensional deformation. The SSA-based estimated GNSS principal strain rate displays less or more participation of contraction perpendicular to the lineament or fault systems in the region. The extensional rates in the available fault system are noted in a negligible amount except a very less amount near the epicenter. The estimated rotation rates at each GNSS from EW to NS direction are very low and statically insignificant. This indicates that the southern peninsula interior region moves as a rigid block.

Keywords Global navigation satellite system · Indian tectonics · Strain and rotation rate · Singular spectrum analysis

1 Introduction

The movement of the tectonic plates is a natural process but sometimes the crustal movement occurs due to processes other than tectonic sources. Identifying and characterizing this kind of ground process because of tectonic movement is an essential task. It is very important to monitor surface deformation to understand the seismotectonic quantitatively. This helps in evaluating the seismic hazards more accurately

K. Ansari (✉)
Integrated Geoinformation (IntGeo) Solution Private Limited, New Delhi 110025, India
e-mail: kdansarix@gmail.com

Department of Geomatics Engineering, Faculty of Engineering, Karadeniz Technical University, Trabzon 61080, Turkey

S. Mukherjee (ed.), *Structural Geology and Tectonics*,
https://doi.org/10.1007/978-981-95-4743-2_11

(Ansari & Park, 2022; Choudhury et al., 2018; Ansari, 2018; Mukul et al., 2014). Ghavri et al. (2022) studied a composite analysis of the geodetic and seismic strain rates of the Kutch paleo-rift basin of western India. Their analysis indicates a high rate of strain build-up on the reactivated major faults in the region that have the potential to generate devastating (Mw $\geq$ 6) earthquakes. In the current study, the studies based on crustal deformation, strain and rotation rate variation in western India measured by the global navigation satellite system (GNSS) measurements are reviewed. The GNSS measuring data has been noted from several previously published research (Dumka et al., 2019, 2022; Jade et al., 2017; Reddy & Sunil, 2008; Appendix 1) and India-fixed velocities are studied.

It has been assumed that the GNSS station shifted linearly or in a periodic way during the data collection but some kind of natural phenomenon such as snow, rainfall, ups, and downs of temperature and partial correction of tropospheric and ionospheric delays may cause random errors in signals. These kinds of errors in signal affect the secular tectonic deformation measurements. The technique of Singular Spectrum Analysis (SSA) is a novel and powerful method of time-series analysis utilized in multidisciplinary subjects e.g., physics, mathematics, social science, oceanography meteorology, and economics in the last decades (Ansari & Park, 2018, 2022; Hassani et al., 2009; Zabalza et al., 2015). All measured velocities at GNSS sites are predicted by SSA. The primary intention of this kind of prediction was to enhance the quality of velocity time-series and study the tectonic crustal deformation without errors. The secondary intention of SSA modeling is to combine the different selected paper's velocities and bring them in a single common platform.

This chapter has tried a comprehensive and critical overview of GNSS results. When a simple observation without any modeling is used, there are high chances of noise error, signal error, and other local deformation errors. To overcome this difficulty, Nocquet (2012) used a combined approach of published results to derive a homogeneous velocity field, consistent at the plate boundary scale of the Mediterranean and the surrounding Alpine belts. Zarifi et al. (2014) used Pearson product-moment correlation to find the correlation between the stress field obtained from the focal mechanism stress inversion and that obtained using the seismic and geodetic strain rates. They assumed that stresses in a continuum are produced by tectonic forces and the consequent deformation on the crustal scale. Therefore, the direction of the stress and strain are ideally to be the same. To overcome such kind of issues, in the current study first obtained GNSS velocity fields have been modeled by a well-known and novel technique known as SSA, and then the India-fixed velocity, strain and rotation rate are analyzed. The SSA technique reduces these kinds of errors and generates secular crustal deformation. This kind of modeled data provides the best accuracy of results and is very close to real observations.

2 Modeling

Consider that the India-fixed velocities of GNSS sites are V (v_1, v_2, v_3, …, v_N) and they are arranged in increasing order (can be arranged in decreasing order as well). SSA method is applied for these arranged velocities and the SSA velocities are predicted (Appendix 1).

The applicability of the SSA method has been divided in four important steps. In the first step named as *embedding,* it is considered that the original time-series of GNSS velocities are V (v_1, v_2, v_3, …, v_N) such that, there exists at least one non-zero term in each series and length of N is greater than 2. The trajectory matrix (T) following Hankel matrix for the SSA method can be written like this (Ansari et al., 2020):

$$T = \begin{bmatrix} v_1 & v_2 & \ldots. & v_{L-1} & v_L \\ v_2 & v_3 & \ldots. & v_L & v_{L+1} \\ \vdots & \vdots & \vdots\vdots & \vdots & \vdots \\ \vdots & \vdots & \vdots & \vdots & \vdots \\ v_{N-L+1} & v_{N-L+2} & \ldots. & v_{N-1} & v_N \end{bmatrix} \quad (1)$$

where L be an integer and can be taken $1 < L < N/2$.

The second step is called *singular value decomposition* (SVD) in which the SVD is applied to the trajectory matrix T. The singular value decomposition (SVD) is a factorization of a real or complex matrix. Suppose $A = TT^T$ is an $n \times n$ order matrix having factorization in the form of $X\sum Y^T$, where X is an $n \times n$ left orthogonal matrix, Σ is an $n \times n$ diagonal matrix with decreasing order eigenvalues ($\lambda_1 \geq \lambda_2 \geq \lambda_3 \geq \ldots. \geq \lambda_{3L}$), and Y is an $n \times n$ right orthogonal matrix. The decomposition of A is given by (Ansari et al., 2020):

$$T = \sum_{i=1}^{d} \sqrt{\lambda_i} X_i Y_i^T \quad (i = 1,\ 2,\ \ldots.,\ d) \quad (2)$$

where d is equal to rank of matrix T and the triple $\sqrt{\lambda_i} X_i Y_i^T$ is called as ith eigentriple of the SVD.

The third step is known as *eigentriple grouping* step which corresponds to splitting the elementary matrices X_i into several groups (less than d) and summing the matrices in each group. If the number of groups is equivalent to d the procedure is called elementary grouping. In the present study, elementary grouping procedure has been followed then:

The SSA-modeled velocities are used and the strain analysis in the western margin of India by applying the zero_strain program of a 2D version (Teza & Galgaro, 2008) is carried out. In this program, it has been assumed that the preliminary coordinates of the geodetic network are given by (x, y). If the coordinates of the original network shift (Δx, Δy) during the repeated measurement then the rectilinear functions shift coordinate (u, v) in the 2D system can be stated by second-order tensor (Teza & Galgaro, 2008):

$$E = \begin{bmatrix} \frac{\partial u}{\partial x} & \frac{1}{2}\left(\frac{\partial u}{\partial y} + \frac{\partial v}{\partial x}\right) \\ \frac{1}{2}\left(\frac{\partial u}{\partial y} + \frac{\partial v}{\partial x}\right) & \frac{\partial v}{\partial y} \end{bmatrix} = \begin{bmatrix} e_{11} & e_{12} \\ e_{21} & e_{22} \end{bmatrix} \text{ and}$$
$$\Omega = \begin{bmatrix} 0 & \frac{1}{2}\left(\frac{\partial u}{\partial y} - \frac{\partial v}{\partial x}\right) \\ -\frac{1}{2}\left(\frac{\partial u}{\partial y} - \frac{\partial v}{\partial x}\right) & 0 \end{bmatrix} = \begin{bmatrix} 0 & \omega \\ -\omega & 0 \end{bmatrix} \tag{3}$$

Here, E is a kind of symmetric matrix representing tensor strain and Ω is a kind of anti-symmetric matrix indicating the rigid body rotational change of motion (Teza & Galgaro, 2008).

3 Result and Discussions

The results are discussed in terms of India-fixed SSA models predicted values. The investigation of strain analysis and rotation rate in western India from these velocities is also raised here.

3.1 GNSS Velocity Field

The SSA predicted horizontal velocity fields in India-fixed reference frame (IREF) from western India provides new constraints of the kinematic motion of the Indian region (Fig. 1). In western India, the Kutch province is categorized as a seismically active zone of the Indian subcontinent, extending ~150 km along N-S and ~250 km along E-W (Arya, 2000). The Bhuj region is one of the major seismogenic terrains of the western India (also see Lohani et al., 2022; Omid et al., 2021; Dhawale et al., 2023 for structural geology and morphotectonics of the Bhuj and the surrounding areas). Recent works on seismicity and structural geology of the Kutch region can be found in Shaikh et al. (2020, 2022) and Maurya et al. (2021).

One of the deadliest earthquakes in this region occurred on 26-Jan-2001, known as the Bhuj earthquake. This earthquake in the Bhuj region is considered to be the second-greatest recorded earthquake in history (Bendick et al., 2001). The epicenter of the Bhuj earthquake with the red star has been shown in Fig. 1. Before this, a damaging earthquake, another earthquake known as the Anjar earthquake (Mw 6) occurred in 1956 with a focal depth of 15 km (Reddy & Sunil, 2008). The GNSS sites DHAM (11.62 ± 1.51 mm yr^{-1}) located in the south of the epicenter and the RATN (5.24 ± 1.49 mm yr^{-1}) located in the north of the epicenter are showing opposite direction velocities with respect to each other (Fig. 1, Appendix 1). This

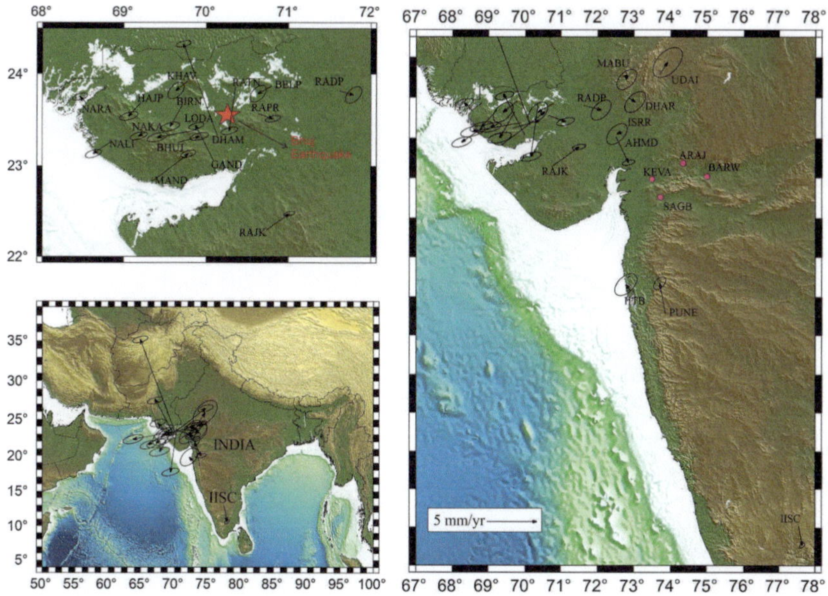

Fig. 1 SSA modeled India-fixed velocity in western India, collected from published research (Dumka et al., 2019, 2022; Jade et al., 2017; Reddy & Sunil, 2008; see Appendix 1) and plotted with Generic Mapping Tools (GMT)

change in velocity from DHAM to RATN GNSS sites across the epicentral track points out a considerable contractional deformation near the Bhuj earthquake and indicates a north–south compressional zone (Reddy & Sunil, 2008). This kind of compression in the epicentral region was expected because of the eastern segment transpressional deformation of the Kutch Mainland Fault (KMF). This KMF interacts with its transfer faults, North Wagad Fault (NWF) in the north (Mandal et al., 2004) and South Wagad Fault (SWF) in the south (Biswas, 2005; Mathew et al., 2006) due to the continuing tectonic uplifts in the Kutch mainland (Wallace et al., 2006). Ravi Shanker (1995) suggested that the crustal structure and differential thermal between RATN and DHAM sites provide the relative movement of the first order. The velocity trend at these sites supports the reverse fault movement with the south-dipping rupture fault plane (Negishi et al., 2002; Wesnousky et al., 2001).

Contrary to the above explanations, the velocity trend of DHAM and GAND sites located in the south of epicenter are parallel to each other support the left lateral strike-slip (Kayal et al., 2002). The GNSS sites RADP, BELP, and KHAV situated in the Kutch basin ~50 to 100 km north from the epicenter of the 2001 Bhuj earthquake, the sites MABU and DHAR are placed around 100 to 200 km further north of the Ahmadabad and the ISRR site is in Gandhinagar near Ahmadabad around 160 km east of the epicenter (Jade et al., 2003). The post-seismic Bhuj earthquake transient deformation was noticed to be very low and reduced rapidly within 3–4 years of the 2001 Bhuj earthquake. The deformation reduced much during 2007–2009 indicating a weak mantle (Chandrasekhar et al., 2009; Choudhury et al., 2013). The change in baseline length indicates an E-W rate of 2.3 ± 2.11 mm yr^{-1} between DHAR-KHAV and an N-S rate of -0.71 ± 2.1 mm yr^{-1} between MABU-ISRR (Jade et al., 2017). Reddy and Sunil (2008) concluded that there was no significant deformation at the GNSS sites located far away form from the epicenters, however previous geodetic studies confirmed the convergence rate across the Kutch Rift Basin (Jade et al., 2003).

Dumka et al. (2022) studied four GNSS sites, viz., BARW (22.04°N; 74.90°E), ARAJ (22.29°N; 74.35°E), SAGB (21.55°N; 73.79°E), and KEVA (21.90°N; 73.67°E) located between the sites ISRR and AHMD in the north and IITB and PUNE in the south (Appendix 1). These locations are shown in Fig. 1 but are not plotted, because their north and east velocity components were unknown. In these cases, the SAGB sites showed a significant SSA modeled maximum intra-plate deformation of ~1.61 ± 0.15 mm yr^{-1}. The site showed NEE-ward displacement (Fig. 4 of Dumka et al., 2019), possibly causing the effect of loading or unloading in the neighboring reservoir (Dumka et al., 2019). The SSA modeled velocity from lowest to highest at the other GNSS sites KEVA, ARAJ, and BARW has been estimated ~0.45 ± 0.15, 0.78 ± 0.15, and 0.91 ± 0.15 mm yr^{-1}, respectively. The movement of these three sites (except SAGB) showed a SE-ward displacement (Fig. 4 of Dumka et al., 2019). Thus, the pattern of deformation on the eastern side differs from that of the western side (Dumka et al., 2022).

The GNSS sites IITB and PUNE are in the Deccan trap region of western India. Refer to Misra et al. (2014), Kaplay et al. (2017a, 2017b, 2021), Misra and Mukherjee (2017), Mukherjee et al. (2017, 2020), Vanik et al. (2018), etc., for tectonics of parts of/surroundings of the Deccan trap. The SSA modeled velocities at these sites showed 0.84 ± 2.32 mm yr^{-1} (IITB; azimuth 231.34°N) and 3.02 ± 1.33 mm yr^{-1} (PUNE; azimuth 258.58°N). The velocity difference between these sites is ~2.18 mm yr^{-1}. Moreover, the GNSS sites of IISC showed 1.92 ± 0.55 mm yr^{-1} horizontal velocity with an azimuth of 169.43°N. The velocity trend at IISC (south direction) is totally opposite to the velocity trend of IITB and PUNE (north direction), indicating an extensional deformation. The Deccan trap is considered as a stable continental seismic region; however, it experienced a significant seismicity level in the past decades, including the 29-Sept-1993 devastating (Mw 6.3) Latur earthquake (Gupta, 1993; Seeber et al., 1996) and September 5, 2000 (Mw 5.2, 17.3°N; 73.8°E) earthquake. This region has complex geophysical signatures and is characterized by numerous hidden tectonic features. The western margin of this region parallel to the west coast fault contains high subsidence as well as uplift, large deformation, high heat flow, hot springs, and magnetism relative to the Dharwar craton (Raval, 1995). Primarily intra-plate earthquakes occurred in this region because of crustal faults and vertical crustal block movements (Kailasam, 1993).

3.2 GNSS Strain and Rotation Rate

The distribution of strain rate assists to evaluate the seismic risk and concludes the causal force of tectonic (Reddy & Sunil, 2008). I used the IREF velocity field in the 2D version of the grid strain (strain_zero) program written by Teza and Galgaro (2008) and estimated spatial strain rate variations in western India. These strain rate variations at the selected GNSS stations have been plotted in red (compression) and blue (extension) in Fig. 2. The positive values are the sign of extension and negative of compression. The estimated principal strain in the southern part of the epicentral region showed the highest compressional rate of strain of ~$-3.86 \times 10^{-7} \pm 1.94 \times 10^{-7}$ yr^{-1} and the lowest extensional rate of ~$6.24 \times 10^{-8} \pm 5.99 \times 10^{-8}$ yr^{-1} at DHAM site (Appendix 2). In the north of the epicenter at the RATN site, the highest compressional strain rate is around $-3.74 \times 10^{-7} \pm 1.77 \times 10^{-7}$ $yr^{-1}$and the lowest extensional rate is around $7.65 \times 10^{-8} \pm 5.00 \times 10^{-8}$ yr^{-1} are estimated. The average compressional and extensional rates in the whole region by using SSA modeled velocities are estimated around $-4.00 \times 10^{-7}$and 7.00×10^{-8} yr^{-1}, respectively, with their visualization trend of NW–SE (compression) and NE-SW (extension) (Fig. 3). The crustal readjustment related to the existing fault system and the main shock jointly contribute to the strain rate anomalies. These kinds of outcomes from the study reveal a continuing transpressional deformation across the region with a blocked embedded structure between the NWF and SWF (Reddy & Sunil, 2008). This deformation might relate to the present-day neotectonic compressive stress regime of the Indian Plate because of its NNE movement against the frontal collision in the north and its proximity to the India, Arabia, and Asia triple junction in the western continental boundary (Biswas, 1987; Biswas & Khattri, 2002; Gowd et al., 1996). These rates significantly exceed the Indian Peninsula shortening rates (~2–3 mm yr^{-1}), which correspond to the rate of strain around 10^{-9} yr^{-1} (Paul et al., 2001). The principal strain rate displays less or more participation of contraction perpendicular to the lineament or fault systems in the region. The extensional rates in the available fault system are noted in a negligible amount except a very less amount near the epicenter (Reddy & Sunil, 2008). The GNSS measurements during the year 2010–2012 showed an extensional rate of ~1.00×10^{-8} at BELP and KHAV sites and their residual velocities were 1.8 ± 1.5 mm yr^{-1} (Jade et al., 2017). These sites are located near the Bhuj earthquake epicenter indicating possible viscoelastic effects of the 2001 earthquake, but the GNSS velocity field current resolution is not able to quantify it (Jade et al., 2017). The details analysis of seismic hazard potential of Kutch paleo-rift basin from seismic and geodetic strain rates can be seen in Ghavri et al. (2022).

The GNSS sites (BARW, ARAJ, SAGB, and KEVA, Fig. 1) studied by Dumka et al. (2022) estimated strain rates in the range of 1.00×10^{-7} yr^{-1} in commonly stable regions (Appendix 2). They estimated a maximum strain rate of 3.00×10^{-7}, a slightly higher rate compared to the stable continental region. These calculated strain rates in this region may be due to the accumulation of slip deficits (Dumka et al., 2022). The SSA modeled India-fixed velocity-based average strain rates in

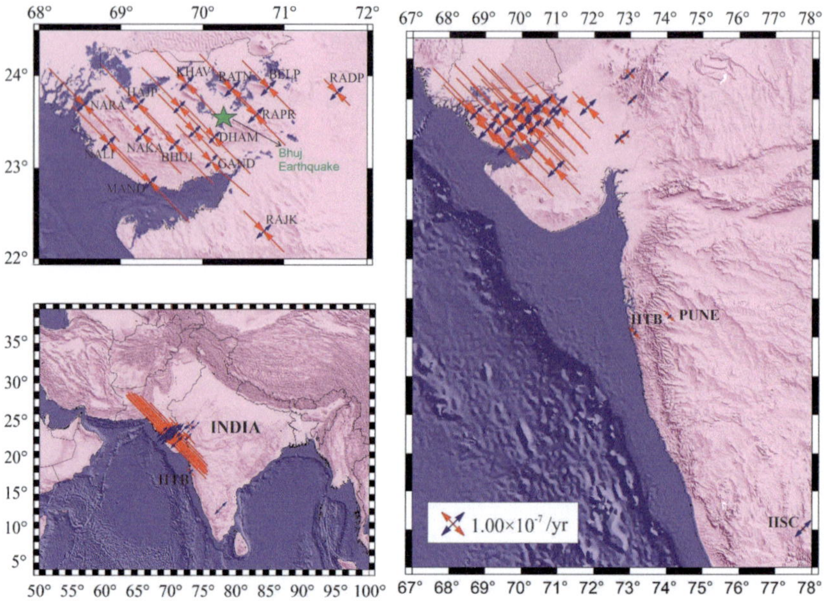

Fig. 2 Strain rate variations at the selected GNSS stations (Appendix 2). The red color arrows show compression and the blue color arrows show extension. The positive values are the sign of extension, and the negative of compression. The figures are plotted with Generic Mapping Tools (GMT)

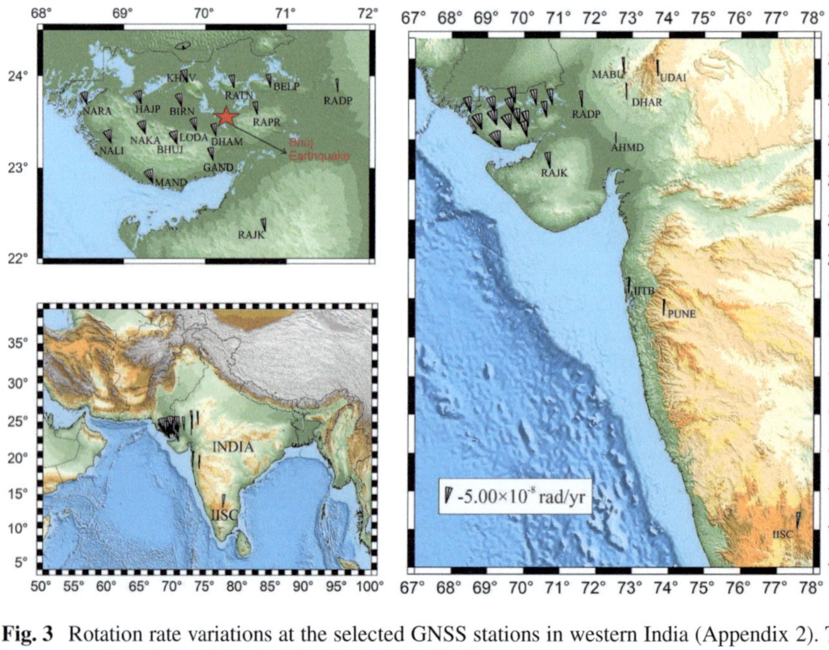

Fig. 3 Rotation rate variations at the selected GNSS stations in western India (Appendix 2). The figures are plotted with Generic Mapping Tools (GMT)

the Deccan tarp region at IITB and PUNE is estimated to be ~$7.78 \times 10^{-9} \pm 2.14 \times 10^{-7}$ yr^{-1} (extension) and $-5.36 \times 10^{-8} \pm 6.41 \times 10^{-7}$ yr^{-1} (compression). The extension rate at the IISC site was estimated around $6.81 \times 10^{-8} \pm 3.52 \times 10^{-8}$ and the compressional rate was $-2.38 \times 10^{-8} \pm 9.11 \times 10^{-8}$ yr^{-1}. These strain rates are quite low. This is possible because its regional part is covered by volcanic Deccan Traps or quaternary deposits, and there are no significant earthquakes that occurred in this region, nor they have clearly identified faults. The small earthquakes that occurred in this region are likely to be related to the small faults corresponding to Mesozoic Narmada trends in the ENE direction or Proterozoic Aravali trends in the NW direction. The hypocenteral depth of these earthquakes was up to 15 km, indicating the concentration of stress also lies within 15 km depth (Rastogi et al., 2013).

The rotation rates at each GNSS station are estimated by using Teza and Galgaro (2008) program and plotted them with black color as shown in Fig. 3. The results showed rotation rates around -1.20×10^{-7} rad yr^{-1} in the Bhuj region swing from EW to NS direction. The tendency of these swings suggests that the north regional area is more influenced by the collision of India and Eurasia. This rotational pattern seems to reflect the transmitting stress field due to this northward continental collision along the Himalayan arc (Reddy et al., 2000). Few N-S trending faults are developed due to the unequal uplift in the western margin during the Quaternary Period (Gubin, 1969). These faults seem to be the site of normal faulting and the source of stress accumulation resulting in EW extension due to the block adjustment along the NS plane (Reddy et al., 2000). The GNSS sites IITB and PUNE located in Deccan Traps showed a rotation rate around $2.16 \times 10^{-8} \pm 3.84 \times 10^{-8}$ rad yr^{-1} and the site IISC in the southern part showed $5.49 \times 10^{-8} \pm 4.45 \times 10^{-8}$ rad yr^{-1}. These rates of rotations from EW to NS direction are very low and statically insignificant (the error is larger than the actual value). The results indicate that the southern peninsula interior region move as a rigid block with the Indian tectonic plate velocity field (Jade, 2004). In these regions there was no significant Intra-plate earthquake/strain accumulation because of regional localized deformation specific to the active dislocations/faults (Jade et al., 2017).

4 Conclusions

This chapter reviews several GNSS sites velocities in the western India. The GNSS sites DHAM (11.62 ± 1.51 mm yr^{-1}) located in the south of the epicenter and the RATN (5.24 ± 1.49 mm yr^{-1}) located in the north of the epicenter are showing opposite direction velocities with respect to each other supports the reverse fault movement with the south-dipping rupture fault plane. The velocity trend of DHAM and GAND sites located in the south of epicenter are parallel and support left lateral strike-slip. The change in baseline length indicates an E-W rate of 2.3 ± 2.11 mm yr^{-1} between DHAR -KHAV and an N-S rate of -0.71 ± 2.1 mm yr^{-1} between MABU-ISRR. The SSA modeled velocity from lowest to highest at the other GNSS sites

KEVA, ARAJ, and BARW has been estimated to be ~0.45 ± 0.15 mm yr^{-1}, 0.78 ± 0.15 mm yr^{-1}, and 0.91 ± 0.15 mm yr^{-1}, respectively, shows the pattern of deformation on the eastern side differs from the pattern of deformation on the western side. The GNSS sites IITB and PUNE SSA modeled velocities showed 0.84 ± 2.32 mm yr^{-1} and 3.02 ± 1.33 mm yr^{-1}. The GNSS sites of IISC showed 1.92 ± 0.55 mm yr^{-1} horizontal velocity is totally opposite to the velocity trend of IITB and PUNE indicating an extensional deformation.

The estimated average compressional and extensional rates in the whole region by using SSA modeled velocities are estimated to be ~−4.00 × 10^{-7} yr^{-1} (compression) and 7.00 × 10^{-8} yr^{-1} (extension), respectively. The GNSS measurement during the year 2010–2012 showed an extensional rate of around 1.00 × 10^{-8} at BELP and KHAV sites and their residual velocities were 1.8 ± 1.5 mm yr^{-1}. The SSA modeled India-fixed velocity-based average strain rates in the Deccan tarp region at IITB and PUNE GNSS sites is estimated around 7.78 × 10^{-9} ± 2.14 × 10^{-7} yr^{-1} (extension) and −5.36 × 10^{-8} ± 6.41 × 10^{-7} yr^{-1} (compression). The results showed rotation rates of ~−1.20 × 10^{-7} rad yr^{-1} (clockwise) in the Bhuj region swing from EW to NS direction. The GNSS sites IITB and PUNE located in Deccan Traps showed a rotation rate around 2.16 × 10^{-8} ± 3.84 × 10^{-8} rad yr^{-1} (anticlockwise) and the site IISC in the southern part showed 5.49 × 10^{-8} ± 4.45 × 10^{-8} ± rad yr^{-1} (anticlockwise). These rates of rotations from EW to NS direction are very low and statically insignificant because there no significant Intra-plate earthquake occurrence or strain accumulation are noticed because of regional localized deformation specific to the active dislocations and faults.

Acknowledgements Soumyajit Mukherjee (IIT Bombay) invited, handled and reviewed this article. An anonymous reviewer provided critical comments that helped to clarify several issues. Mukherjee (2026) summarized this work.

Abbreviations

GMT	Generic Mapping Tools
GNSS	Global Navigation Satellite System
IREF	India-fixed Reference Frame
KMF	Kutch Mainland Fault
NWF	North Wagad Fault
SWF	South Wagad Fault
SSA	Singular Spectrum Analysis
SVD	Singular Value Decomposition

Appendix 1

India-fixed observed and SSA modeled velocity of GNSS sites located in Western India

GNSS sites	Location	Vel (N) (mm yr^{-1})	Vel (East) (mm yr^{-1})	Magnitude (India-fixed)	Magnitude (SSA-predicted)	Azimuth SSA (°N)	References, (data interval)
AHMD	23.04°N; 72.50°E	−2.53 ± 0.40	1.27 ± 0.96	2.83 ± 1.04	2.82 ± 1.04	153.44°N	Reddy and Sunil (2008), (2001–2002)
RAJK	22.29°N; 70.74°E	1.93 ± 0.42	2.747 ± 0.99	3.36 ± 1.06	3.36 ± 1.06	54.87°N	
RAPR	23.57°N; 70.64°E	−0.62 ± 0.59	1.76 ± 1.37	1.87 ± 1.49	1.89 ± 1.49	109.37°N	
RATN	23.86°N; 70.36°E	−5.19 ± 0.67	−0.73 ± 1.55	5.24 ± 1.69	5.24 ± 1.49	187.97°N	
DHAM	23.33°N; 70.14°E	10.87 ± 0.60	−4.18 ± 1.39	11.62 ± 1.51	11.62 ± 1.51	241.97°N	
GAND	23.07°N; 70.10°E	4.13 ± 0.61	−2.20 ± 1.38	4.68 ± 1.51	4.68 ± 1.51	241.97°N	
LODA	23.39°N; 69.89°E	−0.87 ± 0.81	−4.85 ± 1.82	4.92 ± 1.99	4.92 ± 1.99	259.82°N	
BIRN	23.66°N; 69.71°E	−2.56 ± 0.80	−1.52 ± 1.77	2.98 ± 1.94	2.98 ± 1.94	210.71°N	
BHUJ	23.25°N; 69.65°E	0.75 ± 0.70	2.70 ± 1.70	2.80 ± 1.84	2.79 ± 1.84	74.47°N	
MAND	22.83°N; 69.35°E	3.23 ± 0.64	4.33 ± 1.39	5.40 ± 1.53	5.40 ± 1.53	53.34°N	
NAKA	23.36°N; 69.26°E	−0.38 ± 0.69	−0.79 ± 1.58	0.88 ± 1.72	0.88 ± 1.72	244.63°N	
HAJP	23.69°N; 69.21°E	−1.62 ± 0.76	−1.50 ± 1.70	2.21 ± 1.86	2.20 ± 1.86	222.86°N	

(continued)

(continued)

GNSS sites	Location	Vel (N) (mm yr^{-1})	Vel (East) (mm yr^{-1})	Magnitude (India-fixed)	Magnitude (SSA-predicted)	Azimuth SSA (°N)	References, (data interval)
NALI	23.26°N; 68.84°E	−1.41 ± 0.74	−2.15 ± 1.49	2.57 ± 1.66	2.57 ± 1.66	236.80°N	
NARA	23.68°N; 68.54°E	0.78 ± 0.66	−0.79 ± 1.40	1.11 ± 1.55	1.11 ± 1.55	224.60°N	
IISC	13.02°N; 77.57°E	−1.90 ± 0.37	0.36 ± 0.41	1.93 ± 0.55	1.92 ± 0.55	169.43°N	
KHAV	23.92°N; 69.77°E	−1.08 ± 1.47	−1.39 ± 1.48	1.76 ± 2.09	1.75 ± 2.09	232.15°N	Jade et al. (2017), (1996–2015)
RADP	23.82°N; 71.62°E	−0.47 ± 1.48	1.73 ± 1.49	1.79 ± 2.10	1.77 ± 2.10	105.20°N	
BELP	23.87°N; 70.80°E	−0.89 ± 1.49	−1.63 ± 1.49	1.86 ± 2.11	1.85 ± 2.11	241.36°N	
MABU	24.65°N; 72.78°E	−0.75 ± 1.49	0.05 ± 1.50	0.75 ± 2.11	0.71 ± 2.11	176.19°N	
DHAR	24.01°N; 72.85°E	−0.42 ± 1.55	0.65 ± 1.56	0.77 ± 2.20	0.79 ± 2.20	122.87°N	
ISRR	23.16°N; 72.67°E	−0.20 ± 1.53	−0.51 ± 1.54	0.55 ± 2.17	0.53 ± 2.17	248.59°N	
UDAI	24.58°N; 73.71°E	1.19 ± 2.19	0.78 ± 2.21	1.42 ± 3.11	1.41 ± 3.11	33.24°N	
IITB	19.13°N; 72.92°E	0.65 ± 1.63	−0.52 ± 1.65	0.83 ± 2.32	0.84 ± 2.32	231.34°N	
PUNE	18.56°N; 73.88°E	2.97 ± 0.94	−0.60 ± 0.94	3.03 ± 1.33	3.02 ± 1.33	258.58°N	

(continued)

(continued)

GNSS sites	Location	Vel (N) (mm yr^{-1})	Vel (East) (mm yr^{-1})	Magnitude (India-fixed)	Magnitude (SSA-predicted)	Azimuth SSA (°N)	References, (data interval)
BELA	23.87°N; 70.80°E	–	–	2.25 ± 0.12	2.26 ± 0.12	–	Dumka et al. (2019), (2009–2015)
DESA	23.74°N; 70.60°E	–	–	1.53 ± 0.12	1.54 ± 0.12	–	
RAPA	23.56°N; 70.60°E	–	–	0.75 ± 0.14	0.75 ± 0.14	–	
BADA	23.47°N; 70.50°E	–	–	0.64 ± 0.12	0.62 ± 0.12	–	
VAMK	23.43°N; 70.40°E	–	–	1.61 ± 0.12	1.62 ± 0.12	–	
BHAC	23.30°N; 70.30°E	–	–	0.55 ± 0.14	0.59 ± 0.14	–	
KHAV	23.92°N; 69.70°E	–	–	2.91 ± 0.12	2.92 ± 0.12	–	
KUAR	23.99°N; 69.70°E	–	–	2.08 ± 0.10	2.07 ± 0.10	–	
VAND	23.02°N; 69.30°E	–	–	0.67 ± 0.16	0.66 ± 0.16	–	
VKOT	24.20°N; 69.20°E	–	–	1.51 ± 0.12	1.52 ± 0.12	–	
HUBA	23.40°N; 69.80°E	–	–	1.39 ± 0.22	1.37 ± 0.22	–	
CHAD	23.30°N; 70.10°E	–	–	1.07 ± 0.26	1.04 ± 0.26	–	

(continued)

(continued)

GNSS sites	Location	Vel (N) (mm yr^{-1})	Vel (East) (mm yr^{-1})	Magnitude (India-fixed)	Magnitude (SSA-predicted)	Azimuth SSA (°N)	References, (data interval)
DUDH	23.32°N; 70.14°E	–	–	1.16 ± 0.28	1.18 ± 0.28	–	
JNGR	23.36°N; 70.01°E	–	–	1.36 ± 0.22	1.33 ± 0.22	–	
PALN	23.44°N; 69.57°E	–	–	0.69 ± 0.38	0.71 ± 0.38	–	
BARW	22.04°N; 74.90°E	–	–	0.88 ± 0.15	0.91 ± 0.15	–	Dumka et al. (2022), (2009–2016)
ARAJ	22.29°N; 74.35°E	–	–	0.81 ± 0.15	0.78 ± 0.15	–	
SAGB	21.55°N; 73.79°E	–	–	1.62 ± 0.15	1.61 ± 0.15	–	
KEVA	21.90°N; 73.67°E	–	–	0.47 ± 0.15	0.45 ± 0.15	–	

Appendix 2

Spatial strain and rotation rate variation in western India during the interval mentioned in Appendix 1. The extension strain rates are positive while the compression strain rates are negative. The positive rotational rate shows an anticlockwise direction and the negative rotational rate shows a clockwise direction.

GNSS sites	Location	Strain (Emax) Rate (yr^{-1})	Strain (Emin) Rate (yr^{-1})	Rotation rate (rad yr^{-1})
AHMD	23.04°N; 72.50°E	4.24E-08 ± 1.57E-08	−3.72E-08 ± 8.69E-08	−5.61E-09 ± 3.64E-08
RAJK	22.29°N; 70.74°E	7.22E-08 ± 3.02E-08	−2.24E-07 ± 1.21E-07	−8.66E-08 ± 5.39E-08
RAPR	23.57°N; 70.64°E	8.24E-08 ± 4.32E-08	−3.15E-07 ± 1.69E-07	−9.34E-08 ± 8.35E-08
RATN	23.86°N; 70.36°E	7.65E-08 ± 5.00E-08	−3.74E-07 ± 1.77E-07	−1.03E-07 ± 9.19E-08
DHAM	23.33°N; 70.14°E	6.24E-08 ± 5.99E-08	−3.86E-07 ± 1.94E-07	−1.24E-07 ± 1.01E-07
GAND	23.07°N; 70.10°E	6.79E-08 ± 5.51E-08	−3.80E-07 ± 1.86E-07	−1.21E-07 ± 9.82E-08
LODA	23.39°N; 69.89°E	5.79E-08 ± 5.40E-08	−3.98E-07 ± 1.90E-07	−1.39E-07 ± 9.90E-08
BIRN	23.66°N; 69.71°E	5.83E-08 ± 5.09E-08	−4.19E-07 ± 1.80E-07	−1.43E-07 ± 9.60E-08
BHUJ	23.25°N; 69.65°E	5.97E-08 ± 4.32E-08	−4.19E-07 ± 1.85E-07	−1.66E-07 ± 9.33E-08
MAND	22.83°N; 69.35°E	7.26E-08 ± 3.35E-08	−4.33E-07 ± 1.75E-07	−1.92E-07 ± 8.24E-08
NAKA	23.36°N; 69.26°E	6.80E-08 ± 3.75E-08	−4.24E-07 ± 1.76E-07	−1.71E-07 ± 8.85E-08
HAJP	23.69°N; 69.21°E	6.29E-08 ± 3.85E-08	−4.17E-07 ± 1.72E-07	−1.55E-07 ± 8.90E-08
NALI	23.26°N; 68.84°E	7.79E-08 ± 3.29E-08	−4.01E-07 ± 1.72E-07	−1.67E-07 ± 8.48E-08
NARA	23.68°N; 68.54°E	6.33E-08 ± 2.59E-08	−3.87E-07 ± 1.76E-07	−1.33E-07 ± 8.52E-08
IISC	13.02°N; 77.57°E	6.81E-08 ± 3.52E-08	−2.38E-08 ± 9.11E-08	5.49E-08 ± 4.45E-08
KHAV	23.92°N; 69.77°E	6.14E-08 ± 5.00E-08	−3.92E-07 ± 1.75E-07	−1.29E-07 ± 9.34E-08
RADP	23.82°N; 71.62°E	7.00E-08 ± 2.74E-08	−1.36E-07 ± 1.19E-07	−3.94E-08 ± 5.60E-08
BELP	23.87°N; 70.80°E	8.24E-08 ± 3.81E-08	−2.91E-07 ± 1.61E-07	−8.86E-08 ± 7.89E-08

(continued)

(continued)

GNSS sites	Location	Strain (Emax) Rate (yr^{-1})	Strain (Emin) Rate (yr^{-1})	Rotation rate (rad yr^{-1})
MABU	24.65°N; 72.78°E	4.99E-08 ± 2.09E-08	−3.94E-08 ± 7.43E-08	−2.76E-08 ± 3.41E-08
DHAR	24.01°N; 72.85°E	4.26E-08 ± 2.01E-08	−2.11E-08 ± 7.28E-08	−1.45E-08 ± 3.30E-08
ISRR	23.16°N; 72.67°E	3.31E-08 ± 1.84E-08	−2.83E-08 ± 7.90E-08	−2.53E-09 ± 3.43E-08
UDAI	24.58°N; 73.71°E	4.03E-08 ± 8.90E-08	−4.93E-09 ± 2.67E-07	−2.26E-08 ± 2.97E-08
IITB	19.13°N; 72.92°E	7.78E-09 ± 2.14E-07	−5.36E-08 ± 6.41E-07	2.16E-08 ± 3.84E-08
PUNE	18.56°N; 73.88°E	7.79E-09 ± 2.14E-07	−5.37E-08 ± 6.41E-07	2.16E-08 ± 3.84E-08

References

Ansari, K. (2018). Crustal deformation and strain analysis in Nepal from GPS time-series measurement and modeling by ARMA method. *International Journal of Earth Sciences 107*(8), 2895–2905. https://doi.org/10.1007/s00531-018-1633-7

Ansari, K., Panda, S. K., & Jamjareegulgarn, P. (2020). Singular spectrum analysis of GPS derived ionospheric TEC variations over Nepal during the low solar activity period. *Acta Astronautica, 169*, 216–223. https://doi.org/10.1016/j.actaastro.2020.01.014

Ansari, K., & Park, K. D. (2018). Multi constellation GNSS precise point positioning and prediction of propagation errors using singular spectrum analysis. *Astrophysics and Space Science, 363*(12), 258. https://doi.org/10.1007/s10509-018-3479-7

Ansari, K., & Park, K. D. (2022). Geodetic analysis inside the South Korean Peninsula and impact of the 2011 Tohoku–Oki (TO) earthquake. *Acta Geologica Sinica-English Edition, 96*(2), 631–647. https://doi.org/10.1111/1755-6724.14885

Arya, A. S. (2000). Recent developments toward earthquake. *Current Science, 79*(9).

Bendick, R., Bilham, R., Fielding, E., Gaur, V. K., Hough, S. E., Kier, G., Kulkarni, M. N., Martin, S., Mueller, K., & Mukul, M. (2001). The 26 January 2001 "Republic Day" earthquake, India. *Seismological Research Letters, 72*(3), 328–335. https://doi.org/10.1785/gssrl.72.3.328

Biswas, S. K. (2005). A review of structure and tectonics of Kutch basin, western India, with special reference to earthquakes. *Current Science, 88*(10), 1592–1600.

Biswas, S. K., & Khattri, K. N. (2002). A geological study of earthquakes in Kutch, Gujarat, India. *Geological Society of India, 60*(2), 131–142. http://www.geosocindia.com/index.php/jgsi/article/view/83556

Biswas, S. K. (1987). Regional tectonic framework, structure and evolution of the western marginal basins of India. *Tectonophysics, 135*(4), 307–327. https://doi.org/10.1016/0040-1951(87)90115-6

Chandrasekhar, D. V., Bürgmann, R., Reddy, C. D., Sunil, P. S., & Schmidt, D. A. (2009). Weak mantle in NW India probed by geodetic measurements following the 2001 Bhuj earthquake. *Earth and Planetary Science Letters, 280*(1–4), 229–235. https://doi.org/10.1016/j.epsl.2009.01.039

Choudhury, P., Gahalaut, K., Dumka, R., Gahalaut, V. K., Singh, A. K., & Kumar, S. (2018). GPS measurement of land subsidence in Gandhinagar, Gujarat (Western India), due to groundwater depletion. *Environmental Earth Sciences, 77*, 1–5. https://doi.org/10.1007/s12665-018-7966-5

Choudhury, P., Catherine, J. K., Gahalaut, V. K., Chopra, S., Dumka, R., & Roy, K. S. (2013). Post-seismic deformation associated with the 2001 Bhuj earthquake. *Natural Hazards, 65*, 1109–1118. https://doi.org/10.1007/s11069-012-0191-8

Dhawale, M. S., Mukherjee, S., & Biswas, M. (2023). Morphotectonics and paleostress analyses of Kutch area, Gujarat, India. *Results in Earth Sciences, 1*, Article 100002.

Dumka, R. K., Chopra, S., & Prajapati, S. (2019). GPS derived crustal deformation analysis of Kachchh, zone of 2001 (M7. 7) earthquake, Western India. *Quaternary International, 507*, 295–301. https://doi.org/10.1016/j.quaint.2019.01.032

Dumka, R. K., SuriBabu, D., Kotlia, B. S., Kothyari, G. C., & Prajapati, S. (2022). Crustal deformation measurements by global positioning system (GPS) along NSL, western India. *Geodesy and Geodynamics, 13*(3), 254–260. https://doi.org/10.1016/j.geog.2021.05.004

Ghavri, S., Yadav, R. K., & Jade, S. (2022). Seismic hazard potential of Kachchh paleo-rift basin of Indian craton from seismic and geodetic strain rates. *Tectonophysics, 837*, Article 229458. https://doi.org/10.1016/j.tecto.2022.229458

Gowd, T. N., Srirama Rao, S. V., & Chary, K. B. (1996). Stress field and seismicity in the Indian shield: Effects of the collision between India and Eurasia. *Pure and Applied Geophysics, 146*, 503–531. https://doi.org/10.1007/978-3-0348-9200-1_6

Gubin, I. E. (1969). Seismic zoning of the western margin of the Indian peninsula in Maharashtra State. Part 2 UNESCO Report No. 1519/BMS/RD/SCE.

Gupta, H. K. (1993). The deadly latur earthquake. *Science 262*(5140), 1666–1667. https://doi.org/10.1126/science.262.5140.1666

Hassani, H., Heravi, S., & Zhigljavsky, A. (2009). Forecasting European industrial production with singular spectrum analysis. *International Journal of Forecasting, 25*(1), 103–118. https://doi.org/10.1016/j.ijforecast.2008.09.007

Jade, S., Mukul, M., Parvez, I. A., Ananda, M. B., Kumar, P. D., Gaur, V. K., Bendick, R., Bilham, R., Blume, F., Wallace, K., & Abbasi, I. A. (2003). Pre-seismic, co-seismic and post-seismic displacements associated with the Bhuj 2001 earthquake derived from recent and historic geodetic data. *Journal of Earth System Science, 112*, 331–345. https://doi.org/10.1007/BF02709260

Jade, S., Shrungeshwara, T. S., Kumar, K., Choudhury, P., Dumka, R. K., & Bhu, H. (2017). India plate angular velocity and contemporary deformation rates from continuous GPS measurements from 1996 to 2015. *Scientific Reports, 7*(1), 11439. https://doi.org/10.1038/s41598-017-11697-w

Jade, S. (2004). Estimates of plate velocity and crustal deformation in the Indian subcontinent using GPS geodesy. *Current Science*, 1443–1448. https://www.jstor.org/stable/24109220

Kailasam, L. N. (1993). Geophysical and geodynamical aspects of the Maharashtra earthquake of September 30, 1993. *Current Science, 65*(10), 736–739.

Kaplay, R. D., Babar, Md., Mukherjee, S., & Kumar, T. V. (2017). Morphotectonic expression of geological structures in eastern part of south east Deccan volcanic province (around Nanded, Maharashtra, India) In: S. Mukherjee, A. A. Misra, G. Calvès, M. Nemčok (Eds.), Tectonics of the Deccan large igneous province. *Geological Society, London, Special Publications, 445*, 317–335.

Kaplay, R. D., Kumar, T. V., Mukherjee, S., Wesanekar, P. R., Babar, Md., & Chavan S. (2017). E-W strike slip shearing of Kinwat Granitoid at South East Deccan Volcanic Province, Kinwat, Maharashtra, India. *Journal of Earth System Science, 126*, 71.

Kaplay, R. D., Babar, Md., Mukherjee, S., Wable, D., & Pisal, K. (2021). Granitic rocks underlying Deccan trap along the margin of east Dharwar craton, Muntyal (Maharashtra)—Bhaisa (Telengana), India, general description and deformation. In S. Mukherjee (Ed.), *Structural geology and tectonics field guidebook* (Vol. 1, pp. 599–620). Springer Nature Switzerland AG.

Kayal, J. R., De, R., Ram, S., Srirama, B. V., & Gaonkar, S. G. (2002). Aftershocks of the 26 January, 2001 Bhuj earthquake in western India and its seismotectonic implications. *Geological Society of India, 59*(5), 395–417.

Lohani, N., Mukherjee, S., Singh, S., Pawar, A., & Shaikh, M. (2022). Structural geological field guide: Bhuj area (Gujarat, India). In: S. Mukherjee (Ed.), *Structural geology and tectonics field guidebook* (Vol. 2, pp. 227–250). Springer.

Mandal, P., Rastogi, B. K., Satyanaraya, H. V. S., Kousalya, M., Vijayraghavan, R., Satyamurty, C., Raju, I. P., Sarma, A. N. S., & Kumar, N. (2004). Characterization of the causative fault system for the 2001 Bhuj earthquake of Mw 7.7. *Tectonophysics, 378*(1–2), 105–121. https://doi.org/10.1016/j.tecto.2003.08.026

Mathew, G., Singhvi, A. K., & Karanth, R. V. (2006). Luminescence chronometry and geomorphic evidence of active fold growth along the Kachchh Mainland Fault (KMF), Kachchh, India: Seismotectonic implications. *Tectonophysics, 422*(1–4), 71–87. https://doi.org/10.1016/j.tecto.2006.05.009

Maurya, D. M., Shaikh, M., & Mukherjee, S. (2021). Comment on "Structural attributes and paleostress analysis of Quaternary landforms along the Vigodi Fault (VF) in Western Kachchh region". *Quaternary International, 601*, 143–147. https://doi.org/10.1016/j.quaint.2020.07.038

Misra, A. A., Bhattacharya, G., Mukherjee, S., & Bose, N. (2014). Near N-S paleo-extension in the western Deccan region in India: Does it link strike-slip tectonics with India-Seychelles rifting? *International Journal of Earth Sciences, 103*, 1645–1680.

Misra, A. A., & Mukherjee, S. (2017). Dyke-brittle shear relationships in the Western Deccan Strike Slip Zone around Mumbai (Maharashtra, India). In: S. Mukherjee, A. A. Misra, G. Calvès, & M. Nemčok (Eds.), *Tectonics of the Deccan large igneous province* (Vol. 445, pp. 269–295). Geological Society, London, Special Publications.

Mukherjee, S. (2026). Introduction to "Structural Geology & Tectonics". In S. Mukherjee (Ed.), *Structural Geology & Tectonics*. Springer. ISBN: 978-981-95-4743-2.

Mukherjee, S., Dole, G., Yatheesh, V., & Kale, V. (2020). Tectonics of the Deccan trap: Focus on Indian geoscientists' contribution in the last four years. *Proceedings of the Indian National Science Academy, 86*, 237–244.

Mukherjee, S., Misra, A. A., Calvès, G., & Nemčok, M. (2017). Tectonics of the Deccan large igneous province: An introduction. In S. Mukherjee, A. A. Misra, G. Calvès, & M. Nemčok (Eds.), *Tectonics of the Deccan large igneous province* (Vol. 445, pp. 1–9). Geological Society, London, Special Publications.

Mukul, M., Jade, S., Ansari, K., & Matin, A. (2014). Seismotectonic implications of strike–slip earthquakes in the Darjiling–Sikkim Himalaya. *Current Science*, 198–210

Negishi, H., Mori, J., Sato, T., Singh, R., Kumar, S., & Hirata, N. (2002). Size and orientation of the fault plane for the 2001 Gujarat, India earthquake (Mw7. 7) from aftershock observations: A high stress drop event. *Geophysical Research Letters, 29*(20), 10-1. https://doi.org/10.1029/2002GL015280

Nocquet, J. M. (2012). Present-day kinematics of the Mediterranean: A comprehensive overview of GPS results. *Tectonophysics, 579*, 220–242. https://doi.org/10.1016/j.tecto.2012.03.037

Omid, M. W., Mukherjee, S., & Dasgupta, S. (2021). Vein geometry (Bhuj, Gujarat, India). In: S. Mukherjee (Ed.), *Structural geology and tectonics field guidebook* (Vol. 1, pp. 707–714). Springer Nature Switzerland AG.

Paul, J., Bürgmann, R., Gaur, V. K., Bilham, R., Larson, K. M., Ananda, M. B., Jade, S., Mukal, M., Anupama, T. S., Satyal, G., & Kumar, D. (2001). The motion and active deformation of India. *Geophysical Research Letters, 28*(4), 647–650. https://doi.org/10.1029/2000GL011832

Rastogi, B. K., Aggrawal, S. K., Rao, N., & Choudhury, P. (2013). Triggered/migrated seismicity due to the 2001 M w 7.7 Bhuj earthquake, Western India. *Natural Hazards, 65*, 1085–1107. https://doi.org/10.1007/s11069-011-0083-3

Raval, U. (1995). Seismicity of the Koyna region and regional tectonomagmatism of the western margin (India). *Induced Seismicity*, 175–192. https://doi.org/10.1007/978-3-0348-9238-4_15

Reddy, C. D., & Sunil, P. S. (2008). Post-seismic crustal deformation and strain rate in Bhuj region, western India, after the 2001 January 26 earthquake. *Geophysical Journal International, 172*(2), 593–606. https://doi.org/10.1111/j.1365-246X.2007.03641.x

Reddy, C. D., El-Fiky, G., Kato, T., Shimada, S., & Kumar, K. V. (2000). Crustal strain field in the Deccan trap region, western India, derived from GPS measurements. *Earth, Planets and Space, 52*(11), 965–969. https://doi.org/10.1186/BF03352313

Seeber, L., Ekström, G., Jain, S. K., Murty, C. V. R., Chandak, N., & Armbruster, J. G. (1996). The 1993 Killari earthquake in central India: A new fault in Mesozoic basalt flows? *Journal of Geophysical Research: Solid Earth 101*(B4), 8543–8560. https://doi.org/10.1029/95JB01865

Shaikh, M. A., Patidar, A. K., Maurya, D. M., Vanik, N. P., Padmalal, A., Tiwari, P., Mukherjee, S., & Chamyal, L. S. (2022). Building tectonic framework of a blind active fault zone using field and ground-penetrating radar data. *Journal of Structural Geology, 155*, Article 104526.

Shaikh, M., Maurya, D. M., Mukherjee, S., Vanik, N., Padmalal, A., & Chamyal, L. (2020). Tectonic evolution of the intra-uplift Vigodi-Gugriana-Khirasra-Netra Fault System in the seismically active Kachchh Rift Basin, India: Implications for the western continental margin of the Indian plate. *Journal of Structural Geology, 140*, 104124.

Shanker, R. (1995). Fragmented Indian shield and recent earthquakes. *Geological Society of India Special Publications, 27*, 41–48.

Teza, G., & Galgaro, A. (2008). Grid_strain and grid_strain3: Software packages for strain field computation in 2D and 3D environment. *Computers & Geosciences, 34*(9):1142–1153. https://doi.org/10.1016/j.cageo.2007.07.006

Vanik, N., Shaikh, H., Mukherjee, S., Maurya, D. M., & Chamyal, L. S. (2018). Post-Deccan trap stress reorientation under transpression: Evidence from fault slip analyses from SW Saurashtra, western India. *Journal of Geodynamics, 121*, 9–19.

Wallace, K., Bilham, R., Blume, F., Gaur, V. K., & Gahalaut, V. (2006). Geodetic constraints on the Bhuj 2001 earthquake and surface deformation in the Kachchh Rift Basin. *Geophysical Research Letters, 33*(10). https://doi.org/10.1029/2006GL025775

Wesnousky, S. G., Seeber, L., Rockwell, T. K., Thakur, V., Briggs, R., Kumar, S., & Ragona, D. (2001). Eight days in Bhuj: Field report bearing on surface rupture and genesis of the 26 January 2001 earthquake in India. *Seismological Research Letters, 72*(5), 514–524. https://doi.org/10.1785/gssrl.72.5.514

Zabalza, J., Ren, J., Zheng, J., Han, J., Zhao, H., Li, S., & Marshall, S. (2015). Novel two-dimensional singular spectrum analysis for effective feature extraction and data classification in hyperspectral imaging. *IEEE Transactions on Geoscience and Remote Sensing, 53*(8), 4418–4433. https://doi.org/10.1109/TGRS.2015.2398468

Zarifi, Z., Nilfouroushan, F., & Raeesi, M. (2014). Crustal stress map of Iran: Insight from seismic and geodetic computations. *Pure and Applied Geophysics, 171*, 1219–1236. https://doi.org/10.1007/s00024-013-0711-9

Uranium Geochemistry in Phosphate-Bearing Sedimentary Rocks of Masrana and Kimoi Blocks of the Mussoorie Syncline, Uttarakhand, India

Saif A. Khan, Shamim A. Dar, Sarita Patel, and Akhtar R. Mir

Abstract Uranium holds significant global importance due to its crucial role in the energy generation sector. Uranium is present in the Earth's crust with ~1–3 ppm. Furthermore, high concentration is also found in the sediments having rich organic matter content. The average values of uranium in phosphorite, shale, sandstone, and carbonate are 120, 3.5, 0.5, and 2.2 ppm, respectively. The main endeavor of the current work is to discuss the uranium geochemistry of phosphorites of Masrana and Kimoi blocks of Mussoorie syncline to understand the concentration and behavior of uranium with respect to other elements in these phosphorites. The studied phosphorites show dull gray to brown-black color and are hosted by black shale, chert, and minor carbonate rocks of Neoproterozoic to Early Cambrian age, which belong to the lowermost sequence of the Tal Group of the Lesser Himalaya. These phosphorites are rarely bedded and display incrustations of white and orange color. Nodular phosphorite is found in black shales that lie in contact with the phosphorite horizon. Nodules show mostly spheroidal, flattened, plano-convex, and ellipsoidal shapes. Alternate bands of carbonate and phosphate minerals, having a thickness in mm, are seen in laminated phosphorites. In this variety of phosphorites, the phosphatic layer is made of collophane and this class retains the maximum concentration of phosphate than that of other classes. The studied samples are rich in CaO, P_2O_5, SiO_2, Al_2O_3, Fe_2O_3, K_2O, V, Cr, Zn, Sr, Zr, Nb, and Ba but depleted in MgO, MnO, and TiO_2. The uranium content in the studied samples varies from 1.37 to 37.5 ppm with an average of 16.7 ppm. The negative correlation of U with P_2O_5 suggests the removal of uranium from the atomic structure of apatite and substitution of P^{5+}, while the inverse correlation between U and CaO suggests mutual replacement of Ca^{2+} by U^{4+}

S. A. Khan
Department of Geology, Aligarh Muslim University, Aligarh, Uttar Pradesh, India

S. A. Dar (✉) · S. Patel
Department of Geology, Institute of Sciences, Banaras Hindu University, Varanasi, Uttar Pradesh, India
e-mail: sjshamim@gmail.com; sjshamim@bhu.ac.in

A. R. Mir
Department of Earth Sciences, University of Kashmir, Srinagar, Jammu and Kashmir, India

S. Mukherjee (ed.), *Structural Geology and Tectonics*,
https://doi.org/10.1007/978-981-95-4743-2_12

in the apatite crystal structure. Thus, the removal of U from the crystal lattice is responsible for the low contents of U in these phosphorites. The sympathetic/weak positive relationship of U with K_2O indicates partial replacement of Ca^{2+} by K^+ in the apatite atomic structure, which infers the alkaline conditions of the basin of deposition during phosphatization. The positive correlation of U with Ni, Cu, Zn, and Co and the weak negative relation of U with V and Cr, suggest that the phosphate-bearing rocks are deposited under sulfur-reducing/euxinic conditions. Additionally, Ni, Cu, Zn, and Ba are the tracers for the abundance of organic matter (OM), i.e., they are incorporated into the sediments primarily in association with OM. A higher concentration of Ni and Cu suggests (i) a high influx of OM released these elements from the surroundings and concentrated them in the sediments, and (ii) a reducing environment enabled the sedimentary fixation of Ni and Cu. The positive correlation of U with these elements suggests a similar process for U enrichment in the studied phosphate-bearing rocks.

1 Introduction

Uranium (U), being the energy-generating element, has gained its importance in the global market (Mukherjee et al., 2023). Uranium as an incompatible element, is concentrated in the Earth's crust (~1–3 ppm) (Fayek et al., 2011; Hazen et al., 2009). Further, the high concentrations of U are also found in the OM-rich sediments where from which profitable withdrawal of this element is possible (Cumberland et al., 2016). Mafic rocks like basalts have low contents of U on average ~1 ppm (Alloway, 2013), whereas felsic rocks such as granites and rhyolites contain high concentrations of U ~3.6 and 5.0 ppm, respectively (Hobday & Galloway, 1999). Even higher concentrations of U are found in some granite, for example, Proterozoic granites in the Musgrave region of Western Australia contain 50 ppm (Kreuzer et al., 2010). Whereas, in black shales, breccias, clays, limestones, phosphorites, and sandstones, U content averages ~4 ppm (Alloway, 2013). However, in phosphorites, U (~120 ppm) contents are several fold enriched in comparison to these sedimentary rocks (Altschuler, 1980; Krauskopf & Bird, 1995; Slansky, 1986). This enrichment is because of the crystal structure of francolite (carbonate fluorapatite) where U is substituted in place of Ca for having similar ionic radius (Ca = 0.99 Å and U = 0.97 Å McClellan & Van Kauwenbergh, 1990). This enrichment may also be due to the abundance of OM in the accumulation basins of the phosphorites (Burnett, 1990). Some major issues like progression of uranium mineralization in phosphorites during the geological time, uranium forms in phosphorites and controlling main factors of uranium behavior during sedimentation and diagenesis can be addressed with the generation of more geochemical data of phosphorites throughout the globe.

Several groups and/or formations in the Himalayas are known for their higher concentration of uranium. All the known types of uranium occurrences in the Himalayan Belts are listed in Table 1. The table lists all the sources used to collect the information. The main endeavor of the current work is to discuss the uranium geochemistry of phosphorites of the Masrana and Kimoi blocks of the Mussoorie syncline to understand the concentration and behavior of U with respect to other elements in these phosphorites.

2 Geology

The geological division of the Himalayas with modern data sets and ages can be found in Mukherjee (2015). Masrana and Kimoi blocks of Uttarakhand state make a portion of the Lesser Himalaya between the Yamuna Valley on the west and the Ganga Valley on the east (latitudes: 30° 27′ 0″ N and 30° 28′ 0″ N and longitudes: 78° 9′ 0″ E and 78° 8′ 30″ E). The structural geology of the Mussourie syncline was recently revisited by Mahato et al. (2019). The generalized regional and study area geological succession is given in Tables 2 and 3 (modified after Kumar, 2005). The phosphorite deposits of the above-mentioned blocks make up the lowermost sequence of the Tal Group (i.e., the chert unit with thin streaks of phosphate rock and phosphate unit with thin intercalation of shales and chert; see Table 3) of the Lesser Himalayan region of north-western India. The Krol belt in the Lesser Himalaya is made of Neoproterozoic to Early Cambrian rocks of the Blaini, Krol, and Tal Groups (Khan et al., 2016, 2022; Mazumdar & Banerjee, 2001). The Tal Group, having a prospective phosphorite horizon, overlies the Krol Group. Tal Group forms the topmost lithostratigraphic unit of the thick sedimentary sequence of the Krol Nappe. Blaini, Infra Krol, Krol and Tal rocks form a syncline in the Mussoorie area (Fig. 1). Structurally, the region is characterized by Himalayan arc-parallel shear (Biswas et al., 2022) as well as top-to-S/SW foreshear and backshear (Bose & Mukherjee, 2019). The strike of these Groups varies from N10° W to N55° W. Local synforms and antiforms have been made due to refolding at a few places, like in the Pari Tibba-Chamasari area. Steep scarps, shearing, shattering, and drag folding features have been developed in these formations due to faulting. The northern portion of the Krol-Tal sequences along the Krol Thrust is the Tons Thrust, which also overrides the autochthonus Simla slate-Nummulitic zone (Auden, 1934). The Tons Thrust and the Krol Thrust have been found to be independent detachment planes. The Krol and the Giri Thrusts have been found to be the two traces of the same folded thrust plane developed along the plane of unconformity between the Simla Group and Krol belt rocks, and similarly, the Tons and the Chail are the two traces of the same folded thrust plane initiated within the Jaunsar Group along which the Inner Krol belt has moved over the Outer Krol belt (Kumar, 2005). The younger sequence of the Tal Group is found within the Inner Krol belt (Early Cambrian) (Desai et al., 2010).

Auden (1934) classified the Tal Group into lower and upper sequences, whereas Bhargava (1974) divided this group into Lower, Middle, and Upper sequences. The best exposures of these divisions are located in Nigali Dhar and Korgai Synclines. According to Gupta et al. (1993), the Tal Group is made of chert, areno-argillaceous, and arenaceous rocks with subordinate calcareous bands that conformably overlie the Krol Group. Lower Tal is 75–880 m thick and phosphorite mineralization is

Table 1 Types of uranium occurrence in the Himalayan Region

Type of U-occurrence	Host rocks	Age	Locations	References
Sandstone hosted Uranium	Light grey and bluish grey massive sandstone, siltstones, mudstones, and clays	Late Miocene	Surghar-Shingar Range from Dhok Pathan Formation of Middle Siwalik, NW Himalayan foothills, South of MBT	Saraswat and Mahadevan (1989), Ali et al. (2020)
	Greenish grey sandstones, greenish-grey maroon shale, minor red-brown shales	Upper Eocene to Lower Miocene	Basal part of Upper Dharamshala Formation, Mandi district, Himachal Pradesh, bounded by MBT on the NE and Palampur Thrust on the SW	Sharma et al. (2000)
Stratiform remobilized type	Quartzite, phyllite, limestone, boulder bed, ash bed	Precambrian to Lower Paleozoic	Chails/Jaunsar, South of MCT in the Lesser Himalaya	Saraswat and Mahadevan (1989)
The black shale and phosphorite type	Black shale, chert, minor carbonates	Neo Proterozoic to Early Cambrian	Masrana and Kimoi blocks of Lower Tal Group, Mussoorie, Lesser Himalaya	Saraswat and Mahadevan (1989), This Study
	Phosphorites and carbonaceous shale U-hosted in carbonate rocks	Late Neo Proterozoic	Krol D Member, Nainital Syncline, Lesser Himalaya	Sinha et al. (2016)
	Phosphatic carbonaceous phyllite		In Haimanta System, Kinnaur district, Himachal Pradesh of the Greater Himalaya, North of MCT	Saraswat and Mahadevan (1989)

(continued)

Table 1 (continued)

Type of U-occurrence	Host rocks	Age	Locations	References
The structurally controlled hydrothermal type	Quartzite, phyllite, schist, calc-silicate rocks, gneisses, migmatites and vein quartz	Precambrian to Lower Paleozoic	Daling Formation of Sikkim and Arunachal Pradesh, North of MCT, also found in granitic gneisses of Uttar Pradesh	Saraswat and Mahadevan (1989)
Disseminated and vein types	Granites and associated Pegmatites	Tertiary	North of MCT in the Greater and Trans Himalaya	Saraswat and Mahadevan (1989)

MBT: Main Boundary Thrust, MCT: Main Central Thrust

Table 2 Generalized geological succession of the area (after Kumar, 2005)

Group/formation	Age
Alluvium	Recent
Dun gravel/old terraces	Pleistocene to sub-recent
Siwalik group	Middle Miocene to Upper Pliocene
Subathu group	Late Palaeocene to Middle Eocene
Tal group	Cambrian
Infra-Krol/Krol group	Terminal Proterozoic
Blaini group	Neo-Proterozoic
Nagthat group	Meso-Proterozoic
Chandpur group	Late Palaeo-Proterozoic
Simla group	Early Palaeo-Proterozoic

mainly localized to the chert member of this division. The chert member is further divided into a chert horizon and a phosphate horizon. The Chert horizon, made of black cherts plus thin bands of shale, shows nodular characteristics in the upper parts where these nodules are phosphatic in nature. The size of nodules is inversely proportional to phosphate content. The inter-nodular portions of the chert are usually free from phosphatic contamination. Few mm to 10 m thick phosphate horizon is significant economically (Shankar, 1987). This horizon lies in the upper parts of the chert member. At some locations, the phosphorite horizon stands directly on the Krol limestone. In such situations, the Krol limestone is also recognized to hold the bands of phosphorite in its upper parts (Patwardhan & Ahluwalia, 1973; Raha, 1972). The lithological units are shown in Fig. 1. The presence of ripple marks, intraclasts, low-angle cross-bedding, ooids, microbial laminates, and small domal stromatolites indicates an intertidal to shallow subtidal depositional setting for the Lower Tal Group (Mazumdar & Banerjee, 2001). The Upper Tal Group is 70–160 m thick. The lower quartzite member and the upper limestone member are two further

Table 3 Stratigraphic sequence of rock formations in and around the study area. Also see the compiled succession of the Lesser Himalaya in Table 2 of Biswas et al. (2022)

Subathu	Olive shale, shell marl and limestone
Unconformity	
Upper Tal	(i) Limestone member 15–20 m
	(ii) Quartzite member 1300 m
Disconformity	
Lower Tal	(i) Calcareous member 5 m
	(ii) Arenaceous member 300–500 m
	(iii) Argillaceous member 150 m
	(iv) Chert member 150 m
	(a) Phosphate unit (phosphate rock with thin intercalations of shales and chert)
	(b) Chert unit (Bedded black chert with subordinate layers of black shale and thin streaks of phosphate rock)
Disconformity/submarine-diastem (transitional locally)	
Transition zone (developed locally)	Argillaceous limestone often phosphatized interlayered with thin streaks of phosphate rock (also brecciated at places) and chert
Upper Krol	(i) Light grey, argillaceous limestone and purple grey shale
	(ii) Grey to bluish grey dolomitic limestone and dolomites

divisions of this Group. The white to purplish white sandstones/quartzites of the Upper Tal shows rarely cross-bedding and ripple marks.

3 Field Characteristics

The studied phosphorite deposits of Masrana and Kimoi blocks of Uttarakhand show dull gray to brownish-black color. These phosphorites are rarely bedded and display incrustations of white and orange color. These phosphorite deposits are classified as bedded, nodular, lenticular, and laminated (Khan et al., 2011, 2022). Bedded phosphates with clayey and shaly bands (Fig. 2a) are seen in bedded phosphorites. These are made of siliceous or carbonaceous matrix. Nodular phosphorite is found in black shales that occur in contact with phosphorite horizons (Fig. 2b). Nodules show mostly spheroidal, flattened, plano-convex, and ellipsoidal shapes and are feeble to dark gray and black in color. The size of nodules ranges from 4.5 * 2.5 * 1.5 to 27 * 18 * 9 cm^3. Small nodules have higher phosphate content than large nodules. Among the lenticular varieties, dark phosphatic lenticels show alternate associations with light-colored bands. This variety is weakly calcareous and/or has siliceous materials (Fig. 2c). Alternate bands of carbonates and phosphates with mm-scale thickness are seen in the laminated phosphorites (Fig. 2d). In this variety of phosphorites, a

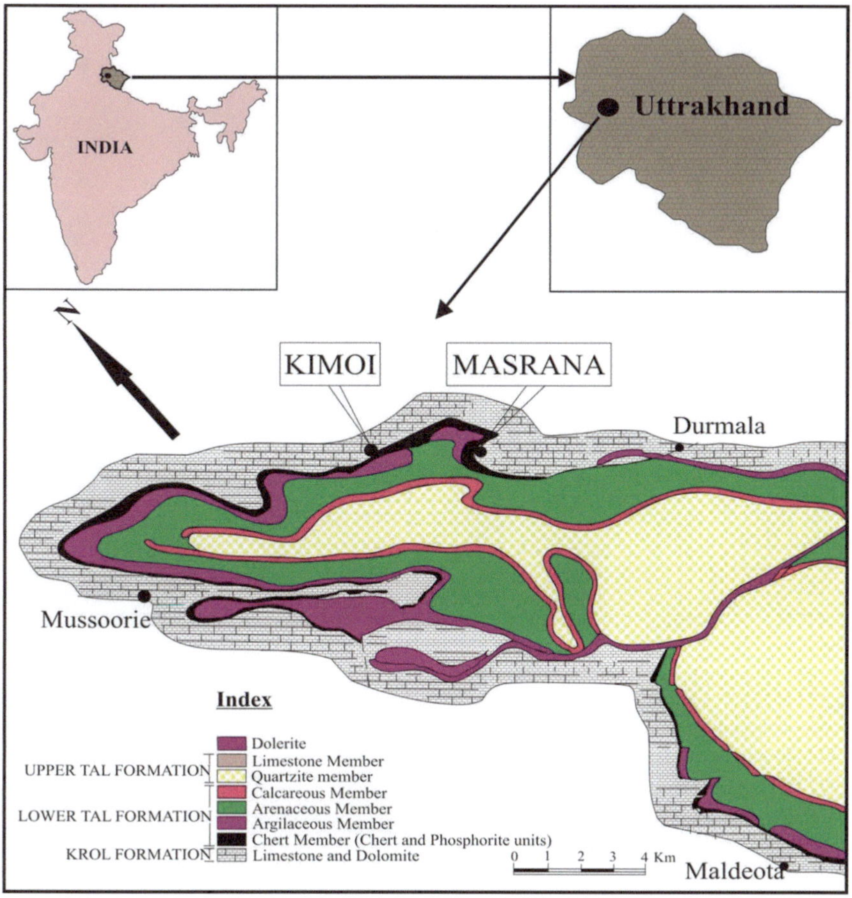

Fig. 1 Geologic map showing the location of the sections examined (modified after Mazumdar & Banerjee, 2001; Khan et al., 2016, 2022)

phosphatic layer is made of collophane (which is a cryptocrystalline variety of apatite mineral, < 1 μm size; Dar et al., 2015) surrounded in a fine phosphate matrix. This variety retains the uppermost concentration of phosphate than that of other classes.

4 Methodology

The present field area falls in the toposheet number 53 J/3. A total of thirty-six fresh representative samples were taken from outcrops and quarries. After petrographic and X-ray diffraction studies of collected samples, only twelve samples were analyzed for major and trace element data by X-ray fluorescence Spectrometry (XRF of the Philips

Fig. 2 **a** Field photograph-bedded phosphorite with layers of bedded phosphate alternate with clayey and shaly bands, **b** Sample photograph showing nodules of collophane of size >2 cm, **c** Sample photograph-lenticles of collophane in siliceous as well as carbonaceous groundmass, **d** Sample photograph-laminations of collophane in a matrix of phosphatic material

Company) and Inductively Coupled Plasma (ICP) using Perkin Elmer, SCIEX, Model ELAN DRC-II, respectively, at National Geophysical Research Institute, Hyderabad, India. The following procedure was followed: first, the selected rock samples were powdered by a jaw crusher to a −200 mesh size at the Department of Geology, Aligarh Muslim University, Aligarh, India. Major element data were generated from the pressed pellets by a Philips MagiX PRO model PW-2440 wavelength dispersive X-ray fluorescence spectrometer (XRF) coupled with an automatic sample changer PW 2540. These pressed pellets for each sample were made of collapsible aluminum cups. These cups were filled with boric acid powder and 1 g of sample was placed over the surface of this boric acid, then these cups were pressed under a hydraulic pressure of 20 tons. The Open acid digestion technique (Roy et al., 2007) was followed for the determination of trace element data. Fifty mg of powder of each sample was put in PTFE Teflon beakers and 10 ml of $HF:HNO_3:HClO_4$ acid mixture in a 7:3:1 proportion was added to each beaker to make the powder of each sample moist. Then these beakers with lids were left overnight for digestion. The next day these beakers were heated on a hot plate with a temperature of ~150 °C, the lids were removed after an hour, and the samples were allowed to dry to form a crystalline paste, 5 ml of the acid mixture was added to each beaker for complete digestion and again these samples were dried up. Ten milliliters of 1:1 HNO_3 solution was added to dissolve residual material in these beakers, and then these beakers were placed on a hot plate for 10 min at a temperature of 70 °C to dissolve all suspended particles. Then, 10 ml of 1 ppm Rhodium solution as an internal standard was added to each beaker and finally, the resulting volume was diluted to 250 ml with the addition of double - distilled water. Then trace elements (including REEs) were determined from this solution by inductively coupled plasma-mass spectrometry techniques (ICP-MS). The precision of ICP-MS data is better than ±6% RSD for all the trace elements and REE. The instrumental and data acquisition parameters are given by Roy et al. (2007). All the available data were standardized against the international reference rock standard NIST-120C for phosphorites. The obtained geochemical data are presented in Table 4.

5 Uranium Geochemistry

Notable concentration of uranium in phosphorites is reported by various workers like Adesanwo et al., 2009 (15–29 ppm in the phosphorites of Ogun state, Nigeria); Al-Hwaiti et al., 2005 (23–42 ppm in the Shidiya Phosphorites, SE Jordan); Dar et al., 2014 (1.67–129.67 ppm in the Paleoproterozoic phosphorites of Lalitpur district, Uttar Pradesh, India); Imamoglu et al., 2009 (0.9–75 ppm in the phosphorites of West Kasrik, SE Turkey); Pitawala et al., 2003 (traces to 0.1 ppm in the apatite bearing Carbonate rocks of Eppawala, Sri Lanka); Rao, 1984 (50–80 ppm in the stromatolitic phosphorites of Udaipur, Rajasthan, India); Rao et al., 2008 (43.6–103 ppm in the phosphorites of Murray Ridge, NW Arabian Sea), Rao et al., 2002 (2.0–194 ppm in the Quaternary phosphorites of the continental margin off Chennai, SE coast of India). The uranium in the studied phosphorite samples ranges from 1.37–37.5 ppm with an average of 16.7 ppm (Table 4). The low U content might be because of elimination from the crystal structure of carbonate-apatite during catagenesis or it may also be due to the low Eh environment in union with plentiful deterioration of OM. Most of the initially incorporated U is generally assumed to be in the reduced (U^{4+}) oxidation state and leached away when it is subaerially exposed to oxidizing conditions or seafloor weathering.

The studied phosphorite samples are rich in CaO, P_2O_5, SiO_2, Al_2O_3, Fe_2O_3, K_2O, V, Cr, Zn, Sr, Zr, Nb, and Ba and depleted in MgO, MnO, and TiO_2 (Table 4). The redox-sensitive elements such as U (1.37–37.5 ppm, average = 16.7 ppm), V (10.9–466 ppm, average = 156 ppm), Cr (45.9–206 ppm, average = 90.5 ppm), Ni (16.1–59.7 ppm, average = 33.4 ppm), Co (2.89–87.1 ppm, average = 22.2 ppm), Cu (11.5–66.2 ppm, average = 31.3 ppm), and Zn (33.9–822 ppm, average = 137 ppm) are present in notable concentration (Table 4).

Several binary plots are shown in Figs. 3, 4, and 5 to understand the behavior of U with different major and trace elements of the studied samples. Uranium has negative correlation with P_2O_5 (Fig. 3a), CaO (Fig. 3b), and SiO_2 (Fig. 3c), whereas it shows weak positive correlation with K_2O (Fig. 3d), Fe_2O_3 (Fig. 3e), and Al_2O_3

Table 4 Uranium geochemistry including major, trace, and REE in phosphate-bearing sedimentary rocks of Masrana and Kimoi blocks of the Mussoorie syncline, Uttarakhand, India

	M6	M12	M21	M25	M27	K41	K38	K43	K47	K51	K58	K64	Max	Min	Average
SiO_2	12.1	10.9	11.9	16.2	11.9	11.6	15.2	11.7	9.88	12.0	16.0	10.4	16.2	9.88	12.5
Al_2O_3	8.09	7.19	8.09	0.88	7.68	8.32	0.22	7.39	8.81	6.14	0.27	0.32	8.81	0.220	5.28
Fe_2O_3	2.41	3.80	2.15	1.64	2.62	1.40	0.56	2.78	4.12	3.16	0.61	0.59	4.12	0.560	2.15
MnO	0.01	0.08	0.01	0.01	0.01	0.01	0.01	0.01	0.07	0.01	0.00	0.01	0.08	0	0.020
MgO	0.40	0.56	0.39	0.29	0.41	0.39	0.28	0.46	0.59	0.39	0.29	1.56	1.56	0.280	0.501
CaO	37.8	38.1	37.8	37.9	37.8	37.8	40.7	38.1	35.4	38.0	40.1	44.8	44.8	35.4	38.7
K_2O	3.76	3.61	3.24	0.06	3.32	3.19	0.01	3.24	4.51	2.96	0.01	0.01	4.51	0.01	2.33
TiO_2	0.31	0.28	0.31	0.10	0.29	0.33	0.08	0.30	0.31	0.26	0.08	0.09	0.33	0.08	0.228
P_2O_5	28.8	28.8	28.8	28.8	28.8	29.8	31.5	28.8	27.2	28.9	31.3	31.8	31.8	27.2	29.4
Sum	93.7	93.3	92.6	85.9	92.8	92.9	88.5	92.8	90.9	91.8	88.6	89.6	93.7	85.9	91.1
Sc	10.1	9.09	10.3	4.28	10.6	11.4	3.98	9.01	9.29	8.66	6.68	4.05	11.4	3.98	8.12
V	114	86.1	136	381	106	210	41.9	109	120	466	88.9	10.9	466	10.9	156
Cr	82.1	59.8	85.0	206	77.0	101	58.9	72.7	94.1	127	76.1	45.9	206	45.9	90.5
Co	19.0	19.7	14.6	30.2	23.4	9.90	12.8	17.1	87.1	17.1	12.2	2.89	87.1	2.89	22.2
Ni	29.2	27.4	26.6	25.7	45.6	48.7	23.9	41.5	59.7	23.0	33.0	16.1	59.7	16.1	33.4
Cu	19.2	24.3	29.3	23.7	25.7	66.2	37.7	24.4	19.3	28.8	65.5	11.5	66.2	11.5	31.3
Zn	66.75	75.45	80.34	33.93	142	822	52.9	63.7	75.7	60.3	130	37.3	822	33.9	137
Ga	120	108	133	13.0	111	136	6.94	109	113	79.0	7.22	4.50	136	4.50	78.3
Rb	117	109	107	10.0	94.6	96.2	1.67	97.6	92.1	67.2	2.98	2.781	117	1.67	66.5
Sr	14.7	43.0	38.9	39.2	28.7	45.2	592.9	24.9	25.5	31.9	432	52.7	593	14.7	114
Y	21.2	16.7	18.3	20.8	18.2	16.2	178	22.9	16.9	45.6	137	1.46	178	1.46	42.8

(continued)

Table 4 (continued)

	M6	M12	M21	M25	M27	K41	K38	K43	K47	K51	K58	K64	Max	Min	Average
Zr	443	278	435	43.9	451	432	27.9	496	457	290	26.5	26.8	496	26.5	284
Nb	256	235	247	29.6	255	269	9.88	257	261	170	17.1	10.0	269	9.88	168
Cs	9.91	11.8	10.8	1.47	9.22	9.64	0.319	11.3	8.75	7.91	0.532	0.343	11.8	0.319	6.83
Ba	553	434	572	121	1290	1117	294	2267	480	1787	410	21.9	2267	21.9	779
La	44.5	326	44.7	11.7	38.9	46.5	56.0	44.7	41.9	66.1	47.4	2.16	326	2.16	64.2
Ce	74.2	63.2	74.6	15.2	66.2	76.3	37.0	76.4	70.7	72.1	28.7	2.57	76.4	2.57	54.8
Pr	9.39	7.84	9.16	2.99	8.47	9.46	10.4	9.79	8.76	14.5	8.83	0.388	14.5	0.388	8.33
Nd	37.5	31.7	35.5	13.8	33.5	37.4	49.5	38.8	34.3	60.8	41.1	1.63	60.8	1.63	34.6
Sm	5.35	4.89	4.72	2.52	4.73	4.49	8.42	5.27	4.75	8.34	6.86	0.248	8.42	0.248	5.05
Eu	1.40	1.26	1.19	0.774	1.46	1.25	2.91	1.86	1.20	2.51	2.69	0.080	2.91	0.08	1.55
Gd	5.47	5.08	4.82	2.92	4.72	4.70	12.6	5.670	4.77	8.83	10.4	0.306	12.6	0.306	5.85
Tb	0.738	0.712	0.611	0.471	0.650	0.578	2.21	0.771	0.644	1.25	1.77	0.045	2.21	0.045	0.870
Dy	4.13	3.94	3.73	3.14	3.85	3.29	16.9	4.65	3.60	8.07	12.7	0.231	16.9	0.231	5.68
Ho	0.903	0.750	0.778	0.679	0.802	0.714	4.303	0.978	0.783	1.897	3.228	0.052	4.30	0.052	1.32
Er	2.68	2.09	2.43	1.88	2.35	2.2	12.8	2.92	2.38	5.74	9.43	0.145	12.8	0.145	3.92
Tm	0.404	0.296	0.368	0.271	0.356	0.323	1.82	0.417	0.364	0.907	1.36	0.017	1.82	0.017	0.575
Yb	3.15	2.20	2.90	1.94	2.64	2.61	12.3	3.17	2.78	7.15	9.44	0.134	12.3	0.134	4.20
Lu	0.391	0.274	0.365	0.229	0.329	0.339	1.56	0.396	0.363	0.986	1.20	0.016	1.56	0.016	0.537
Hf	7.06	4.60	6.93	0.690	6.93	6.98	0.558	8.13	7.18	4.20	0.502	0.388	8.13	0.388	4.51
Ta	3.09	2.63	2.24	2.04	3.57	2.15	1.54	3.30	5.21	2.42	2.49	0.359	5.21	0.359	2.59
Pb	68.5	25.1	66.3	48.3	239	57.2	19.0	53.7	40.0	87.0	78.4	14.0	239	14.0	66.3

(continued)

Table 4 (continued)

	M6	M12	M21	M25	M27	K41	K38	K43	K47	K51	K58	K64	Max	Min	Average
Th	13.2	9.90	13.1	1.02	11.6	9.88	0.469	11.4	12.5	7.29	0.613	0.404	13.2	0.404	7.62
U	9.73	3.11	25.3	3.22	17.6	31.9	20.6	17.4	37.5	10.4	21.7	1.37	37.5	1.37	16.7
ΣREE	190	451	186	58.5	169	190	229	196	177	259	185	8.028	451	8.03	192
REY	211	467	204	79.3	187	206	407	219	194	305	322	9.49	629	9.49	234

(Fig. 3f). The negative correlation between U and P_2O_5 (Fig. 3a) suggests the removal of uranium from the atomic lattice of apatite and substitution of P^{5+}, while the inverse correlation between U and CaO (Fig. 3b) suggests mutual substitution of Ca^{2+} by U^{4+} in the apatite crystal structure (Banerjee et al., 1980; Kajitani, 1970). The better correlation of U with Ca, than U with P, suggests uranium not only exists within the apatite structure but also occupies positions similar to those of calcium within that structure. The negative correlation between U and SiO_2 (Fig. 3c) suggests the substitution of SiO_2 with PO_4 during the early stages of phosphatization. In phosphorites, the SiO_2 might have undergone weathering due to groundwater and marine conditions, leading to an increase in U content over time. This remobilized uranium could have been reprecipitated under appropriate pH and Eh conditions of the basin. Furthermore, a portion of the U content may be present as sorption by gangue minerals (Pirc & Rose, 1981). Overall, this process suggests a complex interplay between phosphatization, weathering, and reprecipitation, as well as the sorption of uranium by gangue minerals. The sympathetic/weak positive relationship of U with K_2O (Fig. 3d) indicates partial replacement of Ca^{2+} by K^+ in the apatite lattice, which points toward alkaline conditions of the basin of deposition during phosphatization. In the studied samples, a weak positive correlation between U and Fe_2O_3 (Fig. 3e) suggests that, although there is some association, iron oxides are not the main factor controlling uranium accumulation, and that U (uranium) is probably associated with other phases or affected by other geochemical processes. The enrichment of Al_2O_3 and its moderate positive correlation with U (Fig. 3f) may indicate the accumulation of U in the basin by weathering and transportation and then attachment of U on the surface of phosphate minerals.

Redox-sensitive trace elements are more soluble under oxidizing conditions than reducing conditions, which makes authigenic enrichments in oxygen-depleted sedimentary facies. Because of this characteristic U, V, Mo, Cr, and Co are precious proxies for the evaluation of paleo-redox conditions. The positive correlation of U with Ni (Fig. 4a), moderate positive correlation with Cu, Zn, Co, (Fig. 4b–d) and weak negative correlation of U with V and Cr (Fig. 5a, b) suggests that the phosphate-bearing rocks are deposited under sulfur-reducing/Euxinic conditions (Tribovillard et al., 2005, 2006). Additionally, Ni, Cu, Zn, and Ba (Figs. 4a–c and 5c) are the tracers for the abundance of OM i.e., they are incorporated into the sediments primarily in association with OM. A higher concentration of Ni and Cu suggests (i) a high influx of OM released these elements from the surroundings and concentrated them in the sediments, and (ii) the reducing environment persisted, enabling the sedimentary fixation of Ni and Cu. For example, elements can be trapped by pyrite if sulfur-reducing conditions prevail (Algeo & Maynard, 2004; Fernex et al., 1992; Nameroff et al., 2004; Tribovillard et al., 2006). The positive correlation of U with these elements suggests a similar process for U enrichment in the phosphate-bearing rocks. Uranium and Th seem to have a positive correlation (Fig. 5d). Concentration of Th resulted in the substitution of Ce in the phosphate-bearing mineral structure (Rao, 1984).

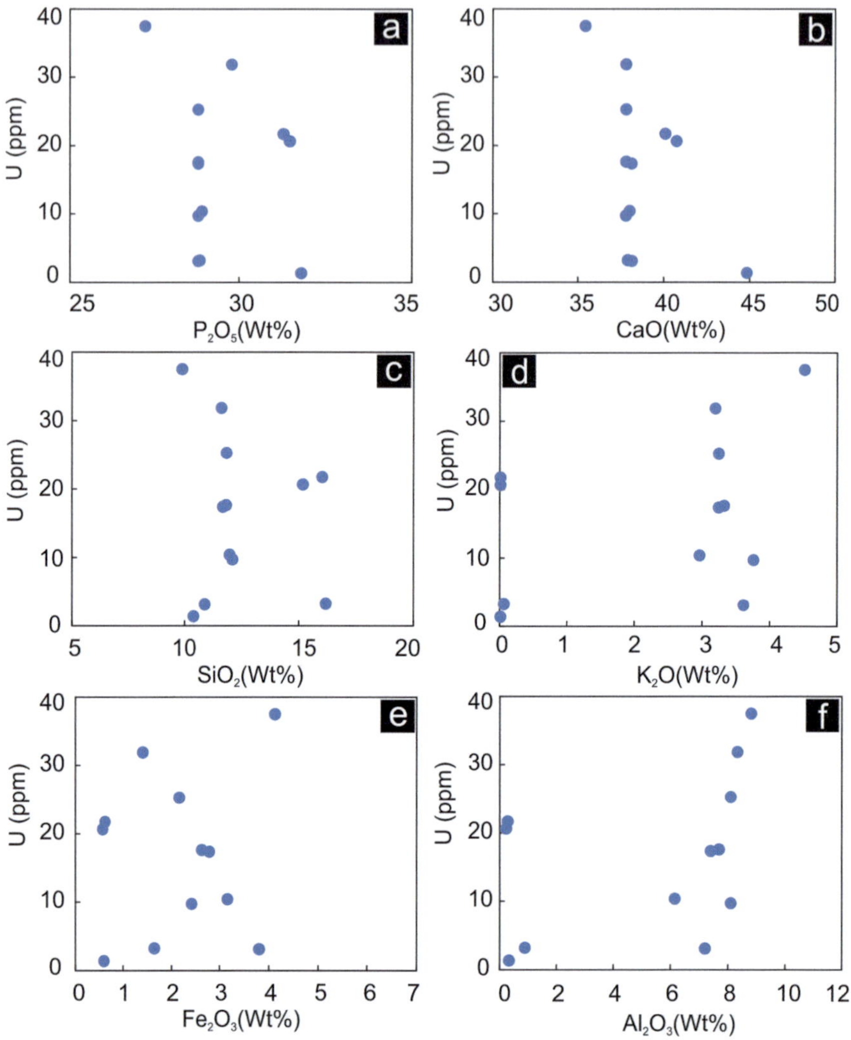

Fig. 3 Cross-plots showing a relationship between U and other major elements (P_2O_5, CaO, SiO_2, K_2O, Fe_2O_3, and Al_2O_3) in the studied samples

6 Conclusions

The studied phosphorite deposits of Masrana and Kimoi blocks of Uttarakhand show dull gray to brownish-black color. These phosphorites are rarely bedded and display incrustations of white and orange color. Bedded phosphate with clayey and shaly bands is seen in bedded phosphorites. These are made of siliceous or carbonaceous matrix.

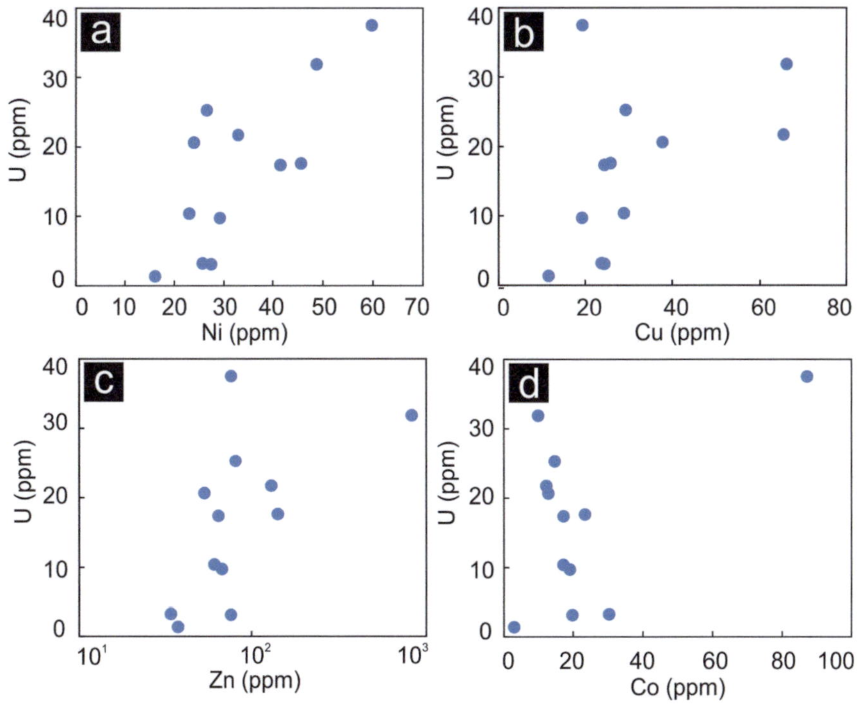

Fig. 4 Cross-plots showing a relationship between U and other trace elements (Ni, Cu, Zn, and Co) in the studied samples

Nodular phosphorite is found in black shales that occur in contact with the phosphorite horizon. Nodules show mostly spheroidal, flattened, plano-convex, and ellipsoidal shapes and are feeble to dark gray and black in color. The size of nodules ranges from 4.5 * 2.5 * 1.5 to 27 * 18 * 9 cm^3. Smaller nodules have a higher content of phosphate than the larger nodules.

Alternate bands of carbonate and phosphate minerals, having a thickness in mm are seen in laminated phosphorites. In this variety of phosphorites, the phosphatic layer is made of collophane surrounded by a fine phosphate matrix. This variety retains the maximum concentration of phosphate than that of other classes.

The low U content in Masrana and Kimoi phosphorites (1.37–37.5 ppm, average 16.7 ppm) might be because of its subtraction from the crystal structure of carbonate-apatite during catagenesis. The negative correlation of U with P_2O_5 suggests the removal of uranium from the atomic lattice of apatite and substitution of P^{5+}, while the inverse correlation between U and CaO suggests mutual substitution of Ca^{2+} by U^{4+} in the apatite crystal structure. The better correlation of U with Ca, than U with P, suggests that uranium not only exists within the apatite structure but also occupies positions similar to those of calcium within that structure.

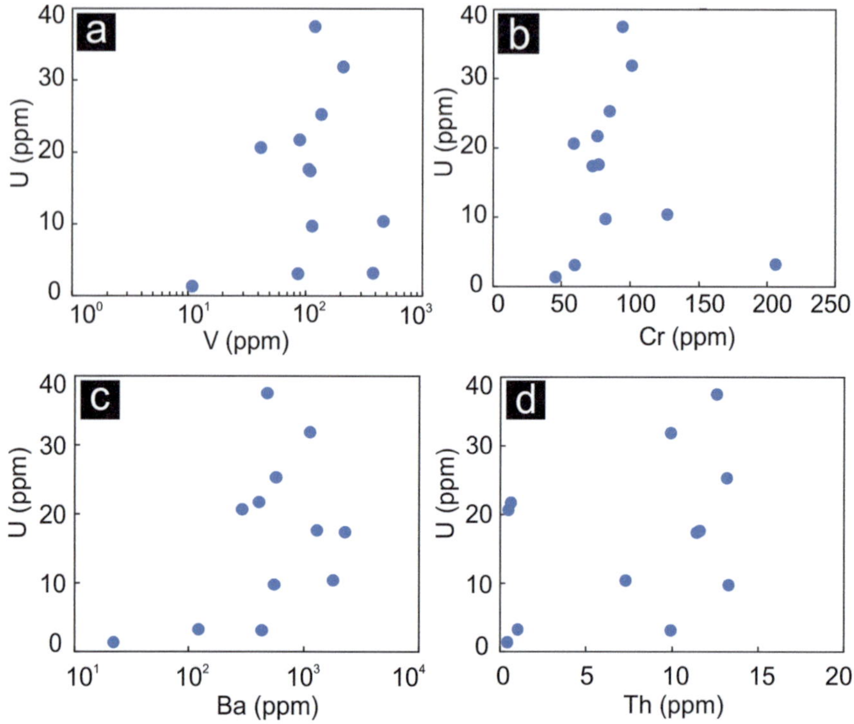

Fig. 5 Cross-plots showing a relationship between U and other trace elements (V, Cr, Ba, and Th) in the studied samples

The sympathetic/weak positive relationship of U with K_2O indicates partial replacement of Ca^{+2} by K^{+1} in the apatite lattice, which points toward alkaline conditions of the basin of deposition during phosphatization.

The positive correlation of U with Ni, moderate positive correlation with Cu, Zn, Co, and weak negative correlation of U with V and Cr suggests that the phosphate-bearing rocks are deposited under sulfate-reducing/Euxinic conditions.

A higher concentration of Ni and Cu in the studied samples suggests that a high influx of OM released these elements from the surroundings and concentrated them in the sediments, and reduced conditions of the basin enabled the sedimentary fixation of Ni and Cu. The influx of OM is also responsible for the enrichment of U in the studied phosphate-bearing rocks based on the positive correlation of U with Ni and Cu.

Acknowledgements We are highly grateful to the chairperson, department of Geology, Aligarh Muslim University, and the Head, Department of Geology, Banaras Hindu University for providing necessary facilities, and support. The authors also extend their sincere thanks to the Director of the National Geophysical Research Institute (NGRI), Hyderabad, for permission to conduct the geochemical analysis of rock samples. Shamim A. Dar is highly thankful to the Institutions of Eminence (IoE), Banaras Hindu University (BHU) (Dev. Scheme No. 6031) and the Department of

Science & Technology, Science and Engineering Research Board (SERB) [Anusandhan National Research Foundation (ANRF)], Government of India, for providing a start-up research grant (M-14/0599; Sanction order no. SRG/2022/001478) for financial assistance. Soumyajit Mukherjee (IIT Bombay) invited, handled and reviewed this article. Mukherjee (2026) summarized this work.

Abbreviations

OM	Organic Matter
XRF	X-Ray Fluorescence Spectrometry
ICP-MS	Inductively Coupled Plasma Mass Spectrometry
REE	Rare Earth Elements

References

Adesanwo, O. O., Adetunji, M. T., Dunlevey, J. N., Diatta, S., Osiname, O. A., & Adesanwo, J. K. (2009). X-ray fluorescence studies of Ogun phosphate rock. *Electronic Journal of Environmental, Agricultural & Food Chemistry, 8.*

Algeo, T. J. & Maynard, J. B. (2004). Trace-element behavior and redox facies in core shales of Upper Pennsylvanian Kansas-type cyclothems. *Chemical Geology, 206*, 289–318. https://doi.org/10.1016/j.chemgeo.2003.12.009

Al-Hwaiti, M., Matheis, G., & Saffarini, G. (2005). Mobilization, redistribution and bioavailability of potentially toxic elements in Shidiya phosphorites, Southeast Jordan. *Environmental Geology, 47*, 431–444. https://doi.org/10.1007/s00254-004-1173-2

Ali, A., Pan, J., Yan, J., & Nabi, A. (2020). Geochemical characteristics and uranium mineralization exploration potential of late Miocene molasse sediments of NW Himalayan foreland basin Pakistan. *Arabian Journal of Geosciences, 13*, 123. https://doi.org/10.1007/s12517-020-5066-9

Alloway, B. J. (2013). Uranium BT—Heavy metals in soils: Trace metals and metalloids in soils and their bioavailability. In B. J. Alloway (Ed.) (pp. 565–577) Springer Netherlands. https://doi.org/10.1007/978-94-007-4470-7_26

Altschuler, Z. S. (1980). The geochemistry of trace elements in marine phosphorites part I. Characteristic abundances and enrichment. *Marine Phosphorites—Geochemistry, Occurrence, Genesis.* https://doi.org/10.2110/pec.80.29.0019

Auden, J. B. (1934). The geology of the Krol belt. *Records of the Geological Survey of India, 67*, 357–454.

Banerjee, D. M., Basu, P. C., & Srivastava, N. (1980). Petrology, mineralogy, geochemistry, and origin of the Precambrian Aravallian phosphorite deposits of Udaipur and Jhabua, India. *Economic Geology, 75*, 1181–1199. https://doi.org/10.2113/gsecongeo.75.8.1181

Bhargava, O. N. (1974). *Geology of the Tal Formation, Nigali Dharsyncline, Sirmur District, India.* Unpublished Ph.D. thesis, Panjab University, pp. 1–134.

Biswas, T., Bose, N., Dutta, D., & Mukherjee, S. (2022). Arc-parallel shears in collisional orogens: Global review and paleostress analyses from the NW Lesser Himalayan Sequence (Garhwal region, Uttarakhand, India). *Marine and Petroleum Geology, 138*, Article 105530.

Bose, N., & Mukherjee, S. (2019). Field documentation and genesis of the back-structures from the Garhwal Lesser Himalaya, Uttarakhand, India. *Geological Society, London, Special Publications, 481*, 111–125.

Burnett, W. C. (1990). Phosphorite growth and sediment dynamics in the modern Peru shelf upwelling system. In *Phosphate deposits of the World. III: Neogene to modern phosphorites* (pp. 62–72).

Cumberland, S. A., Douglas, G., Grice, K., & Moreau, J. W. (2016). Uranium mobility in organic matter-rich sediments: A review of geological and geochemical processes. *Earth-Science Reviews, 159*, 160–185. https://doi.org/10.1016/j.earscirev.2016.05.010

Dar, S.A., Khan, K.F., Khan, S.A., Mir, A.R., Wani, H., & Balaram, V. (2014). Uranium (U) concentration and its genetic significance in the phosphorites of the Paleoproterozoic Bijawar Group of the Lalitpur district, Uttar Pradesh, India. *Arabian Journal of Geosciences, 7*, 2237–2248. https://doi.org/10.1007/s12517-013-0903-8

Dar, S.A., Khan, K.F., Khan, S.A., Khan, S. & Masroor Alam, M. (2015). Petro-mineralogical studies of the Paleoproterozoic phosphorites in the Sonrai Basin, Lalitpur District, Uttar Pradesh, India. *Natural Resources Research, 24*, 339–348. https://doi.org/10.1007/s11053-014-9260-x

Desai, B. G., Shukla, R., & Saklani, R. D. (2010). Ichnology of the Early Cambrian Tal Group, Nigalidhar Syncline, Lesser Himalaya, India. *Ichnos, 17*, 233–245. https://doi.org/10.1080/10420940.2010.534320

Fayek, M., Horita, J., & Ripley, E. M. (2011). The oxygen isotopic composition of uranium minerals: A review. *Ore Geology Reviews, 41*, 1–21. https://doi.org/10.1016/j.oregeorev.2011.06.005

Fernex, F., Février, G., Bénaïm, J., & Arnoux, A. (1992). Copper, lead and zinc trapping in Mediterranean deep-sea sediments: Probable coprecipitation with Mn and Fe. *Chemical Geology, 98*, 293–306. https://doi.org/10.1016/0009-2541(92)90190-G

Gupta, G. R., Kumarupta, G., Mathur, V. K., & Joshi, A. (1993). Stratigraphy of Blaini, Infra Krol, Krol and Tal succession, Krol belt, Lesser Himalaya.

Hazen, R. M., Ewing, R. C., & Sverjensky, D. A. (2009). Evolution of uranium and thorium minerals. *American Mineralogist, 94*, 1293–1311. https://doi.org/10.2138/am.2009.3208

Hobday, D. K., & Galloway, W. E. (1999). Groundwater processes and sedimentary uranium deposits. *Hydrogeology Journal, 7*, 127–138. https://doi.org/10.1007/s100400050184

Imamoglu, M. S., Nathan, Y., Coban, H., Soudry, D., & Glenn, C. (2009). Geochemical, mineralogical and isotopic signatures of the Semikan, West Kasrık "Turkish" phosphorites from the Derik–Mazıdagı–Mardin area, SE Anatolia. *International Journal of Earth Sciences, 98*, 1679–1690.

Kajitani, K. (1970). A geochemical study on the genesis of ningyoite the special calcium uranous phosphate mineral. *Economic Geology, 65*, 470–480. https://doi.org/10.2113/gsecongeo.65.4.470

Khan, K. F., Dar, S. A., & Khan, S. A. (2011). Petrography of phosphorite deposits of Durmala and Maldeota, Uttrakhand. *Journal Indian Association of Sedimentologists, 30*, 33–41.

Khan, S. A., Khan, K. F., & Dar, S. A. (2016). REE geochemistry of Early Cambrian phosphorites of Masrana and Kimoi blocks, Uttarakhand, India. *Arabian Journal of Geosciences, 9*, 456.

Khan, S. A., Dar, S. A., Khan, K. F., & Karim, Y. (2022). Geochemical characteristics of Early Cambrian phosphate bearing sedimentary rocks from the Mussoorie Syncline, India: Implications for paleo-redox conditions. *Geosystems and Geoenvironment, 1*, 100046. https://doi.org/10.1016/j.geogeo.2022.100046

Krauskopf, K. B., & Bird, D. K. (1995). *Introduction to geochemistry*. McGraw-Hill.

Kreuzer, O. P., Markwitz, V., Porwal, A. K., & McCuaig, T. C. (2010). A continent-wide study of Australia's uranium potential: Part I: GIS-assisted manual prospectivity analysis. *Ore Geology Reviews, 38*, 334–366. https://doi.org/10.1016/j.oregeorev.2010.08.003

Kumar, G. (2005). *Geology of Uttar Pradesh and Uttranchal* (p. 383). Geological Society of India. ISBN No: 81-85867-66-6.

Mahato, S., Mukherjee, S., & Bose, N. (2019). Documentation of brittle structures (back shear and arc-parallel shear) from Sategal and Dhanaulti regions of the Garhwal Lesser Himalaya (Uttarakhand, India). *Tectonics and Structural Geology: Indian Context* 411–423.

Mazumdar, A., & Banerjee, D. M. (2001) Regional variations in the carbon isotopic composition of phosphorite from the Early Cambrian Lower Tal Formation, Mussoorie Hills, India. *Chemical Geology, 175*, 5–15. https://doi.org/10.1016/S0009-2541(00)00360-0

McClellan, G. H., & Van Kauwenbergh, S. J. (1990). Mineralogy of sedimentary apatites. *Geological Society, London, Special Publication, 52*, 23–31.

Mukherjee, S. (2015) A review on out-of-sequence deformation in the Himalaya. In Mukherjee S, Carosi R, van der Beek P, Mukherjee BK, Robinson D (Eds.), *Tectonics of the Himalaya. Geological Society, London* (pp. 67–109). Special Publications 412.

Mukherjee, S. (2026). Introduction to "Structural Geology & Tectonics". In S. Mukherjee (Ed.), *Structural Geology & Tectonics*. Springer. ISBN: 978-981-95-4743-2.

Mukherjee, S., Goswami, S., & Zakaulla, S. (2023). Geological relationship between hydrocarbon and uranium: Review on two different sources of energy and the Indian scenario. *Geoenergy Science and Engineering, 221*, 111255. https://doi.org/10.1016/j.petrol.2022.111255

Nameroff, T. J., Calvert, S. E., & Murray, J. W. (2004). Glacial-interglacial variability in the eastern tropical North Pacific oxygen minimum zone recorded by redox-sensitive trace metals. *Paleoceanography, 19*. https://doi.org/10.1029/2003PA000912

Patwardhan, A. M., & Ahluwalia, A. D. (1973). A note on the origin of Mussoorie phosphorite in the Lower Himalaya, India, and its palaeogeographic implications. *Mineralium Deposita, 8*, 379–387. https://doi.org/10.1007/BF00207519

Pirc, S., & Rose, A. W. (1981). Uranium anomalies in paleo-aquifers near sandstone-type uranium deposits in the devonian catskill formation of Pennsylvania. *Journal of Geochemical Exploration, 15*, 219–231. https://doi.org/10.1016/0375-6742(81)90064-9

Pitawala, A., Schidlowski, M., Dahanayake, K., & Hofmeister, W. (2003). Geochemical and petrological characteristics of Eppawala phosphate deposits, Sri Lanka. *Mineralium Deposita, 38*, 505–515. https://doi.org/10.1007/s00126-002-0327-y

Raha, P. K. (1972). Algal stromatolites from the krol formation of the Mussoorie syncline, Uttar Pradesh. *Journal of Palaeosciences, 21*, 227–230. https://doi.org/10.54991/jop.1972.1485

Rao, L. A. K. (1984). Trace-element studies of the Precambrian stromatolitic phosphorites of Udaipur, Rajasthan, India. *Chemical Geology, 45*, 17–31. https://doi.org/10.1016/0009-2541(84)90113-X

Rao, V. P., Michard, A., Naqvi, S. W. A., Böttcher, M. E., Krishnaswamy, R., Thamban, M., Natarajan, R., & Borole, D. V. (2002). Quaternary phosphorites off the southeast coast of India. *Chemical Geology, 182*, 483–502. https://doi.org/10.1016/S0009-2541(01)00336-9

Rao, V. P., Hegner, E., Naqvi, S. W. A., Kessarkar, P. M., Ahmad, S. M., & Raju, D. S. (2008). Miocene phosphorites from the Murray Ridge, northwestern Arabian Sea. *Palaeogeography, Palaeoclimatology, Palaeoecology, 260*, 347–358. https://doi.org/10.1016/j.palaeo.2007.12.003

Roy, P., Balaram, V., Kumar, A., Satyanarayanan, M., & Gnaneshwar Rao, T. (2007). New REE and trace element data on two kimberlitic reference materials by ICP-MS. *Geostandards and Geoanalytical Research, 31*, 261–273. https://doi.org/10.1111/j.1751-908X.2007.00836.x

Saraswat, A. C., & Mahadevan, T. M. (1989). *Genetic aspects of uranium mineralization in the Himalaya*. IAEA.

Shankar, R. (1987). *The Mussoorie phosphate Basin, India* (Vol. 2, pp. 443–448). Cambridge University Press.

Sharma, M., Sharma, Y. C., Basu, B., Chhabra, J., Gupta, R. K., & Singh, J. (2000). Uranium mineralization in the sandstones of Dharamsala, Tileli area, Mandi district, Himachal Pradesh, India. *Current Science, 78*, 897–899.

Sinha, K. K., Ghosh, D., Rai, S. D., & Khandelwal, M. K. (2016). Uranium mineralisation in carbonate rocks from the upper part of Krol formation in Nainital syncline and its significance for uranium exploration in lesser Himalaya. *Journal of the Geological Society of India, 88*, 295–304. https://doi.org/10.1007/s12594-016-0492-8

Slansky, M. (1986). *Geology of sedimentary phosphate* (1st English ed., pp. 59–159). North Oxford Academic Publishers Limited.

Tribovillard, N., Ramdani, A., & Trentesaux, A. (2005). Controls on organic accumulation in Upper Jurassic Shales of Northwestern Europe as inferred from trace-metal geochemistry. Depos. Org. Sediments Model. Mech. Consequences. https://doi.org/10.2110/pec.05.82.0145

Tribovillard, N., Algeo, T. J., Lyons, T., & Riboulleau, A. (2006). Trace metals as paleoredox and paleoproductivity proxies: An update. *Chemical Geology, 232*, 12–32. https://doi.org/10.1016/j.chemgeo.2006.02.012

Structural Style and Deformation of the Gachsaran Formation in the Dezful Embayment of the Zagros Fold Thrust Belt, Chamshir Area, Iran

Ehsan Tavakolian, Raana Razavi Pash, Soumyajit Mukherjee, Ehsan Eghbalpour, Mehdi Ghanavati, Mehdi Khoshnoodkia, Mohammad Mohammadian, Ghodratollah Sadeghi, and Hasan Amiri Bakhtiar

Abstract The Gachsaran evaporate Formation is the caprock of the Asmari oil reservoir in the Zagros oil province. This Formation is a main detachment horizon in the Zagros. As an intermediate detachment horizon, this Formation separates rocks of distinct structural features and folding styles in its overlaying and underlying units. Around 1600 m thick Gachsaran Formation has seven Members (Gs.1–7). The Formation consists of salts, anhydrites, coloured marls, limestones and a single layer of bituminous shale. In this work, we interpret the different Members of the Gachsaran Formation and study the structures in the Chamshir area, located in the lowland basin of the Zagros belt, using surface and subsurface data in 2D sections and 3D models. The presence of a thick layer of the Gachsaran Formation has affected the structural style in the Chamshir area. In the Chamshir area, Members Gs.5–7 crop out. The salt members of the Gachsaran Formation are deformed to several secondary folds and faults. Based on surface and subsurface data (up to 3000 m deep), the area has two types of faults. The faults seen in the exposed part of the studied area often continue to Gs.4, such as the Dezh Soleyman Fault, Chamshir Fault, F4 and F3 faults. The second category of faults developed within the Asmari Formation up to the approximate depth of the Triassic-Jurassic Formations. In other words, a thick layer of the evaporate member of the Gachsaran Formation (Gs.4), as the main detachment layer, has caused the structural style to be separated across it. As a result, the folds above and below the Gachsaran Formation in the Zagros belt

---

E. Tavakolian · E. Eghbalpour · M. Ghanavati · M. Khoshnoodkia · M. Mohammadian · G. Sadeghi · H. A. Bakhtiar
National Iranian South Oil Company (NISOC), Ahwaz, Iran

R. R. Pash (✉)
Department of Earth Sciences, College of Sciences, Shiraz University, Shiraz, Iran
e-mail: ranarazavi@ymail.com

S. Mukherjee
Department of Earth Sciences, Indian Institute of Technology Bombay, Powai, Mumbai, Maharashtra, India

S. Mukherjee (ed.), *Structural Geology and Tectonics*,
https://doi.org/10.1007/978-981-95-4743-2_13

significantly different geometries. As the Gachsaran Formation is the main caprock of the Asmari reservoir in Zagros, understanding the effect of this detachment level on the geometry of the anticlines as oilfields is crucial for future exploration in the region.

Keywords Gachsaran Formation · Structural style · Chamshir area · Zagros belt

1 Introduction

One of the factors controlling the deformation style in the fold-thrust belts is mechanical stratigraphy (e.g. Spratt et al., 2004; Tavakolian & Razavi Pash, 2022). The mechanical stratigraphy of sedimentary cover layers and the presence of incompetent horizons that can act as detachment horizons can be an important parameter controlling the folding style in fold-thrust belts (Cotton & Koyi, 2000; Davis & Engelder, 1985; Misra & Mukherjee, 2015; Razavi Pash et al., 2021a, 2021b; Spratt et al., 2004). The detachment horizons in the middle of the sedimentary cover caused structural separation and the change of the folding style of the upper units, so the geometry of the folds on the surface does not always represent the geometry of those at depth (e.g. O'Brien 1957; Tavakolian et al., 2022).

The Gachsaran evaporate Formation is a main detachment horizon in the foreland of Zagros. In the Dezful Embayment located in the foreland of the Zagros belt (Fig. 1), the Gachsaran Formation has seven Members and consists of salts, anhydrite, marls, limestones and a layer of bituminous shale (Fig. 2). This Formation is as the caprock of Asmari oil reservoirs in Zagros. Several studies have been made on the effect of the Gachsaran Formation on the geometry of petroleum anticlines in Zagros (Table 1).

The Zagros fold-and-thrust belt is one of the prolific petroliferous regions with ~12% of the world's oil reserves (Bordenave & Burwood, 1990). While fold-and-thrust belts have been doubted as suitable for hydrocarbon exploration, the Zagros orogenic belt has never been questioned in this regard (review in Razavi Pash et al., 2023).

Analysis of structural style and deformation in Zagros is crucial for the future exploration and production of hydrocarbons from the reservoirs. Gachsaran evaporate Formation as the main caprock of the Asmari reservoir in Zagros indicates a different geometry in its members in the Chamshir area. Analysis of the structural style of these members aid in hydrocarbon exploration.

Members 5–7 of the Gachsaran Formation are crop out in the Chamshir area. This area includes various types of folds in the different members of the Gachsaran Formation. The performance of the faults in this region has caused the anticlines to be closer to the surface towards the mountain front fault, and for this reason, Gachsaran Formation cropped out.

Member 4 of the Gachsaran Formation has a ductile behaviour in this area and has acted as a detachment layer. Also, evidence of a large salt wall structure can be seen in the Gach Haji anticline. This structure is the result of fault movement and

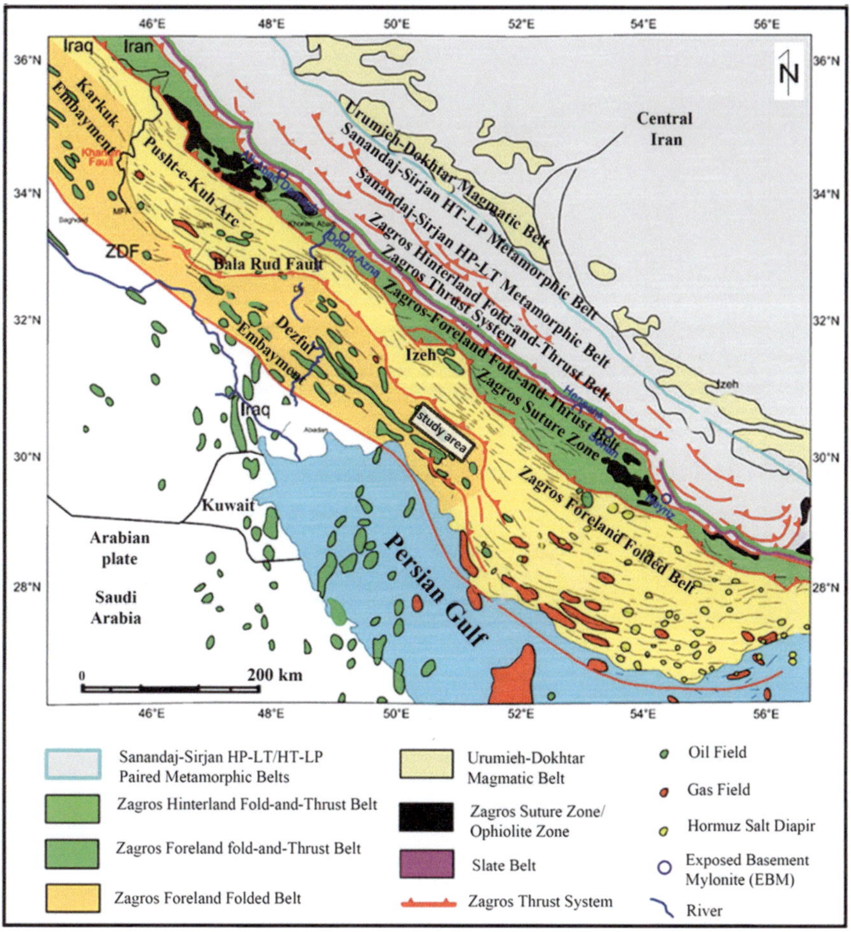

Fig. 1 Location of the Chamshir area (white rectangle) in the foreland of the Zagros Mountain Front Fault (Sarkarinejad & Ghanbarian, 2014)

rising salt from the members of 5 and 6 of the Gachsaran Formation. As a result of the movement of salt, the member of 6 of the Gachsaran Formation has been exposed at the surface.

In this study, the structural style and deformation of the different Members of the Gachsaran Formation around Cheshmir located in the Dezful Embayment have been investigated by performing fieldwork. Seismic and remote sensing images have been used in the work for a better understanding of structures. Twelve seismic profiles within ~2500–3000 m depth have been studied and included in this study. In total, we study folding and faulting styles and deformation of the Gachsaran Formation from surface and subsurface. A 3D structural model of the Chamshir area is presented.

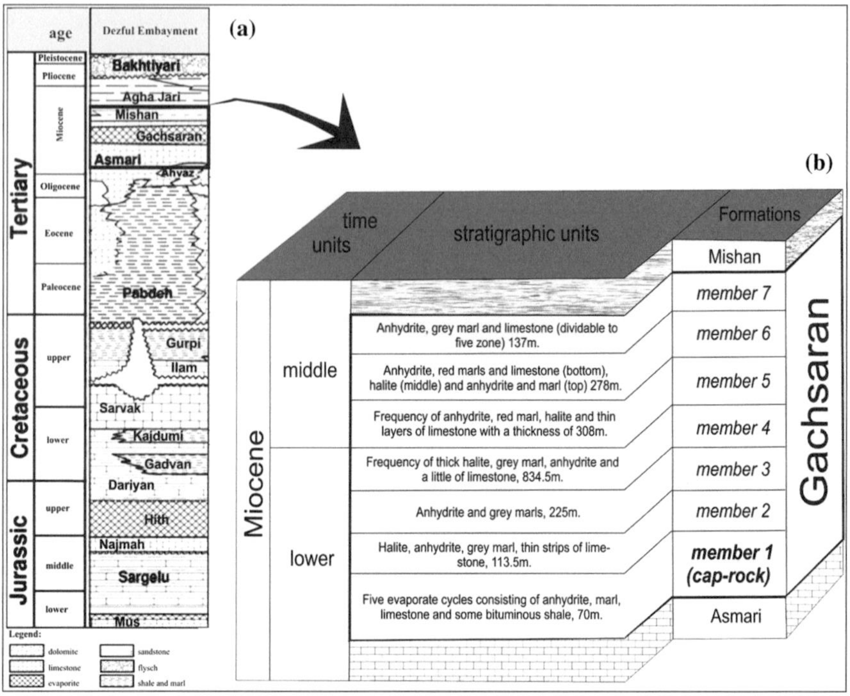

Fig. 2 **a** Stratigraphic column of the Mesozoic and Cenozoic Formations of the Dezful Embayment in the foreland basin of Zagros fold-thrust belt (after Sepehr & Cosgrove, 2004). **b** Lithological units of the Gachsaran Formation (Soleimani & Bahadori, 2015)

2 Geology

The Zagros belt, as a part of the Alpine-Himalayan orogenic belt with a length of ~2000 km, which stretches from northern Iraq to the Strait of Hormuz, is the result of the oblique convergence between the Arabian and Central Iran Plates during the closing of the Neotethys Ocean (Alavi, 1994; Stocklin, 1968).

The Dezful Embayment is located in the southwestern part of the Zagros belt and was considered to be the foreland of Zagros. The Chamshir area is located in the southeastern part of the Dezful Embayment and near the border of the Izeh zone (Fig. 1). The central Zagros is subdivided from NE to SW into the part of High Zagros, Izeh zone and Dezful Embayment. The Dezful Embayment and Izeh zone are separated by the Zagros mountain front fault. The NW and SE boundaries of the embayment coincide with the Balarud and Kazerun faults, respectively. The Zagros mountain front fault is the NE boundary of the Dezful Embayment (Sherkati & Letouzey, 2004).

The Gachsaran Formation is the main detachment horizon in the foreland of Zagros (e.g. Sherkati et al., 2005; Tavakolian & Razavi Pash, 2022). The typical

Table 1 Review on the effect of the Gachsaran Formation on the geometry of anticlines in Zagros

References	Studied terrains	Outcomes
Bahroudi and Koyi (2004)	Zagros foreland basin	Differential propagation of the deformation front above the decollements (especially the Gachsaran Formation) with different mechanical properties (viscous vs. frictional) results in the along-strike irregularity of the Zagros deformation front
Sepehr et al. (2006)	Zagros fold-thrust belt	The folds above and below the incompetent units in the central part of the Zagros Folded Belt have significantly different geometries and wavelengths
Sherkati et al. (2005)	Central and Eastern Zagros fold-belt	Two main detachment levels in the Zagros are the Hormuz basal detachment and the Gachsaran upper detachment. Throughout the Zagros, detachment folds mainly developed during an initial thin-skinned phase of deformation. This was followed by the current thick-skinned stage
Sherkati et al. (2006)	Central Zagros fold-thrust belt	In the Zagros belt, the decollement levels are activated sequentially from deeper horizons to the shallower ones (as the Gachsaran decollement)
Tavakolian et al. (2022)	South Dezful Embayment	The lower–middle Miocene evaporites in the Gachsaran Formation decoupled the folded overlying succession from the underlying competent material, which includes the Asmari Formation reservoir
Tavakolian and Razavi Pash (2022)	Southern Dezful Embayment (Bibi-Hakimeh Anticline)	Asymmetric and disharmonic geometry of the Bibi-Hakimeh anticline is due to the forethrusting and backthrusting and the presence of the Gachsaran Formation as the detachment level
Razavi Pash et al. (2023)	North Dezful Embayment (Qaleh Nar, Lower and Upper Balarud Anticlines)	Changes in overburden pressure, rate of deformation and uplift in the different parts of the subsurface anticlines moved Gachsaran Formation towards both limbs of the anticlines

section adopted from oil drillings in the Gachsaran area indicates that this Formation contains seven members (Fig. 2).

The folds in the Chamshir area are created in three parallel rows (Fig. 3). These three rows of anticlines from north to south are: (i) Se-Qanat–Sarab anticlines, (ii) Dara–Gachsaran anticlines and (iii) Chlingar–Garangan–Bidekarz anticlines.

Geologic Formations at the surface in the Chamshir area, from oldest to youngest are the different Members of the Gachsaran Formation (Early Miocene), Mishan Formation (Early-Middle Miocene), Aghajari Formation (Late Miocene–Pliocene) and alluvial sediments (Fig. 4).

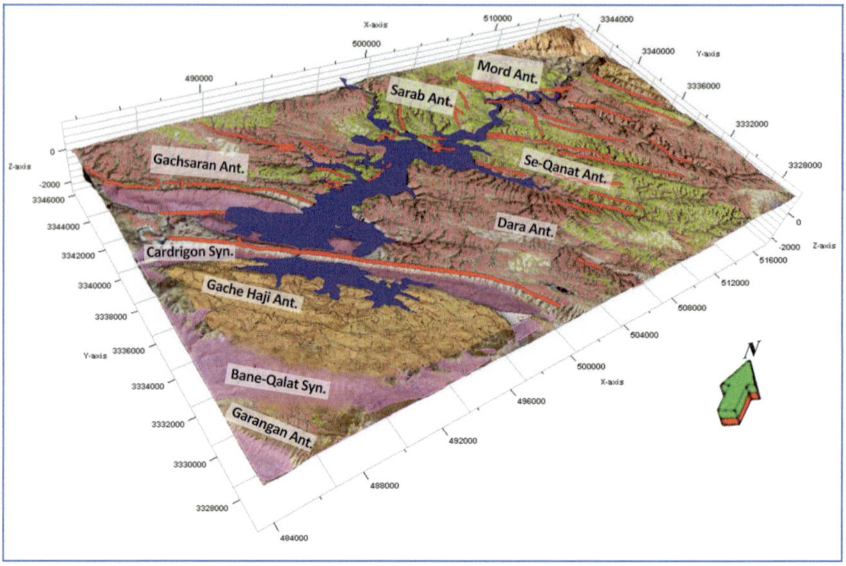

Fig. 3 Location of the studied folds in the Chamshir area

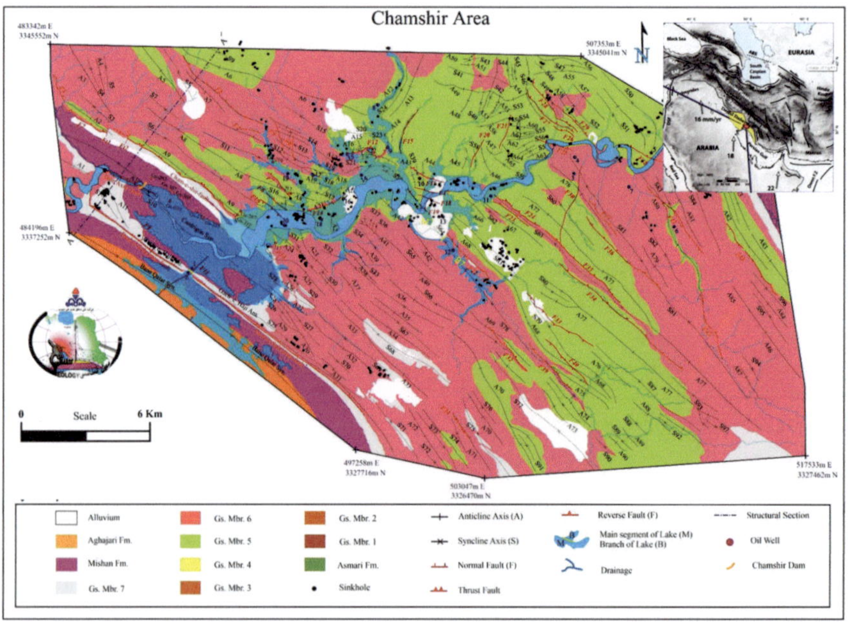

Fig. 4 Geologic map of the Chamshir area. Line AA′ shows the location of cross-section AA′ prepared by the National Iranian South Oil Company (NISOC)

3 Structural Style in the Chamshir Area

To the interpretation of the geometry of the structure, a cross-section AA′ was prepared using surface and ~2500–3000 m deep subsurface data prepared by the National Iranian South Oil Company (NISOC). The AA′ section trends N38E and is ~15 km long (Fig. 4). The northern limb and the central core of the Bane Qalat syncline are located in the southern part of section AA′. The Aghajari and Mishan Formation in this syncline crop out (Fig. 5). On the northern side of the Bane Qalat syncline, the Gach Haji anticline has a low width with two steep and inclined limbs. In both limbs of this anticline, Member 7 of the Gachsaran Formation is exposed. In the hinge area of the anticline, Member 6 of the Gachsaran Formation crops out (Figs. 5 and 6). In this section, along the axial plane of the Gach Haji anticline, the Dezh Soleyman fault passes and uplifted Member 6 of the Gachsaran Formation. This steep fault (NE38/60) dies out in the detachment horizon of the Gachsaran Formation (Fig. 5). In the north of the Gach Haji anticline in section AA′, the Cardrigon syncline is asymmetric. The Mishan Formation crops out in the limbs and central part of this syncline. In the north of the Cardrigon syncline, in the northern part of the AA′ section, several large-scale anticlines and synclines (e.g. the A70 anticline and the S77 syncline) can be seen (Fig. 6c). Within these large-scale folds, called as the folded-faulted zone, disharmonic folds are formed on a smaller scale (Fig. 6d). Member 6 of the Gachsaran Formation has been removed from these anticlines, which shows the structural uplift of the folds in this section. Member 6 crops out as the limbs of the A70 anticline. Member 5 of the Gachsaran Formation is exposed in its central core at the surface. The steep reverse faults F3 and F4 have created the displacements in this folded zone (Fig. 5). The Chamshir fault is located between the Cardrigon syncline and the folded-faulted zone, which thrust Members 5 and 6 of the Gachsaran Formation on Member 7 of the Gachsaran Formation and the Mishan Formation on the northeastern side of the Cardrigon syncline (Fig. 6b).

4 Bane Qalat Syncline

This syncline extends ~17.5 km with a NW trend. Bane Qalat syncline is located between the Gach Haji anticline in the northeast and the Chlingar anticline in the southwest. The Bane Qalat syncline is a double plunging asymmetric fold. The dip of the northeastern limb ranges 30°–60° and the dip of the southwestern limb 15°–25°. Thus the dip of the northeastern limb is more than that of the southwestern limb. The surface outcrop of this syncline consists of the Mishan and Aghajari Formations. This syncline is also a mini basin.

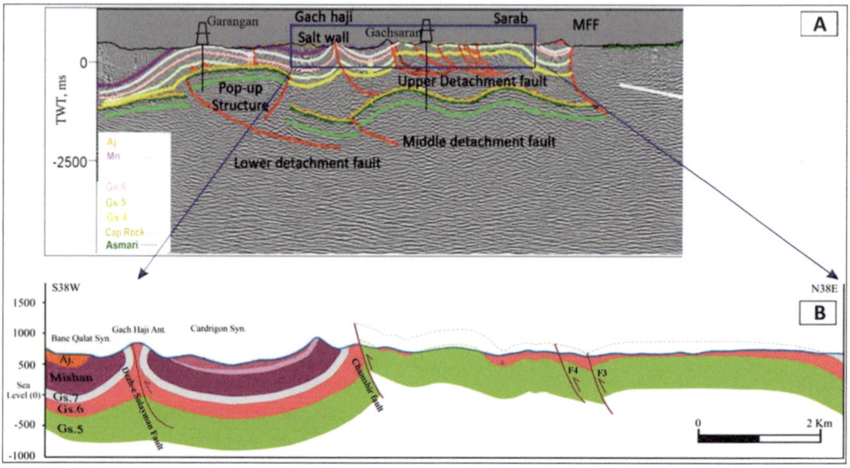

Fig. 5 Regional cross-section AA′ in the Chamshir area. MFF: Zagros Mountain Front Fault

5 Gach Haji Anticline

The Gach Haji anticline is ~18 km long with a NW trend. It is located in the northeast of the Bane Qalat syncline. The Gach Haji anticline is one of the unique features in the Chamshir area. The structure of the Gach Haji anticline is a salt wall affected by the Dezh Soleyman thrust. The anticline is divided into three parts at northwest, middle and the southeast.

The middle part of the Gach Haji anticline can be considered as a salt wall. The southeastern part of the Gach Haji, like the northwestern part, is a salt-cored fold. The core is formed by the salts of the Member 6 of the Gachsaran Formation. In general, the shape of the Gach Haji anticline changes significantly from the surface to the depth, which can be recognized as changes in the structural style of the anticline. The core of the Gach Haji anticline is the salts of Member 6 of the Gachsaran Formation. The anticline is symmetric and fan-shaped. Its northeastern and southwestern limbs are very steep and overturned.

5.1 *The Northwestern Part of the Gach Haji Anticline*

This part of the anticline is ~2.5 km long. It has a single-fold axis and it is a non-cylindrical fold. This part of the anticline actually consists of two anticlines and an intermediate syncline. This part of the anticline is an asymmetric fold. Its southwest limb has dip 40°–50° and its northeast limb has 25°–30° dip. The surface outcrop of this part of the anticline is Members 6 and 7 of the Gachsaran Formation. Member 6 covers the axial zone of this part (Fig. 7a, b). The difference in folding style in the

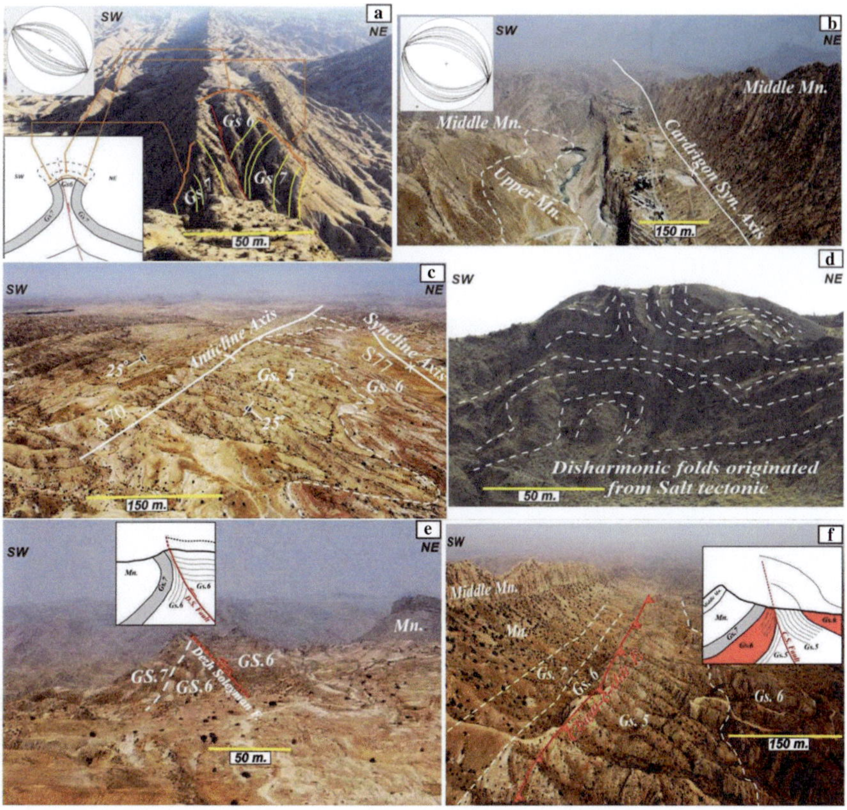

Fig. 6 Satellite images and field photos of the Chamshir area. **a** Satellite image of the Gach Haji anticline, **b** satellite image of the Cardrigon syncline along with, **c** satellite image of the folded and faulted region in the northeastern part of the Chamshir area, **d** field photo of disharmonic folds within the large-scale folds in the folded and faulted region, **e** satellite image of the Dezh Soleyman fault, **f** satellite image of the Chamshir fault. Gs.5-7 are Members 5–7 of the Gachsaran Formation, respectively

core of the Gach Haji anticline (Member 6) and its outer cover (Member 7) is caused by the presence of salt and its rheological property (Fig. 7c, d).

5.2 The Middle Part of the Gach Haji Anticline

This part of the anticline is ~7 km long, has a very narrow axial zone, symmetric and has a fan shape. Its northeastern and southwestern limbs are very steep and overturned. The surface outcrop of this part is formed from Members 6 and 7.

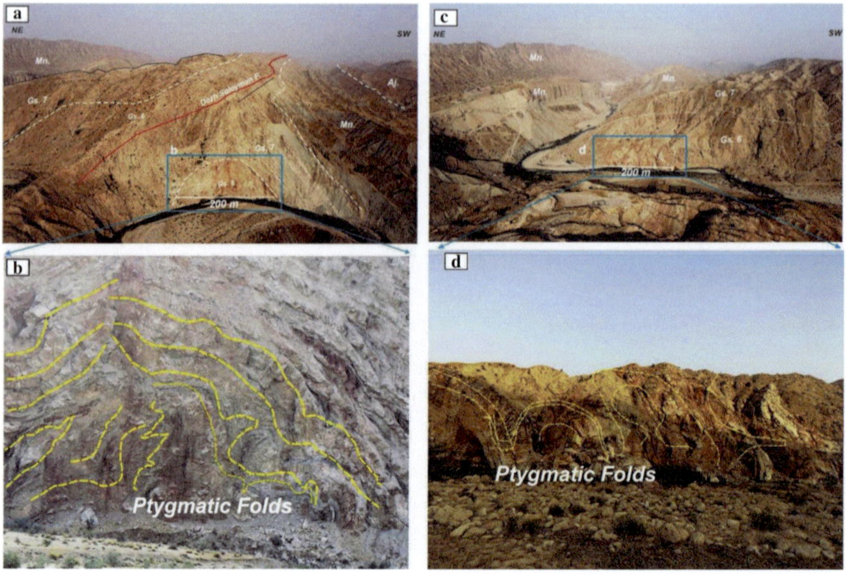

Fig. 7 Gach Haji anticline (view towards the SE). **a** The southern part of the Gach Haji anticline. The core of the anticline is made of Member 6 of the Gachsaran Formation and the Dezh Soleyman fault, the overthrust of Member 6 of the Gachsaran Formation on Member 7 in the axial zone of the anticline happened. **b** The core of the southern part of the Gach Haji anticline and the disharmonic folds were created due to the presence of salt (the layers of Member 6 of the Gachsaran Formation). **c** The northern part of the Gach Haji anticline and Cardrigon syncline. **d** The core of the northern part of the Gach Haji anticline with Member 6 of the Gachsaran Formation and the disharmonic folds in its anhydrite-marl layers of the Gachsaran Formation

5.3 Southeastern Part of the Gach Haji Anticline

In this part, the southwest limb is covered by Member 7 of the Gachsaran Formation, and the axial area is mostly covered by Member 6 of the Gachsaran Formation. Member 7 of the Gachsaran Formation in this part contains secondary folds. Member 6 of the Gachsaran Formation, which contains salt layers, is strongly folded.

6 Dezh Soleyman Fault

It is a reverse fault with 125° trend and ~16 km long along the Gach Haji anticline. The Dezh Soleyman fault passes through almost the hinge zone of the Gach Haji anticline. The dip of the fault plane is high (50°–70°) with a dip direction towards NE. With increasing depth, the dip of the fault gradually decreases. Most likely, it dies out in the salts (Members 2–4) of the Gachsaran Formation. The movement components of this fault are reverse and strike-slip. Its reverse component of this fault has caused

Member 6 of the Gachsaran Formation to be overridden on Member 7 in the axial zone of the Gach Haji anticline (along the entire length of this anticline).

7 Cardrigon Syncline

The syncline is ~11.5 km long and up to ~2.5 km wide. The fold axis trends 128° (NW–SE). The surface outcrop along the entire length of the Cardrigon syncline is the Mishan Formation. A normal fault is observed in the centre of this syncline (Fig. 8). At the southeast end of the Cardrigon syncline is the Mishan Formation exposed, Members 6 and 7 of the Gachsaran Formation (Fig. 9). In this place, Member 7 of the Gachsaran Formation consists of five zones (Z1-5, anhydrite, grey marl, anhydrite, grey marl and anhydrite, respectively).

8 Chamshir Fault

This fault with NW–SE trending can be traced on the northeastern limb of the Cardrigon syncline (Fig. 5). The Chamshir fault with ~65° dip and NE dip direction, like the Dezh Soleyman fault, can be considered as a reverse fault with a strike-slip component. The dip of the fault gradually decreases vertically down so that probably in the salty members of the Gachsaran Formation turns into a flat. On the surface, the effect of this fault is clearly evident, so that the fault has caused Member 5 of the Gachsaran Formation to be overthrusted on Member 6 of the Gachsaran Formation (at other places on Member 7).

To investigate the mentioned anticlines of the Chamshir area, four cross-sections with the northeast-southwest trend were prepared from different parts of the Chamshir area (Fig. 10). In the NE side of the Cardrigon syncline, there are Gachsaran, Sarab, Dara and Se-Qanat anticlines. In general, the thickness of the Gachsaran Formation in the northeast part of the Chamshir area (near the Zagros Mountain front fault) is more than in the southwest. Structural geometry is more complicated in the southwest of the Chamshir than at NE (Fig. 10).

9 Other Structures in Chamshir Area

There are several anticlines and synclines in the Chamshir area (Fig. 4). These folds are named as A1-90 for anticlines and S1-96 for synclines in the geologic map (National Iranian South Oil Company (NISOC) (Fig. 4).

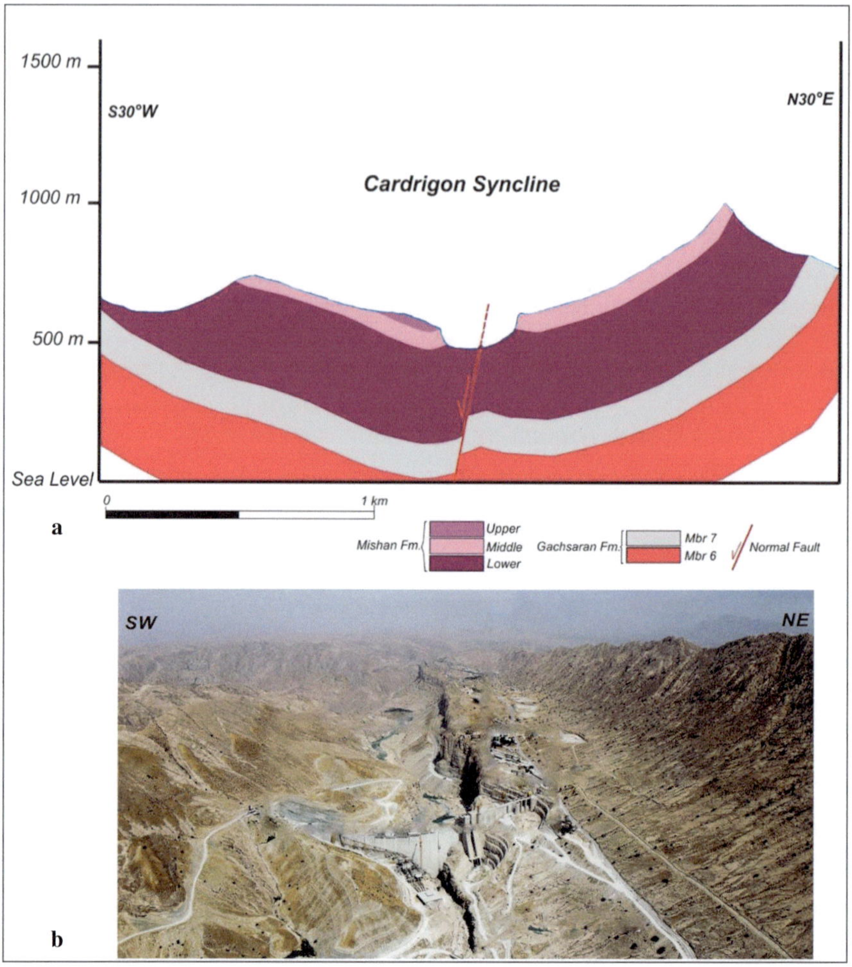

Fig. 8 Cross-section of the Cardrigon syncline (**a**) and satellite image of this syncline (**b**) presented in this study

9.1 Anticline A9

It is located in the vicinity of the northern wall of the Zohreh River. In this anticline, Members 5 and 6 of the Gachsaran Formation are exposed (Fig. 11). Member 6 of the Gachsaran Formation has been completely removed at the axial zone of the fold due to erosion, and Member 5 of the Gachsaran Formation crops out. In this asymmetric fold, the southwestern limb has a greater dip (38° towards the southwest) than the northeast one (20° towards the northeast). The trend of the axis of this anticline is 128E. Several small-scale folds occur in the axial zone and limbs.

Fig. 9 SE end of the Cardrigon syncline (view to north). Here, Mishan Formation (Mn), Member 7 and Member 6 of the Gachsaran Formation are clearly distinguishable. Member 7 of the Gachsaran Formation consists of five zones (Z1 to Z5), anhydrite, grey marl, anhydrite, grey marl and anhydrite, respectively. Gs.6 and Gs.7 are Member 6 and Member 7 of the Gachsaran Formation, respectively. Z1–Z5: Zone 1–Zone 5

9.2 Anticline A18

It is located in the NW part of the Chamshir area. In this anticline, Member 6 of the Gachsaran Formation is exposed (Fig. 12). Both the limbs and the axial zone of this anticline at the surface are composed of Member 6. This anticline is asymmetric that has a greater dip (60° to the northeast) than the southwest one (25° to the southwest). The trend of the axis of this anticline is 111E. The northeastern limb of this anticline is cut by a thrust (F4 with a dip towards the northeast). An unnamed smaller-scale anticline has been cut by this fault on the northeastern limb of the anticline A18.

9.3 Anticline A24

It is located in the Gachsaran area. Members 5 and 6 of the Gachsaran Formation are exposed (Fig. 13). In this asymmetric anticline, the northeast limb has a greater dip (56° to the northeast) than the southwest limb (23° to the southwest). The trend of the axis of this anticline is 133E. In the axial zone and limbs of this anticline, several smaller-scale folds are observed.

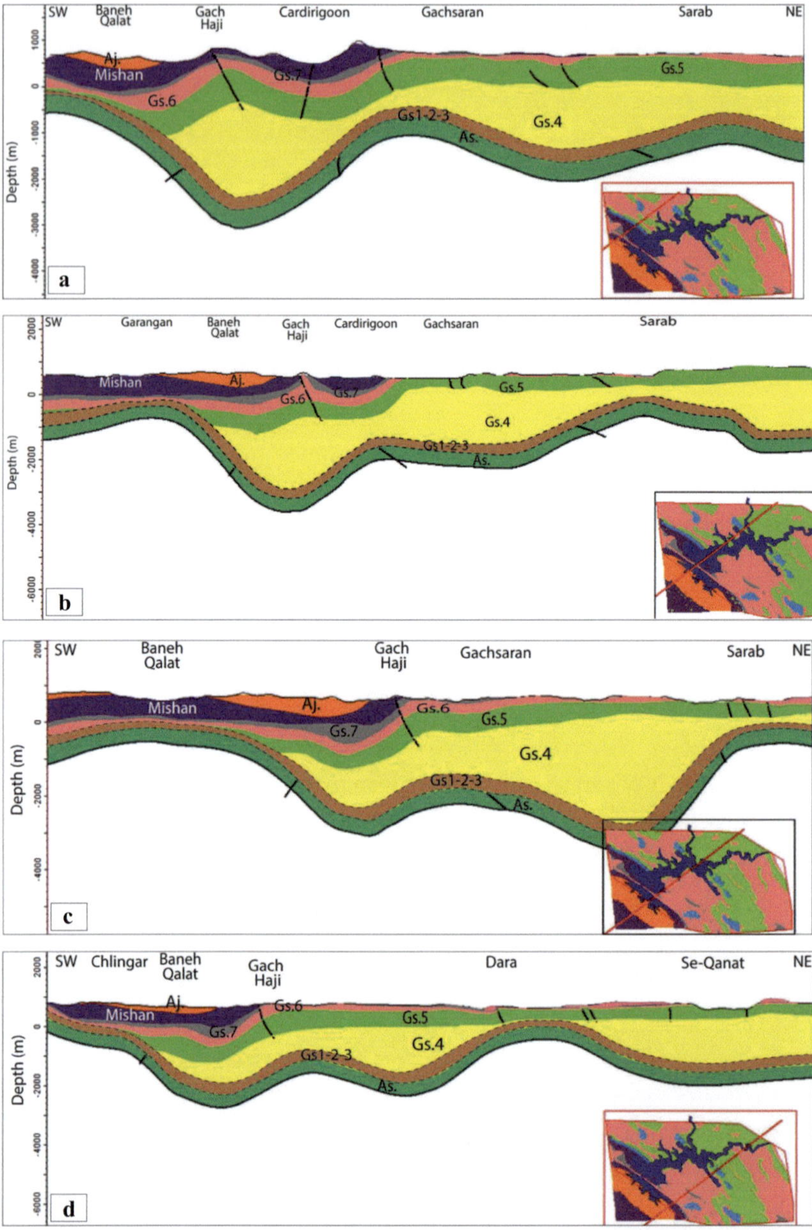

Fig. 10 Cross-sections of the Chamshir area. The location of these cross-sections is shown by the red line in the right corner at the bottom of each cross-section on the geologic map. **a** Cross-section of Bane Qalat–Sarab anticlines, **b** Cross-section of Karangan–Sarab anticlines, **c** Cross-section of Bane Qalat–Sarab anticlines, **d** Cross-section of Chlingar–Se-Qanat anticlines. Aj: Aghajari Formation, Gs.1–Gs.7: members (1–7) of the Gachsaran Formation, As: Asmari Formation. Prepared by using fieldwork data, interpretation of seismic profiles, well data, geologic maps and Petrel software (version 2014). Gs.1-7: Members 1–7 of the Gachsaran Formation

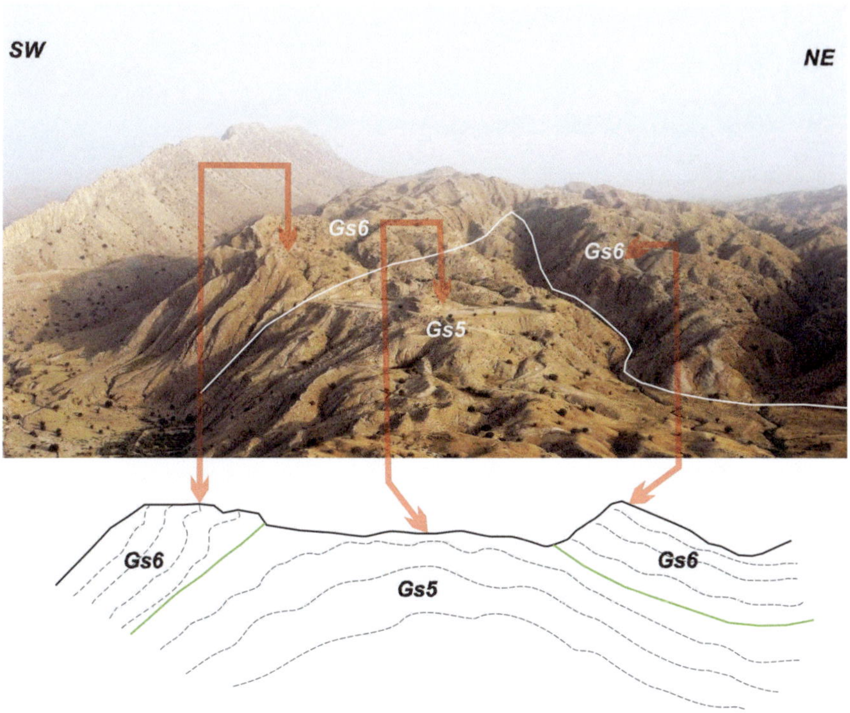

Fig. 11 Anticline A9 in the vicinity of the northern wall of the Zohreh River (view to the northwest). Member 6 of the Gachsaran Formation (Gs.6) has been completely removed from the axial zone of the fold due to erosion, and Member 5 of the Gachsaran Formation (Gs.5) is visible at the surface. In the back side of the image, Member 6 of the Gachsaran Formation also occurs in the axial zone in addition to the limbs. Member 5 is not exposed along the anticline. Gs.5 and Gs.6 are Members 5 and 6 of the Gachsaran Formation

9.4 Anticline A70

It is located in the southern part of the study area. Members 5 and 6 of the Gachsaran Formation are exposed (Fig. 14). At the surface, limbs are formed from Member 6 and the axial zone of the anticline is formed from Member 5 of the Gachsaran Formation. In other words, Member 6 has been removed from the axial zone of the anticline, but it remains on the limbs. In this relatively asymmetric anticline, the northeast limb has a greater dip (29° to the northeast) than the southwest edge (20° to the southwest). The trend of the axis of this anticline is 139E. In the axial zone and limbs of this anticline, several smaller-scale folds are observed.

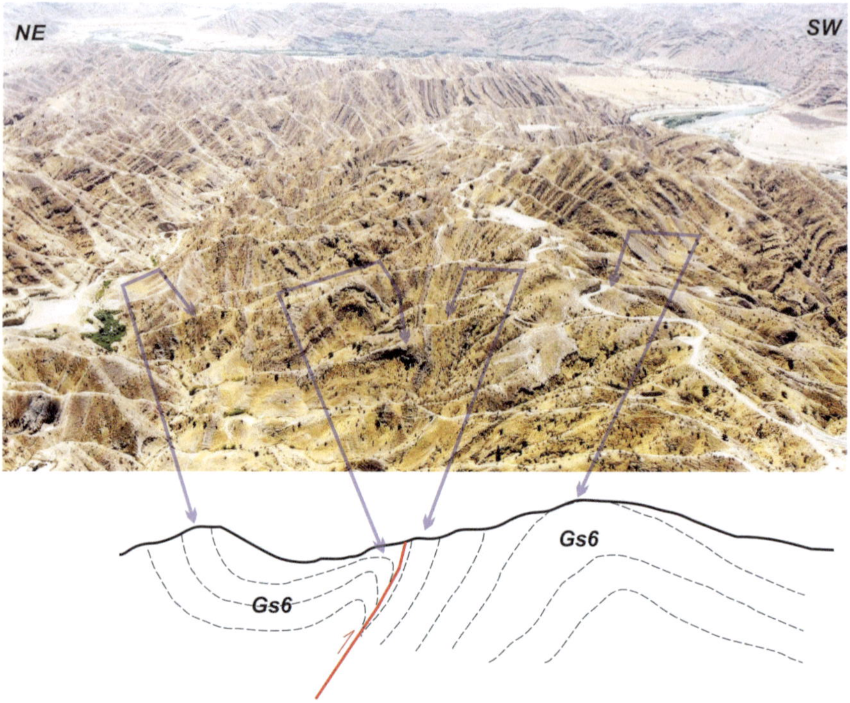

Fig. 12 Anticline A18 (right side of the image), view to the southeast. Gs.6: Member 6 of the Gachsaran Formation

9.5 Anticlines A61 and A62

Folds A61 and A62 are located in the northeast part of the studied area (south of Pakuh village). Only Member 5 of the Gachsaran Formation has been exposed at the surface in these two anticlines (Fig. 15). Both the limbs and the axial zone of both anticlines constitute Member 5. Between these two anticlines, the in between syncline cannot be distinguished, and the anticline (on the northeast side) is overthrust by a fault on the anticline on the southwest side. The trend of the axis of these anticlines is 110E.

9.6 Anticline A58 and Syncline S52

Anticline A58 and syncline S52 are located in the northeast of the study area (south of the Pakuh village). Outcrop in the anticline is the Member 5 of the Gachasaran Formation and in the syncline is the Member 6 of the Gachesaran Formation (Fig. 16). In the anticline, Member 6 of the Gachsaran Formation has been completely eroded. The anticline is an asymmetric fold where the southwest limb is steeper (68° to

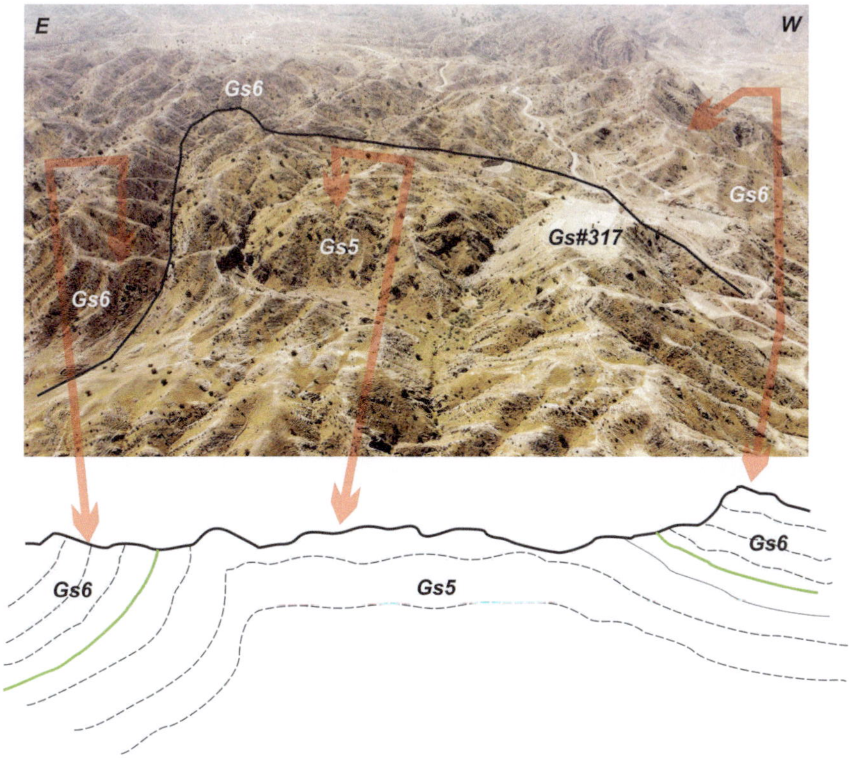

Fig. 13 Anticline A24 in the Gachsaran area (view to the southeast). In front of the image, member 6 of the Gachsaran Formation has been completely removed from the axial zone of the fold due to erosion, and Member 5 of the Gachsaran Formation is visible on the surface, but in the back of the image, Member 6 of the Gachsaran Formation is also present in the axial zone in addition to the limbs, and Member 5 has not been exposed along the anticline. Gs $\neq$ 317 is a drilled well. Gs.5 and Gs.6 are Members 5 6 of the Gachsaran Formation, respectively

the southwest) than the northeast limb (39° to the northeast). The syncline, which is located in the northeast of the anticline, is almost asymmetric, and its southwest limb (32° to the northeast) is steeper than the northeast limb (18° to the southwest). The trend of the axis of this anticline and its adjacent syncline is 130E. In the southwestern limb of the anticline are observed several smaller-scale folds.

9.7 *Anticline A43*

Anticline A43 is located near the intersection of the Delvar Haft Dasht drainage with the Zohreh River. Here, Member 5 of the Gachsaran Formation has a surface outcrop (Fig. 17). The fold is generally asymmetric, where the southwest limb is wavy but

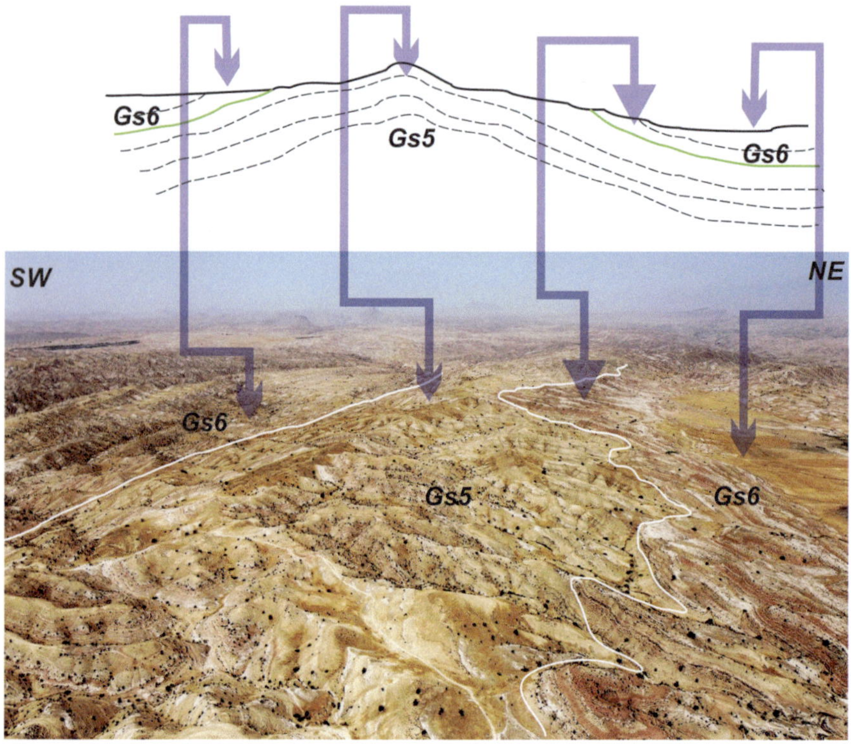

Fig. 14 A70 fold at the southern part of the study area (view to the northwest). Among the stratigraphic units involved in folding, members 6 and 5 of the Gachsaran Formation (Gs.6 and Gs.5) are exposed here at the surface. Gs.5 and 6 are Members 5 and 6 of the Gachsaran Formation, respectively

overall sub-vertical, and the dip of the northeast limb is 48° to the northeast. With a closer look at all parts of this fold, smaller-scale folds can be recognized. In addition, thrust faults can be seen on the northeast limb of the anticline. The trend of the axis of this anticline is about 160E.

9.8 Folded Area in the Se-Qanat Area

In a general view, a folded area in the Se-Qanat area can be considered a large-scale fold that contains several smaller-scale folds. In the northeast and the southwest limbs of this large-scale fold, Member 6 of the Gachsaran Formation is exposed at the surface. The middle part of the anticline is folded in a more intricate way. Member 6 has been completely removed due to erosion, and Member 5 of the Gachsaran Formation is exposed (Fig. 18). In Fig. 18, in the profile drawn, four smaller-scale anticlines can be recognized in the core consisting of the member 5 in the large-scale

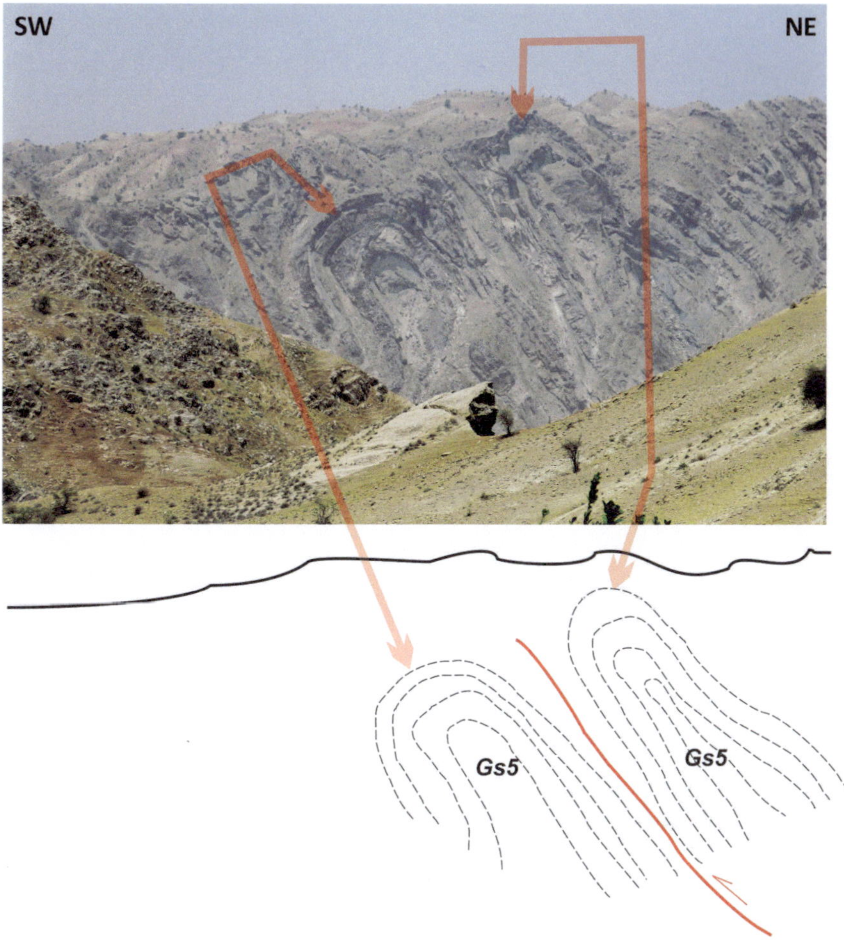

Fig. 15 A photo of the two anticlines A61 and A62 in the northeast of the studied area, view towards the northwest. Gs.5: Member 5 of the Gachsaran Formation

anticlines. The anticline located at the end left (southwest) of the profile corresponds to the anticline A68, and the one located at the extreme right (northeast) of the profile corresponds to the anticline A66.

9.9 Anticline A12

It is located in the north part of the study area around the Sarab drainage. In this anticline, the difference in colour tone between members 5 and 6 of the Gachsaran Formation, which are exposed at the surface, can be seen (Fig. 19). The axis trend

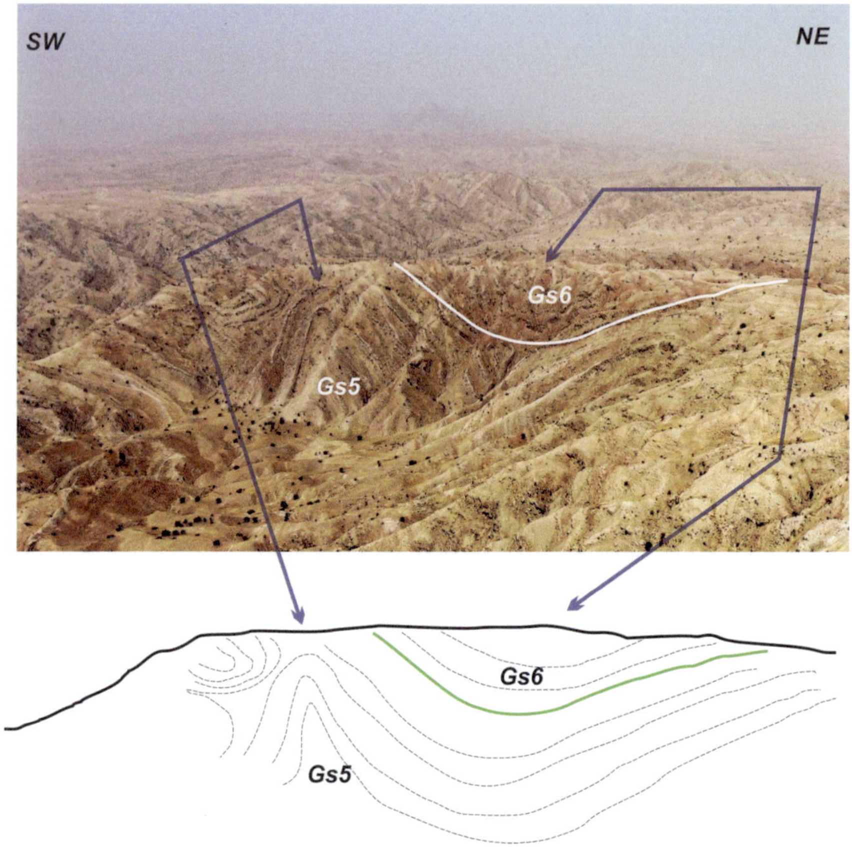

Fig. 16 A photo of the anticline A58 and the syncline S52 in the northeast of the study area, view to the northwest. Gs.5 and Gs.6 are Members 5 and 6 of the Gachsaran Formation, respectively

of this anticline, unlike most of the anticlines of the folded-thrust zone in the study area, is NNE-SSW.

9.10 Small-Scale Faults

The small-scale faults are observed in the northwest part of the study area. Figure 20 shows a normal fault. This normal fault is observed at the successive layers of grey marl, anhydrite and red marl belonging to Member 6 of the Gachsaran Formation. In this sequence, minute fractures can be recognized with the naked eye.

Fig. 17 A43 near the intersection of the Delvar Haft Dasht drainage with the Zohreh River, view towards the northwest

10 Discussions

One of the requirements of the oil industry for the exploration and drilling of hydrocarbon reservoirs is to identify the geometric complexity of the structures and study the behaviour of the intermediate detachment horizons in the fold-thrust belts, such as the Zagros belt in Iran (O'Brien, 1957; Farzipour-Saein et al., 2009; Sherkati et al., 2006) and in Iraq (e.g. Aqrawi et al., 2010a, 2010b), and the Canadian Rocky Mountain (Cooper et al., 2004). The folding style is largely caused by changes in the mechanical characteristics of the sedimentary layers through the folding process (Cotton & Koyi, 2000; Spratt et al., 2004; Vergés et al., 2011).

The Chamshir area located in the Dezful Embayment has many structural complications that have been investigated in this research. In the studied area, the Gachsaran Formation (caprock of the Asmari reservoir) is a detachment horizon. In the Dezful Embayment, (thrust) faults are divided into two categories as per their locations. (i) Faults in the outcrop part of the studied area and often continue to the top of the Member 4 of Gachsaran, such as the Chamshir, Dezh Soleyman, F3 and F4 faults. The faults that have originated from the top of the Asmari Formation to the incompetent Triassic Formations. These deep faults have originated from older detachment surfaces. The same has been observed from the other parts of the Dezful Embayment (Sherkati & Letouzey, 2004; Tavakolian et al., 2022). In general, thicker incompetent units can create more effective structural detachments (Bahroudi & Koyi, 2004; Simpson, 2009; Stewart, 1996). The changes in the thickness of the Gachsaran Formation along the reflection seismic sections indicate the performance of the Gachsaran Formation as a detachment horizon.

Salt diapirs have flowed as salt masses, in a ductile style (Mukherjee et al., 2010), and they seem to penetrate the sedimentary cover disharmonically (Jackson & Talbot,

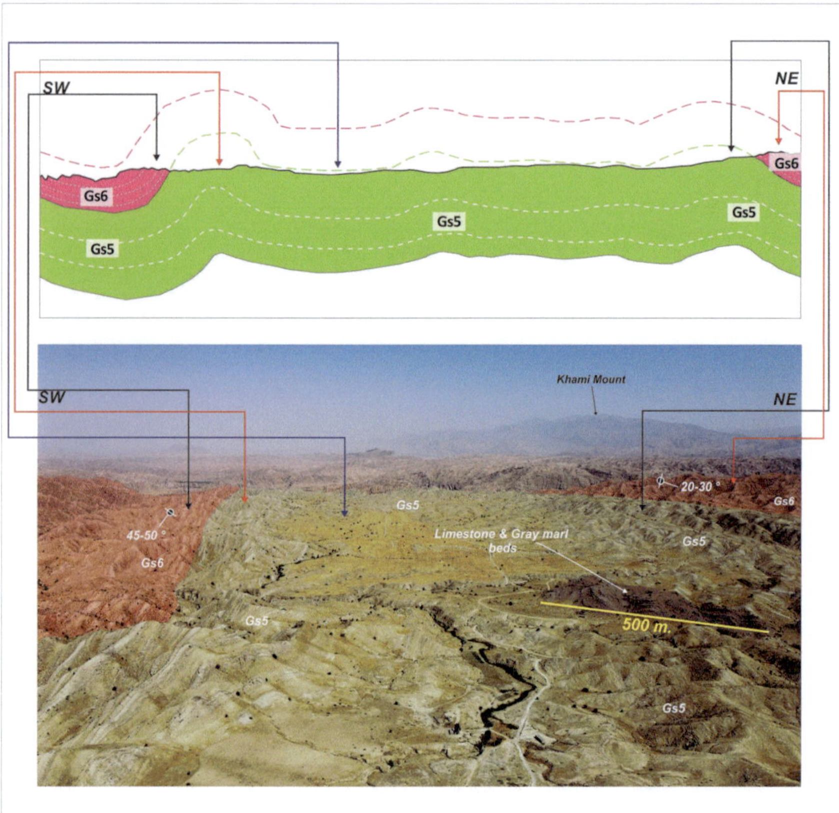

Fig. 18 Folded terrain in the Se-Qanat area in the study area, view towards the northwest. The anticline located at the end left (southwest) of the profile corresponds to the anticline A68, and the one located at the extreme right (northeast) of the profile corresponds to the anticline A66. Gs.5 and 6: Members 5 and 6 of the Gachsaran Formation

1991). Between the two subsurface anticlines of Gachsaran and Garangan, due to compression and the movement of the Gachsaran Formation, the Gach Haji anticline is formed. This anticline is not formed in Formations older than Gachsaran. The Gachsaran Formation in the Gach Haji anticline has been squeezed and Member 6 is raised and therefore crops out. The anticline shows a salt wall structure affected by the Dezh Soleyman thrust fault. The presence of the Cardrigon and Bane Qalat synclines as mini basins on both sides of the Gach Haji anticline has caused insufficient space for salt emplacement of the Gachsaran Formation between these two synclines. This caused upward salt movement along the fault. Thus Member 6 of the Gachsaran Formation moves outward as a salt wall.

Based on the interpretation of seismic profiles, an outcrop of Members 5–7 of the Gachsaran Formation in the Chamshir area was observed at the surface. In the Chamshir, tectonics of the region folded, uplifted and cropped out Members 5 and 6

Fig. 19 Northwest limb of the anticline A12 in the north part of the study area, view towards the north. Gs.5 and Gs.6: Members 5 and 6 of the Gachsaran Formation

of the Gachsaran Formation. Folding in the study area produced a series of anticlines and synclines. Towards the NE, younger portions of the Formation (Members 6, 7 and partly 5) have been removed.

Member 4 of the Gachsaran Formation had insignificant ductility and created a detachment level in the Gachsaran Formation, especially in the synclinal areas. Also, evidence of a large salt wall structure can be seen in seismic profiles in the Gach Haji anticline. This structure is formed by faulting and the rising of Members 5 and 6 salts of the Gachsaran Formation. Member 6 of the Gachsaran Formation was exposed.

In the NE of the Cardrigon syncline, the Gachsaran Formation has been folded and eroded due to shortening. This folding has appeared in a wide area, which includes

Fig. 20 Close-up view of a small-scale (normal?) fault at the NW part of the study area

several large-scale anticlines and synclines and hundreds of disharmonic and small-scale folds, while the Gachsaran Formation in the Garangan anticline, like many anticlines of the Dezful Embayment, is deformed to folds of different shapes.

The faults seen in the exposed part of the studied area often continue to the top of the Member 4 (Gs.4) of the Gachsaran Formation, such as the Dezh Soleyman, Chamshir and along the F4 and F3 faults. A second category of faults developed from the top of the Asmari Formation to the approximate depth of the Triassic-Jurassic Formations. In other words, a thick layer of the evaporate Member of the Gachsaran Formation (Gs.4), as a detachment layer, has caused the structural style to be separated in the upper and lower layers (Fig. 21).

Based on the interpretation of seismic profiles and their correlation to the geological map and field data of the region, there is a good match between the surface and subsurface information in the Chamshir area, especially in relation to the spreading of different Members of the Gachsaran Formation at the surface (Fig. 22).

11 Conclusions

The Gachsaran Formation is the main detachment horizon in the Foreland basin of Zagros. As a detachment horizon, this Formation separates structural features and folding style in its overlaying and underlying units. The presence of a thick layer of Gachsaran Formation has affected the structural style in the Chamshir area. In the Chamshir area, Members Gs.5-7 of the Gachsaran Formation are exposed at the surface. The salt Members of the Gachsaran Formation are ductile and deformed to several secondary folds. Based on surface and subsurface data, the area has two

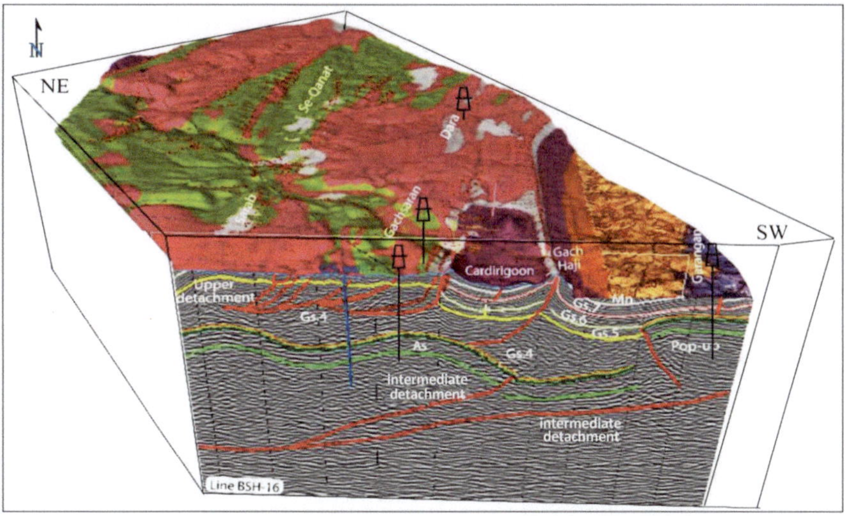

Fig. 21 3D model of the Chamshir area based on surface and subsurface data using Petrel software (v. 2014). Two categories of faults are shown—(1) those extending from the surface to the top of Member 4 (Gs.4) of the Gachsaran Formation and (2) those developed from the top of the Asmari Formation to the approximate depth of the Triassic-Jurassic Formations. Gs.4 to Gs.7 are Members 4 to 7 of the Gachsaran Formation, respectively

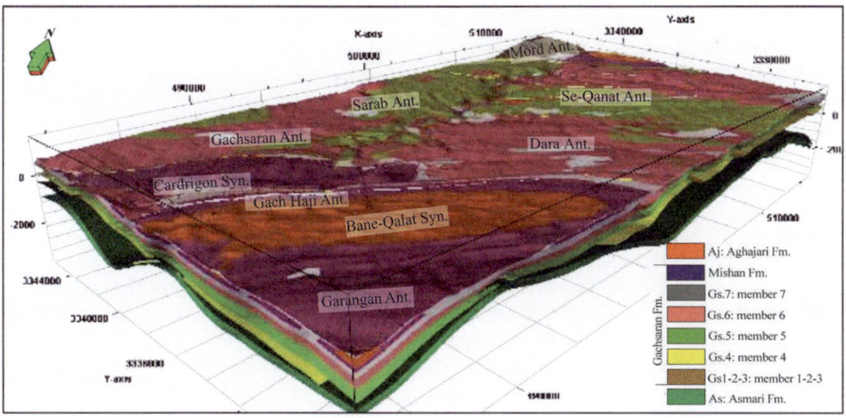

Fig. 22 3D model of the studied area in the Chamshira region from the base of the Asmari Formation to the surface

types of faults. The faults in the exposed part of the studied area often continue to the top of Member 4 (Gs.4) of the Gachsaran Formation, such as the Dezh Soleyman, Chamshir, F3 and F4 faults. The second category of faults developed atop the Asmari Formation to the approximate depth of the Triassic-Jurassic Formations. In other words, a thick layer of the evaporate member of the Gachsaran Formation (Gs.4), as a main detachment layer, has caused the structural style to be separated in the upper and lower layers in the Chamshir area.

Acknowledgements The authors would like to thank the proofreaders and the publisher. The authors are especially indebted to the National Iranian South Oil Company (NISOC) for providing the wells data and seismic profiles for this study. This research was supported by the National Iranian South Oil Company (NISOC) funding which is gratefully acknowledged. Mukherjee (2026) summarized this work.

References

Alavi, M. (1994). Tectonics of the Zagros orogenic belt of Iran: New data and interpretations. *Tectonophysics, 229*(3–4), 211–238.

Aqrawi, A. A. M., Goff, J. C., Horbury, A. D., & Sadooni, F. N. (2010a). *The petroleum geology of Iraq* (pp. 1–424). Scientific Press Ltd.

Aqrawi, A. A. M., Mahdi, T. A., Sherwani, G. H., & Horbury, A. D. (2010b). Characterisation of the Mid-Cretaceous mishrif reservoir of the southern Mesopotamian Basin, Iraq. In *American Association of petroleum geologists conference and exhibition* (Vol. 7, pp. 7–10).

Bahroudi, A., & Koyi, H. A. (2004). Tectono-sedimentary framework of the Gachsaran Formation in the Zagros foreland basin. *Marine and Petroleum Geology, 21*(10), 1295–1310.

Bordenave, M. L., & Burwood, R. (1990). Source rock distribution and maturation in the Zagros orogenic belt: Provenance of the Asmari and Bangestan reservoir oil accumulations. *Organic Geochemistry, 16*, 369–387.

Cooper, M., Brealey, C., Fermor, P., Green, R., & Morrison, M. (2004). Structural models of subsurface thrust-related folds in the foothills of British Columbia, Case studies of sidetracked gas wells. In K. R. McClay (Eds.), *Thrust tectonics and hydrocarbon systems: AAPG Memoir* (Vol. 82, pp. 579–597).

Cotton, J. T., & Koyi, H. A. (2000). Modeling of thrust fronts above ductile and frictional detachments: Application to structures in the Salt Range and Potwar Plateau, Pakistan. *Geological Society of America Bulletin, 112*(3), 351–363.

Davis, D. M., & Engelder, T. (1985). The role of salt in fold-and-thrust belts. *Tectonophysics, 119*(1–4), 67–88.

Farzipour-Saein, A., Yassaghi, A., Sherkati, S., & Koyi, H. (2009). Basin evolution of the Lurestan region in the Zagros fold-and-thrust belt, Iran. *Journal of Petroleum Geology, 32*(1), 5–19.

Jackson, M. P. A., & Talbot, C. J. (1991). *A Glmwry of 'Su/r tectonics*. Bureau of Economic Geology, University of Texas at Austin, Geological Circular 91-4.

Misra, A. A., & Mukherjee, S. (2015). *Tectonic inheritance in continental rifts and passive margins* (pp. 1–86). Springer International Publishing.

Mukherjee, S. (2026). Introduction to "Structural Geology & Tectonics". In S. Mukherjee (Ed.), *Structural Geology & Tectonics*. Springer. ISBN: 978-981-95-4743-2.

Mukherjee, S., Talbot, C. J., & Koyi, H. A. (2010). Viscosity estimates of salt in the Hormuz and Namakdan salt diapirs, Persian Gulf. *Geological Magazine, 147*, 497–507.

O'Brien, C. A. E. (1957). Salt diapirism in south Persia. *Geologie en Mijnbouw, 19*(9), 357–376.

Razavi Pash, R., Sarkarinejad, K., Sherkati, S., & Motamedi, H. (2021a). Analogue model of the Bala Rud Fault, Zagros: An oblique basement ramp in a fold-and-thrust belt. *International Journal of Earth Sciences, 110*, 741–755.

Razavi Pash, R., Sarkarinejad, K., Zarehparvar Ghoochaninejad, H., & Motamedi, H. (2021b). Application of 3D structural modeling to analyze the structural geometry and kinematic evolution: A case study from Lab-e-Safid and Qale Nar subsurface anticlines in the Northern Dezful embayment, Iran. *Geotectonics, 55*, 261–272.

Razavi Pash, R., Seraj, M., Mukherjee, S., & Radmehr, A. (2023). Structural relationship between subsurface oil fields in the North Dezful Embayment: Qaleh Nar, Lower and Upper Balarud Anticlines (central Zagros, Iran). *Bulletin of the Mineral Research and Exploration*. https://doi.org/10.19111/bulletinofmre.1344433

Sarkarinejad, K., & Ghanbarian, M. A. (2014). The Zagros hinterland fold-and-thrust belt in sequence thrusting, Iran. *Journal of Asian Earth Sciences, 85*, 66–79.

Sepehr, M., & Cosgrove, J. W. (2004). Structural framework of the Zagros fold–thrust belt, Iran. *Marine and Petroleum Geology, 21*(7), 829–843.

Sepehr, M., Cosgrove, J., & Moieni, M. (2006). The impact of cover rock rheology on the style of folding in the Zagros fold-thrust belt. *Tectonophysics, 427*(1–4), 265–281.

Sherkati, S., & Letouzey, J. (2004). Variation of structural style and basin evolution in the central Zagros (Izeh zone and Dezful Embayment), Iran. *Marine and Petroleum Geology, 21*(5), 535–554.

Sherkati, S., Molinaro, M., de Lamotte, D. F., & Letouzey, J. (2005). Detachment folding in the Central and Eastern Zagros fold-belt (Iran): Salt mobility, multiple detachments and late basement control. *Journal of Structural Geology, 27*(9), 1680–1696.

Sherkati, S., Letouzey, J., & Frizon de Lamotte, D. (2006). Central Zagros fold-thrust belt (Iran): New insights from seismic data, field observation, and sandbox modeling. *Tectonics, 25*, TC 4007.

Simpson, G. D. H. (2009). Mechanical modeling of folding versus faulting in brittleductile wedges. *Journal of Structural Geology, 31*, 369–381.

Soleimani, B., & Bahadori, A. (2015). The Miocene Gachsaran formation evaporite cap rock, Zeloi oilfield, SW Iran. *Carbonates and Evaporites, 30*, 287–306.

Spratt, D. A., Dixon, J. M., & Beattie, E. T. (2004). Changes in structural style controlled by lithofacies contrast across transverse carbonate bank margins—Canadian Rocky Mountains and scaled physical models.

Stewart, S. A. (1996). Influence of detachment layer thickness on style of thin-skinned shortening. *Journal of Structural Geology, 18*(10), 1271–1274.

Stocklin, J. (1968). Structural history and tectonics of Iran; a review. *American Association of Petroleum Geologists Bulletin, 52*(7), 1229–1258.

Tavakolian, E., & Razavi Pash, R. (2022). The Impact of Gachsaran evaporate formation on the structural style of folds: Bibi-Hakimeh Anticline, Zagros Belt (Iran). *Geotectonics, 56*(5), 654–662.

Tavakolian, I., Yassaghi, A., & Najafi, M. (2022). Structural style in the south Dezful embayment, SW Iran: Combined influence of the Zagros frontal fault system and the detachment in the Miocene Gachsaran formation. *Journal of Petroleum Geology, 45*(3), 303–323.

Vergés, J., Saura, E., Casciello, E., Fernandez, M., Villaseñor, A., Jimenez-Munt, I., & García-Castellanos, D. (2011). Crustal-scale cross-sections across the NW Zagros belt: Implications for the Arabian margin reconstruction. *Geological Magazine, 148*(5–6), 739–761.

Folding of Post-bijawar Dykes of Central India Around Hirapur

Gaurav K. Singh, Rajendra K. Trivedi, Ankit K. Jhariya, and Ankit Jain

Abstract The dykes form a significant marker to comprehend geodynamics provided the post emplacement structural modifications are considered. Most works on dykes are mainly confined to their petrology and geochemistry without much emphasis placed on their deformation. The dominant trend of dykes documented so far in Bundelkhand craton is NW-SE. In the present contribution, we report post-Bijawar dykes and their folding for the first time. Dyke planes and bedding planes of host rock have been measured at three locations along with 18 samples thin sectioned for petrography. It has been found that the dykes have suffered polyphase folding and are slightly metamorphosed. This folding has modified the dyke geometry and that NW-SE trend is attributed to deformation driven orientation.

Keyword Post-Bijawar dykes; Bundelkhand granite; Deformation

1 Introduction

Granite-greenstone belt comprises a significant part of basement rocks in the Bundelkhand craton. It consists of Archean granitoids emplaced in three pulses between 2492 ± 10 to 2515 ± 7 Ma (Gopalan et al., 1990; Wiedenbeck & Goswami, 1994; Wiedenbeck et al., 1996; Mondal et al., 1997, 1998, 2002; Malviya et al., 2004, 2006) as a major phase of felsic magmatism. These granitoides enclose various enclaves of Archean gneisses, amphibolites, ultramafics, BIF, tonalite-trondhjhemite gneisses, marbles, calc-silicate rocks, fuchsite quartzites and other metasediments. This granite-greenstone ensemble is intersected by many minor intrusions including pegmatitic veins, dykes and felsic units (Rao et al., 2005; Pati et al., 2008; Pradhan et al., 2012). The veins are dominantly aligned along ENE-WSW direction whereas the majority of dykes are oriented NW-SE with trends swinging from ENE-WSW

G. K. Singh (✉) · R. K. Trivedi · A. K. Jhariya · A. Jain
Department of Applied Geology, Dr. HarisinghGour Vishwavidyalaya, Sagar, Madhya Pradesh, India
e-mail: gauravks@dhsgsu.edu.in

S. Mukherjee (ed.), *Structural Geology and Tectonics*,
https://doi.org/10.1007/978-981-95-4743-2_15

to NE-SW. These dykes and ductile shear zones in the host granitoides represent a prominent tectono-thermal event in the Archean Bundelkhand craton (Pati, 1999; Senthiappan, 1981) around Palaeoproterozoic to Mesoproterozoic time. In this craton, volcano-sedimentary succession is found along three main ENE-WSW trending tracts viz. Gwalior, Bijawar and Sonrai. No significant report of dyke intrusion however, has so far been reported from the sedimentary rocks of these basins.

Here, we present the first report of the yet undocumented Precambrian mafic–ultramafic/felsic dykes intruding Bijawar sedimentaries exposed around Hirapur, Central India. These dykes are of three different types as identified in field namely felsic, basic and doleritic, which intrude dolomite and fine ferruginous clastics. In all these sections, it is found that both host rock and the intrusion are folded and therefore, gives different trends besides their principal NW-SE orientation. These dykes are affected by buckle folding which explains why the dykes in this and in previous reports show variable trends. The work also offers an insight in to post-Bijawar activity of dyke intrusion and subsequent deformation due to which the original trend seems to have been considerably modified.

2 Geology

Hirapur is a town in Sagar district, Madhya Pradesh (Fig. 1) and is ~80 km northeast of Sagar city toward Kanpur. The area essentially forms a part of the southern fringe of the Bundelkhand craton, which comprises of older TTG gneisses and supracrustal enclaves including Banded Iron Formation (BIF), fuchsite quartzite, dunite-peridotite pods exposed around Girar, Madaora, and Lalitpur district as Mehroni group is considered to be an event ~3200–3300 Ma (Ramakrishnan & Vaidyanandan, 2008). The Bundelkhand Granite is younger than the entire aforesaid ensemble and is related to the granite plutonism contributing to cratonization phenomena in Bundelkhand craton ~2500 Ma and can be related to granite bloom (Davis & Bleeker, 1999). It forms a basement for the Palaeoproterozoic and Mesoproterozoic sedimentary successions viz. Bijawar Group and Vindhyan Supergroup, respectively. The granites therefore form the oldest litho-unit in the study area because no part of Mehroni Group of rocks has been found in the study area during mapping. The contact between granite and Bijawar Group is a non-conformity as seen near Kawar and is defined by a quartz pebble conglomerate (QPC) followed by a succession of mafic lavas known as Bhusar Basalt. It is followed by a horizon of dolomite called as Bajno Dolomite which is stromatolitic at some places. These rocks show slumping as well as intra-formational conglomerate with fragments of dolomitic chert as seen near Bheemkund. Bajno dolomite is intruded by a basaltic sill called as Dargawan Trap that has been dated as ~1691 Ma (Haldar & Ghosh, 2000) giving Paleoproterozoic age of the Bijawar Group. The exact nature of contact between dolomite and the sill is not yet fully established which is why the two has so far been treated as

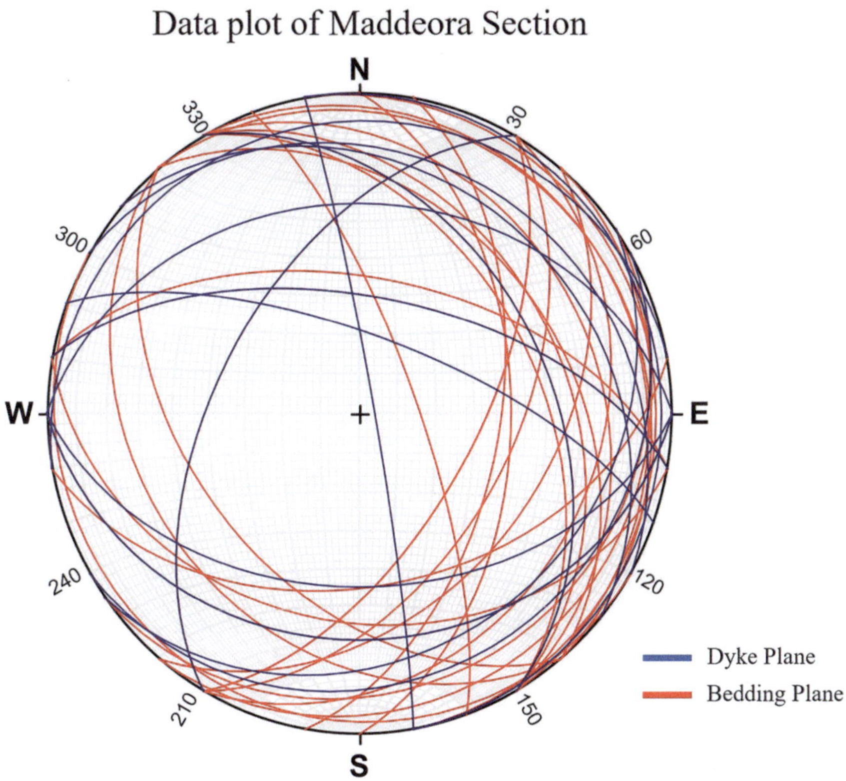

Fig. 1 The geographical location map of Sagar and adjoining areas showing accessibility to Hirapur

separate stratigraphic units. The dolomites are overlain by a succession of ferruginous sandstone known locally as Pukhra Sandstone grading into chert breccia also called as Malehra Chert Breccia. The exposures of sandstone and chert breccia are described as Karri Formation while Hirapur Phosphorite includes ferruginous, carbonaceous and phosphatic shales and sandstones. The observed rock types in the study area are dolomite, ferruginous sandstone, chert breccia and phosphoritic rocks with sandstone-shale intercalations (Fig. 2). The Bijawar sediments have higher dips where they are influenced by folds and faults otherwise it is usually ~15–20°.

3 Field Description of Dykes

Some characteristic physical properties have been used to differentiate dykes from the surrounding rocks including color, composition, tenacity and resistance to erosion. Post-Bijawar intrusives are well exposed mainly at three locations viz. Maddeora, Luhani and Taura (Fig. 2).

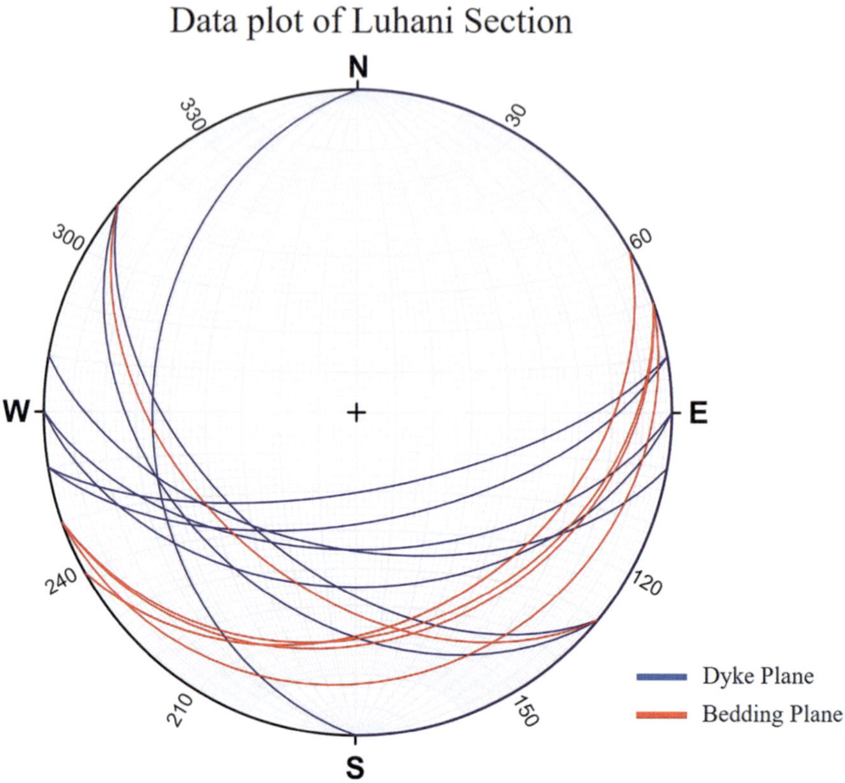

Fig. 2 The geological map of Hirapur marked with Maddeora, Luhani and Taura as study locations (modified after Roday et. al., 1989)

3.1 Maddeora Section

Maddeora section is located near Maddeora High School ~77 km NE of Sagar town toward the Chattarpur road. The hillock (24° 20′54″N–79° 11′00″ E) shows dolomites interlaminated with chert resting over the basement granite. Bajno dolomites have caught up patches of granite at the contact with granite (Figs. 3 and 4). A NW-SE trending felsic dyke intrudes Bajno dolomite (Fig. 5a, b) besides numerous quartz veins with varying trends. It also shows vesicles filled with secondary materials (Fig. 5c). The measurements of bedding and dyke planes give NW-SE to NE-SW trends indicating folded nature of the dyke and dolomites (Fig. 6). This felsic dyke can be traces all along the contact of granite and Bijawar sediments for several km.

Fig. 5a Aplitic dyke intrusion within dolomite observed ~1 km southeast of Maddeora. Scale bar: 50 cm b Dolomite intruded by aplitic dyke seen ~2.5 kms east of the Maddeora High School. Scale bar: 15 cm c Filled vesicles in aplitic rock found 1.5 kms southeast of the Maddeora High School. Scale bar: 10 cm d Doleritic rock having intrusive relation with ironstone found along the first order stream south

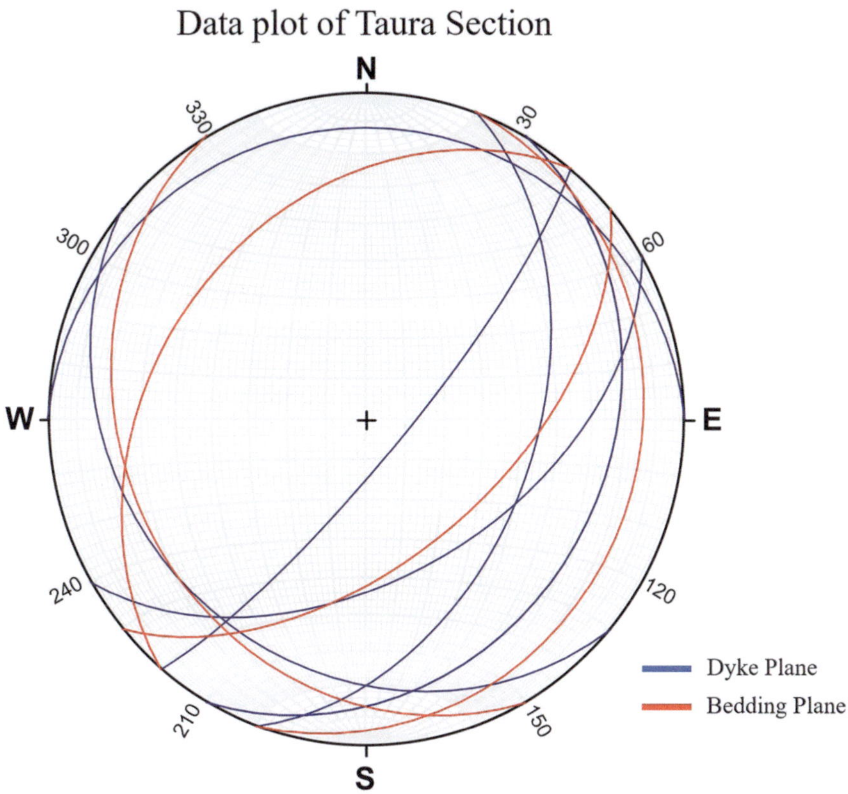

Fig. 3 Caught up patches of granite (arrow) within dolomite exposed near Maddeora. Scale bar is 20 cm

Fig. 4 Granite-Bijawar contact exposed near Luhani. The scale bar is 30 cm

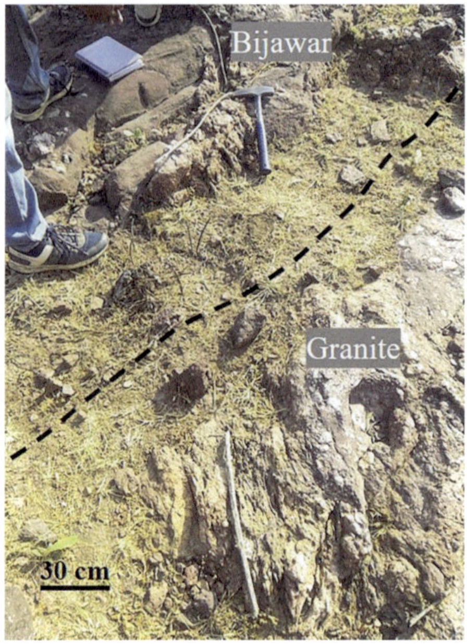

Fig. 5 **a** Aplitic dyke intrusion within dolomite observed ~1 km southeast of Maddeora. Scale bar: 50 cm **b** Dolomite intruded by aplitic dyke seen ~2.5 kms east of the Maddeora High School. Scale bar: 15 cm **c** Filled vesicles in aplitic rock found 1.5 kms southeast of the Maddeora High School. Scale bar: 10 cm **d** Doleritic rock having intrusive relation with ironstone found along the first order stream south of Taura. Scale bar: 20 cm **e** Dyke showing hook-shaped refolding near Luhani. Scale bar: 50 cm **f** S-fold seen in basic dyke near Luhani. Scale bar: 20 cm

of Taura. Scale bar: 20 cm e Dyke showing hook-shaped refolding near Luhani. Scale bar: 50 cm f S-fold seen in basic dyke near Luhani. Scale bar: 20 cm

3.2 Taura Section

A doleritic rock (Fig. 5d) intruding ironstone is exposed (24°21′ 21″N–79° 14′0.5″ E) along NW-SE deflection of a first order stream trending N-S. This stream merges with the easterly flowing Kalidhar River almost at right angle which indicates that the drainage is tectonically influenced. Along the same nala section, a major contact between Vindhyan sandstone and ironstone of Bijawar Group is exposed. No effect of this intrusion is found in the Vindhyan sandstone which means that the intrusion is Pre-Vindhyan. The measurement of the dyke planes gives many orientations swinging from NW-SE to NE-SW indicating that the dyke is folded (Fig. 7).

3.3 Luhani Section

The section is located SW of Luhani village (24°22′ 54.6″N–79° 09′6.7″ E) where a contact between granite and Bijawar rocks is exposed (Fig. 4). Dolomites are overlain by Karri sandstone with minor occurrence of laminated chert. Two dykes are intruded

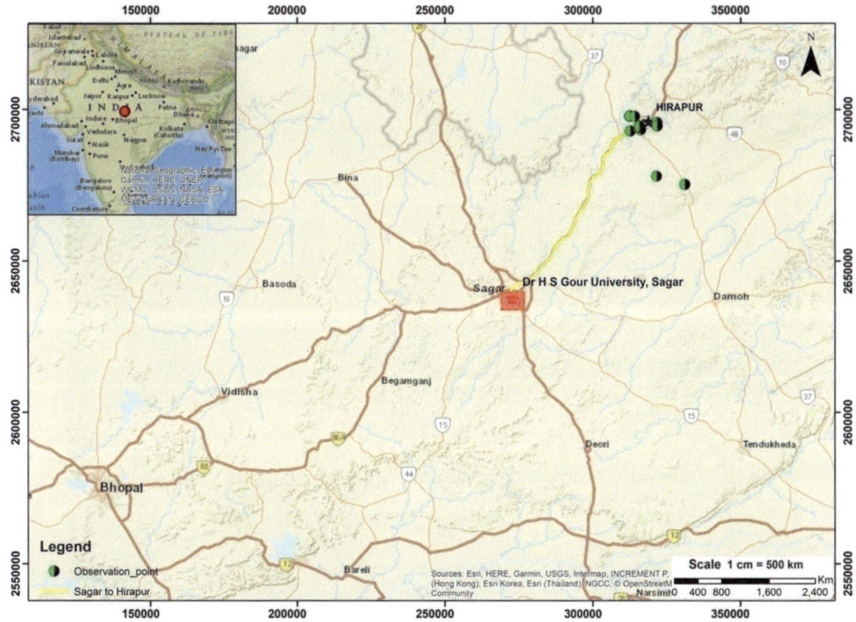

Fig. 6 The felsic dykes of Maddeora section plotted as great circles on equal area stereonet along with bedding planes of host sediment. It shows folds plunging mostly toward NE, SE and SW

in to dolomites and chert and are folded (Fig. 5e, f). It also gives orientation of dykes (Fig. 8) ranging from NE-SW to NW-SE similar to Maddeora and Taura sections indicating a folded nature of dykes.

4 Petrography

Post-Bijawar dykes around Hirapur are mainly of three principal types namely felsic, metadoleritic and metabasic. The felsic intrusive rocks are fine grained with quartz as the most dominant component (Fig. 9a) along with subordinate feldspars and mica. Pyrite and magnetite are observed in the reflected light and (Fig. 9b) occur as accessory minerals. The overall composition is granitic but since the rock is dominantly fine grained, therefore can be petrographically termed as aplite. Another variant of dykes around Hirapur is the metabasic rocks of Luhani section. It has hornblende as the most prominent mineral (Figs. 9c, d) along with chlorite and insignificant feldspar. Effects of metamorphism are seen as foliation and syn-kinematic rotation of mineral grains. Intrusive rock of the Taura section are doleritic with metamorphic imprints, however, the melt supported fabric is still discernible as sub-ophitic texture. These rocks are medium to coarse grained (Fig. 9e, f) with feldspars, amphiboles and chlorite. It seems that all the pyroxenes have been converted in to amphiboles.

Fig. 7 Doleritic dyke planes exposed south of Taura plotted on equal area net together with bedding planes of ferruginous sandstones in which they are intruded. Folds show plunge toward NE, SE and SW

5 Discussion and Conclusions

Palaeoproperozoic Bijawar sediments are intruded by dykes of three principal compositions- felsic, metadoleritic and metabasic. Due to the lack of geochronological data, their emplacement timing is not known. Dykes along with their host

Fig. 8 Planes of the dyke of Luhani section plotted on equal area net and that of host sediments. Folds plunge toward NE, SE and SW

rocks have suffered polyphase folding. First generation folds are tight to nearly isoclinal (Fig. 5f), which was refolded subsequently producing hook-shaped interference (Fig. 5e). The third folding event has produced dome-basin structure affecting the overall outcrop pattern of the area. It seems that the mineral transformation from pyroxenes to amphiboles would have enabled the dykes to be relatively incompetent and therefore, more prone to folding (Ramsay & Huber, 1987, pp. 401). This effect of folding of dykes has led to the modification of the original orientation in which they were emplaced. It can therefore be concluded that the dominant NW-SE attitude

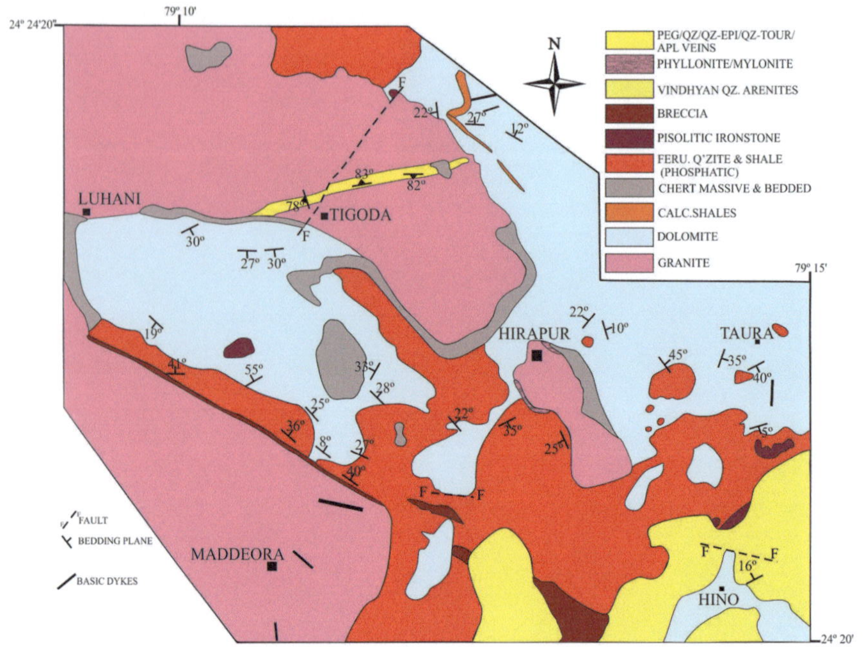

Fig. 9 **a** Maddeora aplite under cross-polarized light showing quartz feldspar and opaque minerals. **b** Aplite under reflected light with magnetite and pyrite. **c** and **d** altered amphiboles of Luhani section. **e** doleritic rock south of Taura showing altered feldspar. **f** Altered feldspars with secondary minerals

or their variations proposed so far (Rao et al., 2005; Pati et al., 2008; Pradhan et al., 2012) requires to be re-investigated especially in terms of folding.

Acknowledgements We thank Prof. Soumyajit Mukherjee (IIT Bombay) for inviting, handling and reviewing this work. Dr. Vibha Katiyar and Mr. Ansar Iqbal are thanked for assisting in preparing the maps. Mukherjee (2026) summarized this work.

References

Davis, W. J., & Bleeker, W. (1999). Timing of plutonism, deformation and metamorphism in the yellowknife domain, Slave Province, Canada. *Canadian Journal of Earth Sciences, 36*, 1–38.

Gopalan, K., MacDougall, J. D., Roy, A. B., & Murali, A. K. (1990). Sm–Nd evidence for 3.3 Gaold rocks in Rajasthan, Northwestern India. *Precambrian Research, 48*, 287–297.

Haldar, D., & Ghosh, R. N. (2000). Eruption of Bijawar lava: An example of Precambrian volcanicity under stable cratonic conditions. *Geological Survey of India Special Publication, 57*, 151–170.

Malviya, V. P., Arima, M., Pati, J. K., & Kaneko, Y. (2004). First report of metamorphosed basaltic pillow lava from central part of Bundelkhand craton, India: An Island arcsetting of possible Late Archaean age. *Gondwana Research, 7*, 1338–1340.

Malviya, V. P., Arima, M., Pati, J. K., & Kaneko, Y. (2006). Petrology and geochemistry ofmetamorphosed basaltic pillow lava and basaltic komatiite in Mauranipur area: Subduction related volcanism in Archean Bundelkhand craton, Central India. *Journal of Mineralogical and Petrological Sciences, 101*, 199–217.

Mondal, M. E. A., Deomurari, M. P., Goswami, J. N., Rahman, A., & Sharma, K. K. (1997). 207Pb/206Pb zircon ages of samples from Bundelkhand massif, Central India. In *Abstract in International Conference on Isotopes in Solar System* (pp. 80–81).

Mondal, M. E. A., Sharma, K. K., Rahman, A., & Goswami, J. N. (1998). Ion microprobe207Pb/206Pb zircon ages for gneiss–granitoid rocks from Bundelkhand massif: Evidence for Archean components. *Current Science, 74*, 70–75.

Mondal, M. E. A., Goswami, J. N., Deomurari, M. P., & Sharma, K. K. (2002). Ion microprobe 207Pb/206Pb ages of zircons from the Bundelkhand Massif, northern India: Implications for crustal evolution of the Bundelkhand–Aravalli supercontinent. *Precambrian Research, 117*, 85–100.

Mukherjee, S. (2026). Introduction to "Structural Geology & Tectonics". In S. Mukherjee (Ed.), *Structural Geology & Tectonics*. Springer. ISBN: 978-981-95-4743-2.

Pati, J. K. (1999). Study of granitoid mylonites and reef/vein quartz in parts of Bundelkhand Granitoid Complex(BGC). *Records of the Geological Survey of India, 131*(8), 95–96.

Ramakrishnan, M., & Vaidyanandan, R. (2008). Geology of India. *Geological Society of India., 1*, 254–255.

Ramsay, J. G., & Huber, M. I. (1987). Modern structural geology, volume 2: Folds and fractures. Academic Press, London.

Roday, P. P., Chourasia, L. P., & Shailendra, C. (1989). Structural analysis of shear zones in basement granite and their relationship with folding, shearing and faulting in the cover sediments near Hirapur, district Sagar, Central India. *Journal of Geological Society of India, 33*, 387–408.

Senthiappan, M. (1981). Geology of the area along the Raksa Shear zone, Jhansi district, Uttar Pradesh. *Special Publication, Geological Survey of India, 3*, 73–76.

Wiedenbeck, M., & Goswami, J. N. (1994). High precision 207Pb/206Pb zircon geochronology using a small ion probe. *Geochimica Cosmochimica Acta, 58*, 2135–2141.

Wiedenbeck, M., Goswami, J. N., & Roy, A. B. (1996). Stabilization of the Aravalli Craton of northwestern India at 2.5 Ga: An ion microprobe zircon study. *Chemical Geology, 129*, 325–340.

An Updated Geological Map of Hirapur and Adjoining Areas of Madhya Pradesh, Central India

Gaurav K. Singh, Ankit Jain, and Rajendra K. Trivedi

Abstract Hirapur is mainly known in India and elsewhere for its phosphorite reserves preserved exclusively in fine ferruginous clastic facies of Palaeoproterozoic age. These rocks belong to the Bijawar Group and have been mapped several times by the Geological Survey of India and others along with the granitic basement. In some of the maps, granites have been shown to be intersected by basic dykes, yet post-Bijawar dykes have not been mapped so far. Here, we present an updated map of the Hirapur and the adjoining areas where the post-Bijawar dykes are mapped.

Keywords Geologic map · Dykes · Bijawar group · Geologic mapping

1 Introduction

The rocks of the Bijawar basin exposed around Hirapur represent the southern fringe of Bundelkhand craton and the SW closure of the basin. This ENE-WSW trending intra-cratonic isolated volcano-sedimentary succession is sandwiched between the Bundelkhand Granite in the north and the Vindhyan basin toward the south. The Bijawar sediments have been primarily investigated by stalwarts of Geological Survey of India (GSI) including H. B. Medlicott (1959), A. F. Wilson (1873–1877), G. Mani (1969) and S. M. Mathur (1970). The geological map was prepared by Kumar (1984) especially pertaining to the Phosphorite investigation and also by Kumar et al. (1990). The stratigraphy of Bijawar succession is worked out by R. N. Srivastava and his team given in GSI (2004). Other significant contributions include Pant and Banerjee (1990), Haldar and Ghosh (2000), Rajarajan (1978) and Ramakrishnan and Vaidyanadhan (2008). The structural map of Hirapur and adjacent localities has been prepared by Roday et al. (1989). The map by Sharma and Rahman (2000) depicts the orientation of quartz veins, dykes, etc. in the Bundelkhand craton.

G. K. Singh (✉) · A. Jain · R. K. Trivedi
Department of Applied Geology, Dr. HarisinghGour Vishwavidyalaya, Sagar, Madhya Pradesh, India
e-mail: gauravks@dhsgsu.edu.in

S. Mukherjee (ed.), *Structural Geology and Tectonics*,
https://doi.org/10.1007/978-981-95-4743-2_16

Though the older pre-Bijawar dykes have been documented in some of the works mentioned above, no map or any account is found related to the post-Bijawar dykes. In this work, dykes of felsic and basic composition intersecting dolomites and ferruginous clastic rocks of Bijawar Group have been mapped which updates the pre-existing geological map of the area.

2 Geology

Geologically, the area is a Precambrian terrain with Late Archaean Bundelkhand Granite forming the oldest lithology known so far (~2550 Ma, Crawford, 1970). It is intruded by dykes of felsic to basic composition. The granite forms the basement for the Palaeoproterozoic Bijawar basin (Haldar & Ghosh, 2000) and shares a tectonic contact with them in the study area. Locally, the Bijawar succession can be differentiated in to lower carbonate facies represented by Bajno dolomite interlaminated with chert and upper fine ferruginous clastic facies referred generally as the Karri Formation. This Formation constitutes ferruginous sandstones, shales, chert breccia, etc. The rocks of the Bijawar Group are followed by Vindhyan succession exposed toward south-east. They include basal sandstone unit overlain by olive green shale and limestone.

3 Post-Bijawar Dykes

So far, dykes were only described from the Bundelkhand Granite whereas the Bijawar succession is also intersected by felsic and basic dykes (Fig. 1). These dykes have been encountered while mapping the Granite-Bijawar contact from Mardewrato Luhani. Along these tracts, the dykes are mainly felsic in composition as revealed from field observation. They intrude both the Bajno Dolomite and rocks of Karri Formation as well. Since the dykes are folded along with their host sediments, they show variable orientations (Fig. 1) and hence, do not represent the true zone in which they were emplaced. Another short stretch of a dyke intruding ironstones of the Karri Formation having basic composition is exposed close to the Kalidhar River (Fig. 1). Additionally, pre-Bijawar dykes within granites have also been marked at several new locations.

4 Conclusions

The following conclusions can be drawn from the recent mapping of the Pre-Cambrian rocks around Hirapur:

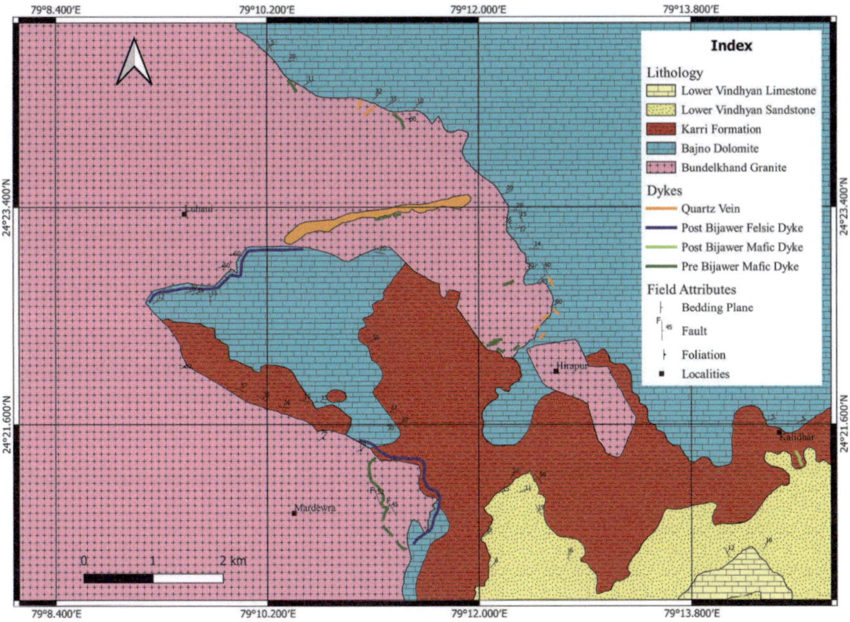

Fig. 1 The modified geological map of Hirapur and adjoining areas added with post-Bijawar dykes. They intrude both Bajno Dolomite and rocks of Karri Formation

- The dykes are not only confined to the granites as previously mapped but also affect the Bijawar succession and therefore, are post-Bijawar in nature.
- They range in composition from felsic to basic and intrude the rocks of Bajno Dolomite and Karri Formation as well.
- The dykes are folded together with the host sediments and hence, show variable trends. It can therefore be said that they were certainly not emplaced in their present trend.

Acknowledgements We thank Prof. Soumyajit Mukherjee (IIT Bombay) for inviting, handling and reviewing this work. Mukherjee (2026) summarized this chapter.

References

GSI (2004). Manual of geology of India. v.1: Precambrian. Pt IV (N & NW part of penninsula) (4th ed., p. 257). Geological Survey of India Special Publication No. 77.

Haldar, D., & Ghosh, R. N. (2000). *Eruption of Bijawar lava: an example of Precambrian volcanicity under stable cratonic conditions* (pp. 151–170). Geological Survey of India Special Publication No. 57.

Kumar, B. (1984). *Final Report on Hirapur Phosphorite Investigation, Sagar and Chhattarpur Districts, Madhya Pradesh (1977–1983).* Geological Survey of India Project M.P. II.

Kumar, B., Srivastava, R. K., Jha, D. K., Pant, N. C., & Bhandari, B. K. (1990). A revised stratigraphy of the type area of Bijawar Group in central India. *Indian Minerals, 44*, 303–314.

Mukherjee, S. (2026). Introduction to "Structural Geology & Tectonics". In S. Mukherjee (Ed.), *Structural Geology & Tectonics*. Springer. ISBN: 978-981-95-4743-2.

Pant, N. C., & Banerjee, D. M. (1990). *Patterns of sedimentation in the type Bijawar basin of central India* (pp. 156–166). Geological Survey of India Special Publication No. 28.

Rajarajan, K., (1978). Geology of Sagar district and Western part of Damoh district, M.P. *Memoirs of the Geological Survey of India, 109*, 99.

Ramakrishnan, M., & Vaidyanandan, R. (2008). Geology of India. *Geological Society of India, 1.*

Roday, P. P., Chourasia, L. P., & Shailendra, C. (1989). Structural analysis of shear zones in basement granite and their relationship with folding, shearing and faulting in the cover sediments near Hirapur, district Sagar, Central India. *Journal of Geological Society of India, 33*, 387–408.

Sharma, K. K., & Rahman, A. (2000). The early Archaean-Palaeoproterozoic crustal growth of the Bundelkhand craton, northern Indian shield. In: M. Deb (Ed.), *Crustal Evolution and Metallogeny in Northwestern Indian Shield* (pp. 51–72). Narosa Publication House, New Delhi.

Field Observations of Dykes and Geometry of Igneous Enclaves in Ladakh Batholith Granitoids Around Taru, Leh, Ladakh, North India

Akhtar R. Mir, Soumyajit Mukherjee, A. Shaida, H. Nazir, and Shamim A. Dar

Abstract The Ladakh Batholith is part of Andean-type Trans-Himalaya (Gangdese) plutonic belt that extends for 2500 km from Afghanistan in the west to east of Lhasa in Tibet. Dykes of varied composition e.g. doleritic, andesitic, basaltic andesite with 24–54 Ma age intrude Ladakh batholith along the southern margin at a number of places like Taru, Umla, Saboo, Thikse, Hemis-Shangpuchan etc. Ladakh batholith is also intruded by pegmatitic and aplite veins throughout its length and breadth. In the present work, dykes intruding Ladakh batholith at Taru village of Leh district of the Union Territory Ladakh have been studied to understand their mineralogical characteristics and stress regimes. The studied dykes are post-plutonic and extend from a few metres to several metres along with 1–3 m width. Nine dykes exist within ~5 km^2 area around Taru. Mostly these are composed of plagioclase, pyroxenes, hornblende, k-feldspars, quartz and biotite. The phenocrysts have sizes between 1 to 10 mm. The studied dykes around Taru village have dominant NE-SW trend. Rose diagram of joints of the studied dykes shows that NE is the dominant trend of joints followed by NW, SW and SE as the other trends. The dominant strikes of the studied dykes and those of the joints is NE-SW. Igneous enclaves occur at places as grey to blackish blobs in the Ladakh granitoids around the study area. These enclaves are mafic, felsic and hybrid, fine to medium grained, melanocratic to mesocratic, without or with phenocrysts and vary in size significantly. The enclaves have varied shapes like schlieren, spindle, star, batman, tear drop, elongated, fish etc. The presences of these igneous enclaves indicate that the mingling process of mafic

A. R. Mir (✉)
Department of Earth Sciences, University of Kashmir, Srinagar, JK, India
e-mail: mirakhtar@uok.edu.in

S. Mukherjee
Department of Earth Sciences, Indian Institute of Technology Bombay, Mumbai, Maharashtra, India

A. Shaida · H. Nazir
Department of Geology, University of Ladakh, Leh, Ladakh, India

S. A. Dar
Department of Geology, Banaras Hindu University, Uttar Pradesh, Varanasi, India

S. Mukherjee (ed.), *Structural Geology and Tectonics*,
https://doi.org/10.1007/978-981-95-4743-2_17

and felsic composition magmas had occurred within these granitic plutons present around Taru village.

Keyword Ladakh batholith · Dykes · Igneous enclaves

1 Introduction

The Himalayan Mountain range is formed due to subduction of Tethys Ocean and subsequent collision between the Indian and the Eurasian tectonic plates (Aitchison et al., 2009; Mukherjee et al., 2013, 2015). Trans-Himalaya is situated north of the main Himalayan belt. The Ladakh batholith (LB) is part of the Trans-Himalaya. LB is made up of multiple calc-alkaline magmatic bodies, extends NW to SE for ~600 km with 30–50 km width. LB is delimited by two suture zones- the Shyok and the Indus Suture zones at north and south, respectively (Fig. 1). The rock types of LB are norites, gabbros, granodiorites, diorites and leco-granite (Honegger et al., 1982). Biotite-bearing granites and granodiorites are the chief rock types. Crystallization ages of few granitoids of LB are 102-60 Ma (Honegger et al., 1982). LB is intruded by dykes at several places in west of Leh, such as around Phyang, Taru, Umla, Tunglung, Likir, Yangthang and Hemis-Shugpachan, and also east of Leh (e.g. Nyoam) (Ahmad et al., 1998). These dykes are 54–24 Ma old (Ravikant & Guha, 2002; Hari et al. 2015). LB also encloses igneous enclaves of varied shapes (rounded, elongated, lenticular etc.). The study of dykes is important for understanding the geodynamic processes of the area, in addition to their significant role for scrutinizing the geochemical evolution of mantle (Mir et al., 2010; Tarney, 1992). The host rock type of all studied dykes is Ladakh granitoids. The present work focuses on mineralogy, structural data of joints of dykes and geometry of igneous enclaves, which are present in the LB around the Taru village (Fig. 2). The location of studied dykes is shown in Google Earth map (Fig. 3) and geographical coordinates of each studied dyke is given in Table 1. The studied area is near the Leh-Srinagar highway. Structural data (including dip and trend) of joints within dykes were recorded in the field with the help of Brunton Compass and is given in Table 2. Data were plotted stereographically using the STEREONET software, version 9.96 of 2017. Field observations like colour, nature and contact relationship of igneous enclaves with host rock and geometry of these enclaves were documented. Dykes have been taken as reliable deformation indicators to decipher tectonics (e.g. Babar et al., 2017; Misra & Mukherjee, 2017).

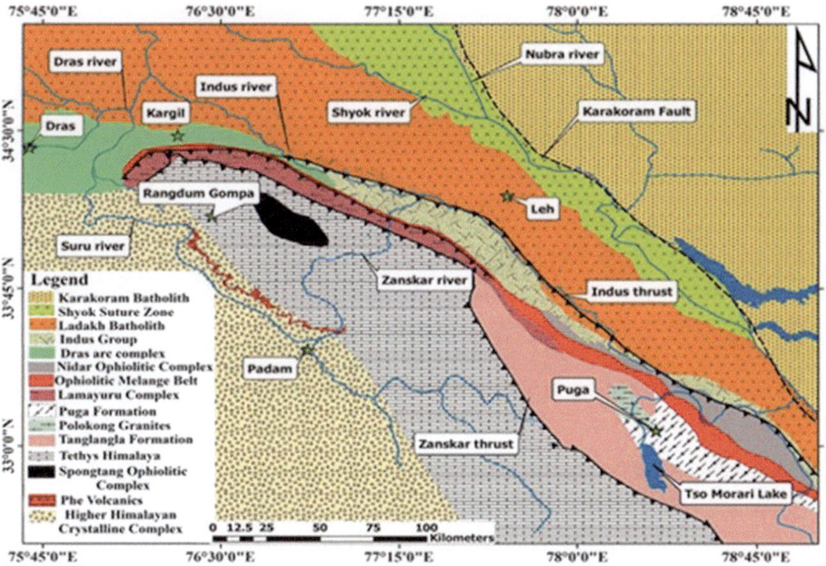

Fig. 1 Simplified geological map of the northern Ladakh–Zanskar Himalaya (after Maheo et al., 2004)

Fig. 2 Geological map around Taru village (after, Heri et al., 2015), showing study area in red circle

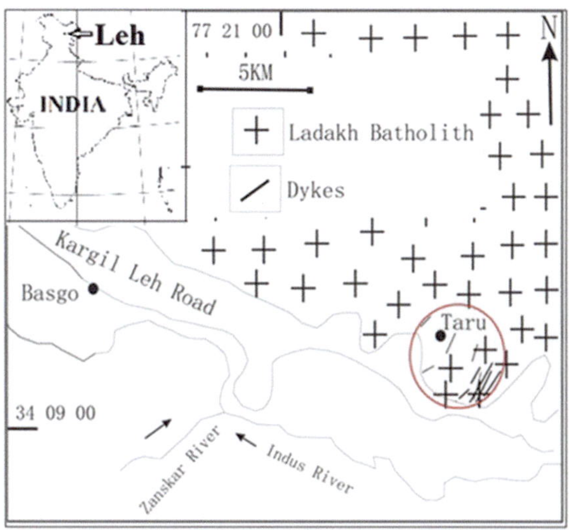

Fig. 3 Google earth map showing location of studied dykes near Taru village, Leh, Ladakh

Table 1 Geographical coordinates of studied dykes around Taru village, Leh, Ladakh

S. no.	Dyke number	Latitudes	Longitudes
01	Dyke-1	34°11′ 23″N	77°24′ 09″E
02	Dyke-2	34°10′ 49″N	77°25′ 43″E
03	Dyke-3	34°11′ 59″N	77°25′ 60″E
04	Dyke-4	34°10′ 25″N	77° 25′ 55″E
05	Dyke-5	34°10′ 33″N	77°26′ 14″E
06	Dyke-6	34°10′ 0.5″N	77° 26′ 55″E
07	Dyke-7	34° 10′ 45″N	77° 26′ 08″E
08	Dyke-8	34°10′ 19″N	77°26′ 51″E
09	Dyke-9	34°10′ 42″N	77°26′ 51″E

2 Field Characteristics of Dykes

Dyke 1

It is situated ~1 km away from the Leh campus, University of Ladakh near the Gurdawara Pathar Sahib. Here, post-plutonic dyke exists adjacent to ~10 m wide syn-plutonic dyke (Fig. 4). The host rock of both these dykes is the Ladakh Granite. The strike of post-plutonic dyke is N50°W. Dip amount and dip direction of this dyke are 45° and N40°E, respectively. Three sets of joints are observed in post-plutonic dyke. These are oblique, cross and longitudinal joints. The cross joint also cuts through the host rock. Spheroidal weathering pattern is noticed in parts of this dyke (Fig. 5).

Table 2 Structural data of joints of dykes intruding Ladakh Batholith around Taru, Leh, Ladakh

Parameters	Strike (in degrees)	Dip amount (in degrees)	Dip direction (in degrees)
Dyke 1, Taru, West of Taru Campus, UoL, Near Gurduwara			
Dyke 1	310 (N50°W)	45	40 (N40°E)
Oblique joint-1	275 (N85°W)	80	05 (N5°E)
Oblique joint-2	220 (S40°W)	75	130 (S50°W)
Oblique joint-3	55 (N55°E)	58	145 (S35°E)
Cross joint-1	240 (S60°W)	45	330 (N30°W)
Cross joint-2	N55°E	50	325 (N35°E)
Cross joint-3	N45°E	76	135 (S45°E)
Longitudinal joint-1	340 (N20°W)	70	70 (N70°E)
Longitudinal joint-2	115 (S65°E)	71	205 (S25°W)
Dyke 2, Taru, East of Taru Campus, UoL			
Dyke 2	24 (N24°E)	90	Vertical
Oblique joint-1	95 (S85°E)	71	5 (N5°E)
Oblique joint-2	95 (S85°E)	84	Vertical
Pegmatite vein	315 (N45°W)	60	45 (N45°E)
Dyke 3, Taru, East of Taru Campus, UoL			
Dyke 3	75 (N75°E)	80	165 (S15°E)
Oblique joint	70 (N70°E)	60	340 (N20°W)
Cross joint	15 (N15°E)	30	105 (S75°E)
Longitudinal joint	80 (N80°E)	60	170 (S10°E)
Dyke 4, Taru, East of Taru Campus, UoL			
Dyke 4	75 (N75°E)	80	165 (S15°E)
Oblique joint	10 (N10°E)	45	100 (S80°E)
Cross joint-1	130 (S50°E)	50	40 (N40°E)
Cross joint-2	340 (N20°W)	80	250 (S70°W)
Dyke 5, Taru, East of Taru Campus, UoL			
Dyke 5	80 (N80°E)	90	Vertical
Oblique joint	50 (N50°E)	60	140 (S40°E)
Cross joint	330 (N30°W)	85	60 (N60°E)
Dyke 6, Taru, East of Taru Campus, UoL			
Dyke 6	75 (N75°E)	75	165 (S15°E)
Oblique joint	105 (S75°E)	45	195 (S15°W)
Cross joint-1	130 (S50°E)	50	40 (N40°E)
Cross joint-2	190 (S10°W)	55	100 (S80°E)
Longitudinal joint	290 (N70°W)	54	200 (S20°W)
Dyke 7, Taru, East of Taru Campus, UoL			
Dyke 7	85 (N85°E)	90	Vertical

(continued)

Table 2 (continued)

Parameters	Strike (in degrees)	Dip amount (in degrees)	Dip direction (in degrees)
Cross joint	190 (S15°E)	80	280 (N80°W)
Longitudinal joint-1	85 (N85°E)	80	175 (S5°E)
Longitudinal joint-2	90 E	45	360 N
Oblique joint-1	315 (N45°W)	65	225 (S 45°W)
Oblique joint-2	305 (N55°W	70	35 (N35°E)
Dyke 8, Phyang (NW of Phyang Petrol pump)			
Dyke 8	55 (N55°E)	90	Vertical
Oblique joint-1	10 (N10 °E)	40	100 (S80°E)
Oblique joint-2	295 (N65°W)	20	25 (N25°E)
Oblique joint-3	45 (N45°E)	25	145 (S35°E)
Cross joint-1	135 (S45°E)	80	225 (S45°W)
Cross joint-2	160 (S20°E)	65	70 (N70°E)
Dyke 9, Phyang (NW of Phyang Petrol pump)			
Dyke 9	70 (S70°E)	80	340 (N20°W)
Cross-1	20 (N20°E)	55	110 (S70°E)
Cross-2	20 (N20°E)	65	290 (N70°W)
Cross-3	140 (S40°E)	85	230 (S50°W)
Longitudinal joint-1	235 (S55°W)	65	145 (S35°E)
Longitudinal joint-2	70 (N70°E)	80	160 (S20°E)
Oblique joint-1	210 (S30°W)	65	300 (N60°W)
Oblique joint-2	295 (N65°W)	82	25 (N25°E)
Oblique joint-3	120 (S60°E)	60	210 (S30°W)

Dyke 2

This dyke is situated ~1 km from the University of Ladakh Leh Campus towards Phyang village along NE (Fig. 6). The strike of this dyke is N24°E and dips vertically. Extension of this dyke is ~150 m. Aplite vein also intrudes in Ladakh granitoids in the vicinity of this dyke (Fig. 7). Attitude data of joint sets as documented in field (Fig. 8) is given in Table 1.

Dyke 3

This dyke is situated in Taru village and ~1.5 km away from the University of Ladakh, Leh Campus towards north. The strike of this dyke is N75°E with dip amount of 80° and dip direction of S15°E. The dyke is oblique-slip faulted at two places (Fig. 9). The dyke is also cut by pegmatite vein. Three sets of joints- oblique, cross and longitudinal are documented at this dyke.

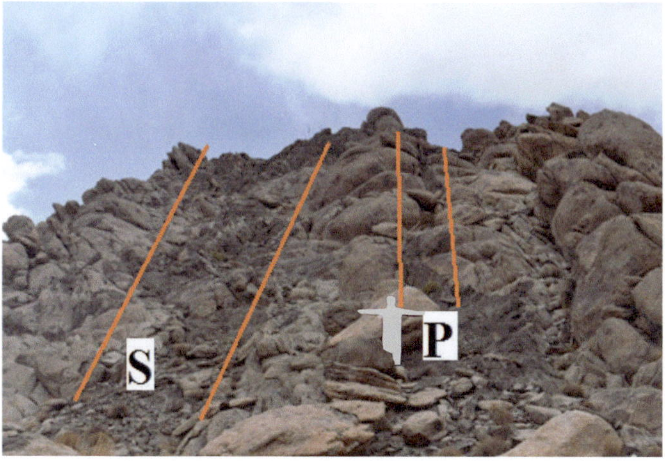

Fig. 4 Syn (S) and post-plutonic (P) dykes intruded in the Ladakh granitoids near Gurduwara Pathar Sahib, ~1 km from the Leh campus, UoL. Marker: Akhtar R. Mir (5.6 feet height)

Fig. 5 Spheroidal weathering on dyke 1. Marker: Geological hammer (1.9 feet length)

Dyke 4

This dyke is situated at a distance of nearly 1 km from UoL, Leh Campus in South-east direction towards Phyang village (Fig. 10). The orientation of dyke is N 75° E. It dips 80° and towards S15°E. The dyke contained 3 sets of joints they are oblique joint, cross joint-1, and cross joint-2.

Dyke 5

This dyke is situated towards east of dyke 4. The dip of this dyke is vertical having N80°E strike. This is a fine-grained dyke (Fig. 11). Width of this dyke is ~3 m and is ~300 m long. The dyke contains two sets of joint (Fig. 12).

Fig. 6 Dyke 2 having strike NE-SW and having almost vertical dip intruding in Ladakh granitoids (LG). Marker: Nazir A. (5.2 feet height)

Fig. 7 Aplite vein (AV) intruding in the Ladakh granitoids (LG) in the vicinity of dyke-2

Dyke 6

This dyke is east of dyke 5 towards Phyang village. It is andesitic in nature and shows porphyritic texture (Fig. 13). The width of this dyke is ~5 m and extends for more than 100 m. The strike of the dyke is 75^0, the dip amount is 75 and the dip direction is S15 °E. The dyke contains four sets of joints (oblique, cross and longitudinal joints). This dyke shows zoned phenocrysts of k-feldspars embedded in fine groundmass (Fig. 14).

Fig. 8 Joints within dyke 2. Marker: Ball pen (5.5 inches long)

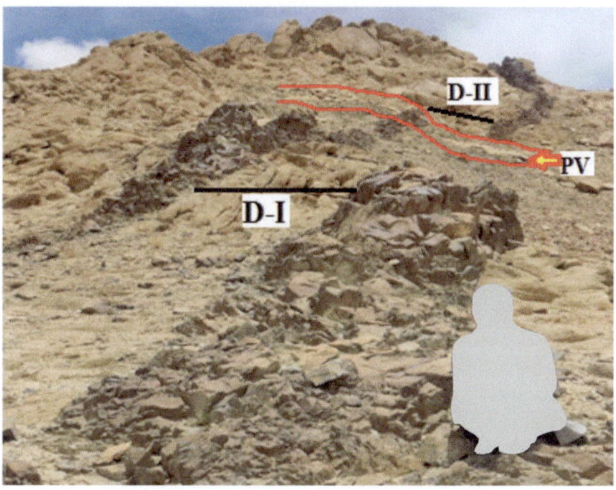

Fig. 9 Displacement of dyke-3 at two places marked as D-I and D-II. Pegmatite vein (PV) also cuts through this dyke. Marker: Shaida (3 feet- waist to head height)

Dyke 7

This dyke is situated east of Taru towards the Phyang village (Fig. 15). This vertical dyke strikes N85ºE. The dyke contains four sets of joints (oblique joints-1, cross and longitudinal-1, 2) (Fig. 16). Exposed parts of this dyke are more weathered than the other dykes in the surrounding area. A hand specimen sample was collected that shows the contact between dyke and host granite (Fig. 17).

Fig. 10 Dyke-4 intruding in the Ladakh granitoids. Marker: Shaida (5 feet height)

Fig. 11 Dyke-5 intruded in the Ladakh granitoids. Marker: Shaida (5 feet height)

Dyke 8

This dyke is situated north-west of the Phyang petrol pump at ~2 km distance towards Taru village (Fig. 18). The strike of this dyke is N55ºE and dips vertically. This dyke contains oblique, cross, and longitudinal joints.

Dyke 9

This dyke is situated towards north of dyke 8 at a distance of ~150 m (Fig. 19). The strike of this dyke is N70ºE having dip = 80° and dip direction = 340° N. The width

Fig. 12 Close view of joints within dyke 5. Marker: Ball pen (5.5 inch long)

Fig. 13 Dyke 6 intruded in the Ladakh granitoids. Marker: Shaida (5 feet height)

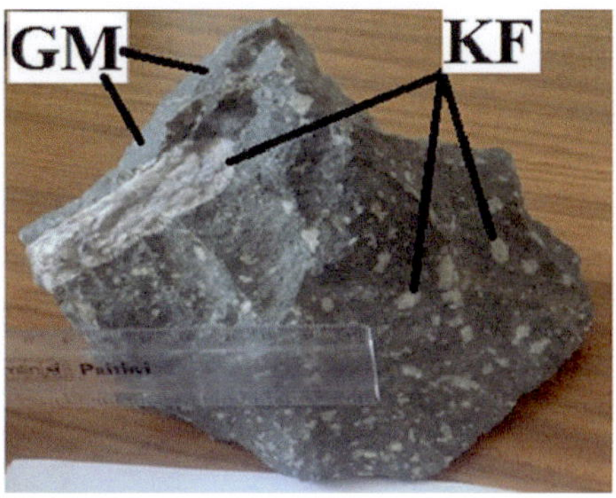

Fig. 14 Handspecimen of andesite dyke 6 showing porphyritic texture wherein phenocrysts of K-feldspars (KF) are embedded in a fine groundmass (GM)

Fig. 15 Dyke 7 intruding in the Ladakh granitoids. Marker: Shaida (5 feet height)

Fig. 16 Joint sets of dyke 7. Marker Ball pen (5 inch long)

Fig. 17 Handspecimen showing contact between dyke 7 (D) and host granite (G)

Fig. 18 Dyke 8 intruding in the Ladakh granitoids. Marker: Shaida (5 feet height)

Fig. 19 Dyke 9 intruding in the Ladakh granitoids. Marker: Shaida (5 feet)

and length of this dyke is 2 m and ~500 m respectively. The structural data of joints within this dyke is given in Table 1.

From the attitude data of dykes and joints (within dykes) of study area (Table 1), stereo-projections and rose diagrams have been drawn (Fig. 20). Figure 20a shows that NE is the dominant trend of the dykes within the study area, whereas NW is the second dominant direction. Further, rose diagram of joints (Fig. 20b) inside the dykes shows that NE is the dominant trend followed by NW, SW and SE directions.

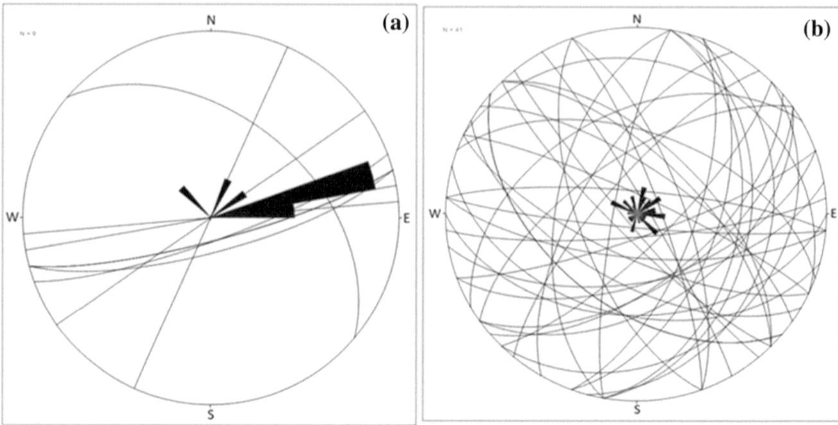

Fig. 20 **a** Showing rose diagram of dykes and **b** rose diagram of joints of dykes around Taru village, Leh, Ladakh

3 Geometry of Igneous Enclaves

Location 1

Near the Taru streamlet, at ~1.5 km from the University campus at Taru towards south-east, mafic schlieren have been noticed within the Ladakh granitoids (Fig. 21). Schlieren are dark, fine-grained, biotite-rich segregations within the host granitoid. They vary in length (several cm to >1 m) and width (<1 cm to 10 cm). Schlieren are most common near the margins of enclave swarms and appear texturally identical to the biotite-rich parts associated with enclaves. Most of these schlieren show diffusive contact with the host granite.

Location 2

Near location 1, a nearly spindle-shaped enclave has been noted. This enclave shows defused contact with the host granitoids (Fig. 22).

Location 3

Felsic enclave including the xenoliths of host granite is seen on the left bank of the Taru streamlet (Fig. 23). This site is ~1.5 km from the Ladakh University Leh campus.

Location 4

Numerous trails of oriented elliptical mafic enclaves showing possibly a magmatic flow structure within the host Ladakh Granite have been documented near the Taru Phyang village boundary at ~2.5 km from the Ladakh University Leh campus towards SE (Fig. 24). Some of these enclaves are tear-drop and deformed star-shaped.

Fig. 21 Schlieren of mafic composition within the Ladakh granitoids

Fig. 22 A spindle-shaped enclave in the Ladakh granitoids

Location 6

Near Taru-Phyang village boundary, elongated mafic enclave has been seen with crenulated contact with host granite. Some K-feldspar crystals of host granite are seen fully and partially enclosed within this enclave as xenocrysts (Fig. 25). From the same location, a mafic enclave quadrilateral is also noted with crenulated contacts (Fig. 26).

Location 7

Two adjacent mafic enclaves were seen in the Ladakh batholith at Taru village (Fig. 27). These enclaves show sharp contact with the host granite.

Fig. 23 Felsic enclave inside the Ladakh granitoids

Fig. 24 Mafic enclave swarm inside the Ladakh granitoids

Location 8

Near Chamba of Ladakh University, Leh Campus (Taru), ~2 feet long fine-grained mafic enclave was noted in the Ladakh granites (Fig. 28).

Fig. 25 Elongated mafic enclave with crenulated contact with the host granite

Fig. 26 A quadrilateral mafic enclave with crenulated contact with the host granite

4 Summary and Conclusion

The dominant strike of the studied dykes is NE followed by subordinate directions of NW and E-W. The dip amount of these dykes varies from 45 to 90°. These dykes extend for ~50–300 m with a width of ~0.5–6 m. In some cases, cross joints inside the dykes extend and cut through the host granite body. Igneous enclaves occur as grey to black blobs throughout the Ladakh batholith granitoids around Taru village. These enclaves are mafic, felsic and hybrid in nature. These enclaves are fine to medium grained, and melan- to mesocratic. While some mafic enclaves have sharp boundaries with the host rock, others have diffuse contacts. Further, these enclaves have different shapes like schlieren, spindle, star, batman, tear drop, elongated, fish etc. The presences of these igneous enclaves indicate that the mingling process of

Fig. 27 Adjacent mafic enclaves in the Ladakh granitoids, near Taru village

Fig. 28 Mafic enclave near the Chamba, University of Ladakh, Leh Campus

mafic and felsic composition magmas had occurred within the granitic plutons near Taru area.

Acknowledgements This work was a part of SA's Master's thesis under ARM and SM and part of NA's Master's thesis under ARM. We thank the Director, Leh Campus University of Ladakh, UT Ladakh for providing necessary facilities. We acknowledge Dr. Rakesh Chandra, Dr. Malik Zubair and Dr. Farooq Ahmad for field assistance. Mukherjee (2026) summarized this work.

References

Ahmad, T., Thakur, V. C., Islam, R., Khanna, P. P., & Mukherjee, P. K. (1998). Geochemistry and geodynamic implications of magmatic rocks from the Trans Himalayan arc. *Geochemical Journal, 32*, 383–404.

Aitchison, J., Jason, R., Chan, A., & Davis, A. (2009). Tectonic implications of felsic tuffs with in the lower Miocene Gangrinboche conglomerates, Southern Tibet. *Journal of Asian Earth Science, 34*, 287–292.

Babar, M., Kaplay, R. D., Mukherjee, S., & Kulkarni, P.S. (2017). Evidences of deformation of dykes from central deccan volcanic province, Aurangabad, Maharashtra, India. In S. Mukherjee, A. A. Misra AA, Calvès G, Nemčok M. (Eds.), *Tectonics of the Deccan Large Igneous Province* (Vol. 445, pp. 337–353). London: Geological Society, Special Publications.

Heri A. R., Arrchison J. C., King, J. A., & Villa, I. M. (2015). Geochronology and isotope geochemistry of Eocene dykes intruded in Ladakh batholith. *Lithos 212–215*,111–121.

Honegger, K., Dietrich, V., Frank, W., Gansser, A., Thoni, M., & Trommsdorf, V. (1982). Magmatic and metamorphism in the Ladakh Himalayas (the Indus-Tsangpo suture zone). *Earth Planetary Science Letters, 60*, 253–292.

Maheo, G., Bertrand, H., Guillot, S., Villa, I. M., Keller, F., & Capiez, P. (2004). The south Ladakh ophiolites (NW Himalaya, India): An intra-oceanic tholeiitic origin with implication for the closure of the Neo-Tethys. *Chemical Geology, 203*, 273–303.

Mir, A. R., Alvi, S. H., & Balaram, V. (2010). Geochemistry of mafic dikes in the Singhbhum Orissa craton: Implications for subduction-related metasomatism of the mantle beneath the eastern Indian craton. *International Geology Review, 52*(1), 79–94.

Misra, A. A., & Mukherjee, S. (2017). Dyke-brittle shear relationships in the Western Deccan Strike Slip Zone around Mumbai (Maharashtra, India). In S. Mukherjee, A. A. Misra, G. Calvès, N. Nemčok (Eds.), *Tectonics of the Deccan Large Igneous Province* (Vol. 445, pp. 269–295). London: Geological Society, Special Publications.

Mukherjee, S. (2026). Introduction to "Structural Geology & Tectonics". In S. Mukherjee (Ed.), *Structural Geology & Tectonics*. Springer. ISBN: 978-981-95-4743-2.

Mukherjee, S., Carosi, R., van der Beek, P. A., Mukherjee, B. K. & Robinson, D. M. (2015). Tectonics of the Himalaya: An introduction. In S. Mukherjee, R. Carosi, P. van der Beek, B. K. Mukherjee & D. Robinson (Eds.), *Geological Society* (Vol. 412, pp. 1–3). London: Special Publications.

Mukherjee, S., Mukherjee, B., & Thiede, R. (2013). Guest Editors. Geosciences of the Himalaya-Karakoram-Tibet Orogen. *International Journal of Earth Sciences 102*, 1757–1758.

Ravikant, V., & Guha, D. (2002). Ultra-potassic post collisional dyke from the Leh batholith, Northwest Himalaya. *Journal of the Geological Society of India, 59*, 473–476.

Tarney, J. (1992). Geochemistry and significance of mafic dyke swarms in the Proterozoic. In K. C. Condie (Ed.), *Proterozoic Crustal Evolution* (pp. 151–179). Elsevier.

Hydrocarbon Prospectively of the Ganga Basin (India) Based on Plate Tectonic Reconstruction

Subhobroto Mazumder and Sushil Kumar

Abstract The Ganga basin in the northern part of the Indian plate is characterized by a thick sequence of Neoproterozoic Cambrian sediments that are similar in age to hydrocarbon producing Oman Salt basins in Arabian plate and Bikaner Nagaur basin in western India. Though a number of exploratory wells have been drilled throughout the Ganga basin, commercial success in terms of hydrocarbon discovery still remains elusive. The present study attempts to re-evaluate the prospectivity of the Ganga basin by tectonic reconstruction of the Indian plate during Neoproterozoic-Cambrian. Correlations with similar producing basins in Arabian plate and western part of Indian plate are made. Based on the tectonic reconstruction, the strike-slip Najd Fault system in the Arabian plate is envisaged to continue on to the Indian plate extending across the Aravalli Delhi Fold Belt on to the Ganga basin in its eastern part. The constituent faults of the Najd Fault system are deciphered by gravity, seismic data and geomorphology to have the present day *en-echelon* E-W trend. Strike-slip activity along these trends developed two broadly N-S oriented rift systems correlating with the Sarda and the Gandak depressions. These serve as areas of deposition of Neoproterozoic Ujhani, Tilhar and Karanpur Formation that are coeval stratigraphically to their counterparts in western Indian plate and the Arabian plate.

Keywords Ganga basin · Najd Fault · Tectonics

1 Introduction

The Ganga basin (GB) hosts a vast extent of sedimentary succession (areal extent of ~3,00,000 km^2) ranging in age from Proterozoic to Recent that has been tectonically categorized as an internal sag basin overlain by a foredeep basin (Fig. 1).

S. Mazumder (✉)
Western Offshore Basin, Oil and Natural Gas Corporation, Mumbai, Maharashtra, India
e-mail: subhom007@gmail.com

S. Kumar
KDMIPE, ONGC, Dehradun, Uttarakhand, India

S. Mukherjee (ed.), *Structural Geology and Tectonics*,
https://doi.org/10.1007/978-981-95-4743-2_18

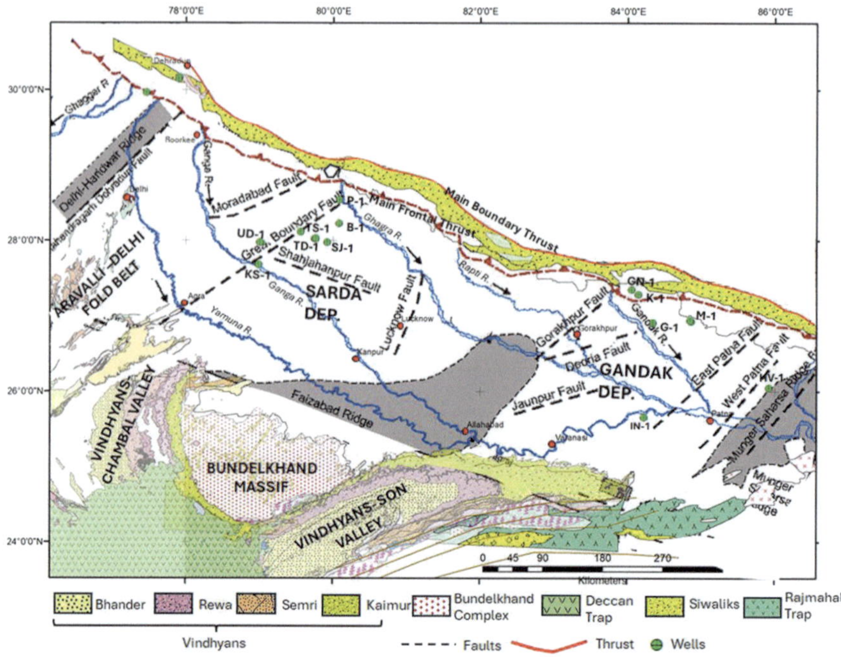

Fig. 1 Disposition of tectonic elements, boundary litho-units and subcrops of Neoproterozoic in the GB (after Azmi et al., 2008; Dutta et al., 2023; Goswami et al., 2009; Montemagni et al., 2016; Patel et al., 2020). Sarda Dep.: Sarda Depression, Gandak Dep.: Gandak Depression

The thickness of the sediments hosted by the GB is found to reach a magnitude of ~8000 m and have been probed by several deep wells most of which are found to show hydrocarbon indications during drilling implying the existence of active petroleum systems. In terms of maturity of hydrocarbon resources, the Ganga basin is classified as a Category-III Basin, characterized with prospective resources awaiting discovery (128 MMToE prognosticated hydrocarbon resources: Dutta et al., 2023) and is considered a frontier basin in terms of exploration. Attempts in exploratory drilling for hydrocarbon by multiple companies, e.g., Oil and Natural Gas Corporation, Oil India Limited and Cairn Energy India Limited for more than 6 decades is yet to achieve commercial success in the GB. However, renewed exploratory efforts in rapidly changing hydrocarbon exploration scenario has placed GB as one of the thrust areas and calls for a reassessment of its geologic model.

The present study is an attempt to re-evaluate the prospectivity of the Ganga basin by tectonic reconstruction of the western part of India during Neoproterozoic-Cambrian based on the integration of subsurface seismic and gravity data with remote sensing and geomorphic data. Additionally, the study also aims to delineate the eastern extent of the Najd Fault (NF). Implication of NF in forming prospective corridors in Ganga basin is also envisaged.

During the end of Neoproterozoic and Early Cambrian, the Gondwana assembly reached its final shape with its north-east margin extending from the northwest Arabian-Nubian Shield, through the Indian craton extending further eastwards to western Australia (Domeier, 2018). This assembly resulted from a prolonged phase of tectonics, magmatism and metamorphism known as the *Pan-African Orogeny*. Initiating at ~650 Ma and lasting till the earliest Palaeozoic, the oceans that split Rodinia began to close (Hallett & Clark-Lowes, 2016). This consolidated continental domains (Hefferan et al., 2014) and formed several orogens. In the resultant assembly, the Indian plate was contiguous with Madagascar, Africa and Arabia in the west and the Antarctic and Australian plate on its east forming the Eastern Gondwana (Fig. 2).

In the Arabian plate constituting the Eastern Gondwana, the Arabian-Nubian Shield is characterized by transpressive suturing between East and West Gondwana with the closure of intervening oceans. The final cratonization of the terranes post-closure of oceans was predominated by a major period of intra-continental rifting and wrenching (Husseini and Husseini, 2014) that resulted in anastomosing strike-slip faults in the northern part of the Arabian Shield formed between 630 to 530 Ma. These faults entitled as NF form a system of NW–SE trending sinistral strike-slip fault system that either are aligned or cut across the margins of the earlier Pan-African structures (Nehlig et al., 2002). The Najd Fault System is constituted of three principal and parallel fault zones mapped for a length of ~1100 km, with 5 to

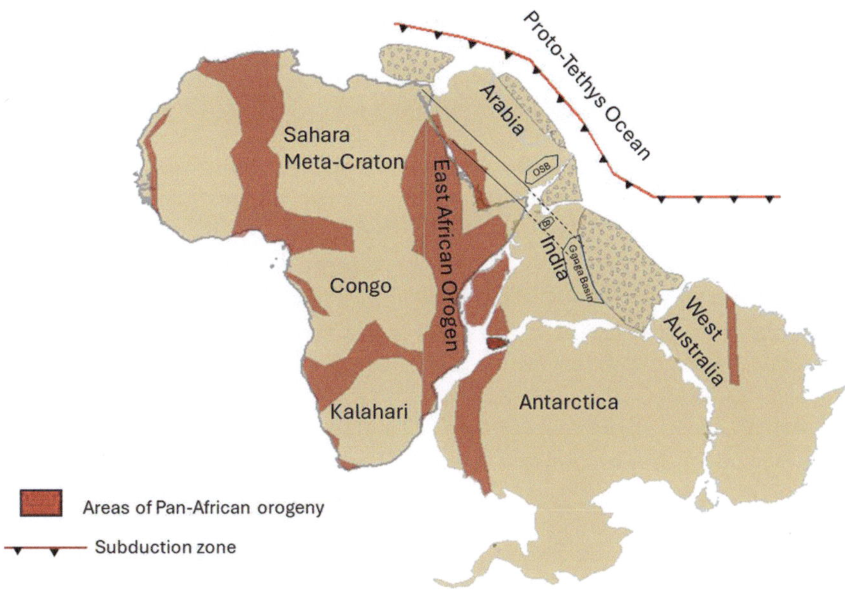

Fig. 2 Tectonic reconstruction at the end of Cambrian depicting areas of Pan-African orogeny (after Kroner and Stern, 2004) and layout of the Najd fault system in the Arabian plate. B: Bikaner Nagaur Basin. OSB: Oman Salt Basins

10 kms average individual width and a total width of ~300 km (Fig. 2a) (Al-Husseini, 2000). The fault system has been interpreted to extend up to Egypt (Mohamed & Wahed, 2010), which defines its western extremity (Stern, 1985; Mohamed et al., 2010). The eastern extent of NF has been mapped up to NW India with effects of strike-slip are observed in the Bikaner-Nagaur basin (Fig. 2; also see Biswas et al., 2024). Studies based on aeromagnetic data indicate that the Najd System to be much broader and continuous at depths than its counterpart exposed in the outcrops (Stern, 1985).

Differential left lateral slips along the NF as well as faults parallel/sympathetic to them like the Zagros Line marking the eastern depositional boundary of the Infracambrian Hormuz Formation, developed several ~N-S grabens (after Husseini, 1988) in the cratonic extent from Africa, Arabia to India that marks the NE margin of the Eastern Gondwana. These grabens in most cases are sedimentary basins hosting Neoproterozoic and Cambrian sequences, some of which are prolific producers of hydrocarbon, e.g., the Huqf basin in southern Oman, and the Hormuz basins in Eastern Iran and Arabian/Persian Gulf (e.g., Mukherjee et al., 2010).

2 Neoproterozoic-Cambrian Facies Distribution, Their Correlatives and Structural Framework/Genesis of the Basins

A review of a few of the Neoproterozoic-Cambrian basins in the NE margin of the Eastern Gondwana has been carried out to delineate the similarity within the basins in terms of facies correlatives and basin framework.

2.1 Ghaba Salt Basin and South Oman Salt Basin

The Oman salt basins are situated on the southeastern margin of Arabian plate adjacent to the boundaries of the Indian, Iranian and the African plates. These occur as three NE aligned almost discrete basins where the South Oman Salt basin is separated from the Ghaba basin in North Oman by the Central Oman High. The Mabrouk Makarem High separates the Ghaba basin from the Fahud Basin (Fig. 3) (Heidorn et al., 2018). These highs are part of a series of N-S to NE-SW trending basement highs that broadly parallels the salt basins separating them and constitutes the dominant gravity grain of the terrain (Loosveld et al., 1996).

These salt basins in the eastern part of the Arabian peninsula hosts the Huqf Supergroup (Table 1) with 725–540 Ma rock record resting unconformably on a Pan-African basement of 800 Ma constituted of older igneous-metamorphic rocks (Allen, 2007). The Huqf Supergroup is classified into three groups. In younging direction, these are the Abu Mahara Group, the Nafun Group and the Ara Group.

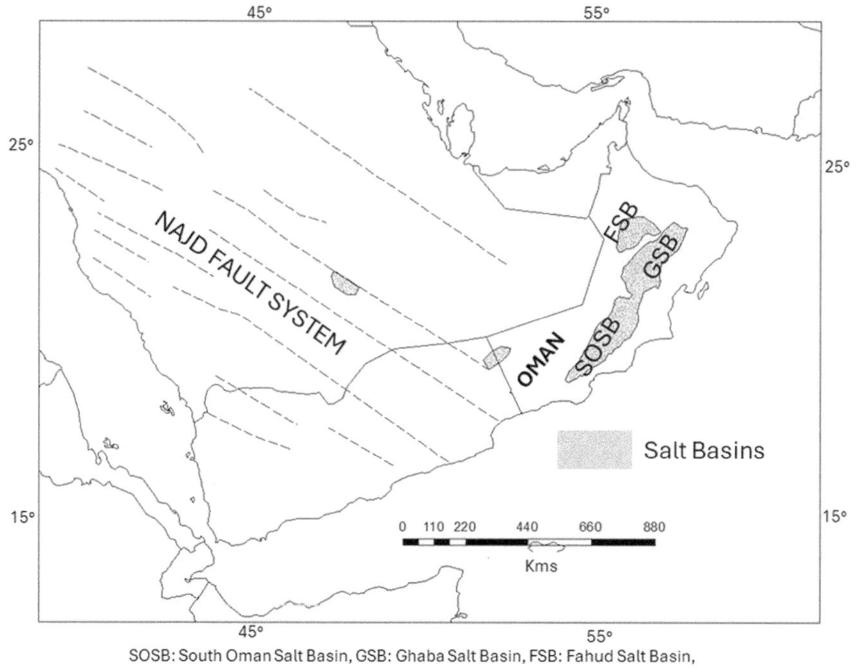

Fig. 3 Disposition of the Najd fault system in the Arabian peninsula and Neoproterozoic-Cambrian salt basins in Oman (after Kusky & Matsah, 2003; Schroder et al., 2004; Meyer et al., 2014)

The Abu Mahara Group has an age range of 725–640 Ma and is constituted of glacial and volcanogenic clastics. The Nafun Group is Ediacaran and in the Huqf area consists of ramp-peritidal carbonate facies of the Khufai Formation overlain by marine, storm-generated shelf clastics and carbonates of the Shuram Formation. The overlying Buah Formation is composed of limestones and dolomites with stromatolites, which are deposited in a carbonate ramp. These imply a facies change from deep to shallow water (Yue & Vandeginste, 2020). The Ediacaran–Cambrian Ara Group represents a thick package of platform dolomitized microbial carbonates interlayered with evaporites with at least six cycles (Nicholas & Gold, 2012) deposited in a supratidal environment.

In terms of hydrocarbon exploration, the Huqf Supergroup hosts one of the oldest petroleum systems with silicilyte and carbonate stringer plays in the Ara Group in South Oman and the gas-bearing plays of the Buah Formation of Nafun Group in North Oman being the most significant (Cozzi et al., 2012).

Tectonically, the salt basins in Oman are considered to be subsiding rift basins developed in Neoproterozoic and lower Cambrian (~600–540 Ma) with their origin attributed to the rifting and wrenching due to sinistral strike-slip of the Najd transform fault system (Pollastro, 1999). The salt basins are bound by NNE-SSW normal faults whereas a dominant NW-aligned lineament system sympathetic to the Najd strike-slip fault system cuts them at high-angles (Fig. 2a). Strike-slip creates isolated highs

Table 1 Stratigraphy, broad lithology and environment of deposition for Huqf Supergroup in Oman (After Allen, 2007)

Age		Formation	Lithology	Environment
Cambrian	Nafun Group	Ara Group	Microbial carbonates interlayered with evaporites	Supra-tidal
Neoproterozoic		Buah Formation	Limestones and dolomites with stromatolites	Shelfal/change from deep to shallow marine
		Shuram Formation	Siltstone/shale/ carbonates	Storm generated shelfal deposits
		Khufai Formation	Carbonates/ dolomites	Carbonate Shelf/ Peritidal
		Abu Mahara Group	Extensive sandstone beds with thin limestone	Glacial and volcanogenic

as is found in the Ghawar field. The axis the individual highs are found to offset NW with multiple occurrences of them associated with NW-trending faults, implying their genesis due to the Najd faulting or their reactivation (Perinçek et al., 2006).

2.2 *Punjab and Thar Platform, Indus Basin*

The Neoproterozoic-Cambrian sequences in the Indus Basin (Pakistan) are encountered in exposures along ~E-W Salt Range Thrust that lies north of the Sargodha Ridge. The thrust and the associated outcrops are crossed by the Indus river with its western part referred to as Trans Indus Ranges. Further south of the Salt Range Thrust, Neoproterozoic to Cambrian sequences are encountered as subcrops in the Punjab platform and in the Thar platform (Siddiqui, 2016) (Fig. 4).

The Punjab platform in the easternmost part of the Middle Indus basin is located ~the NW edge of Indian plate and defines a tectonically very stable area. It shows a monoclinal geometry, dipping very gently at ~1° toward west/the Sulaiman foredeep. The platform is bounded by the Lahore-Sargodha High in the north, Sulaiman depression and the Sulaiman fold belt in the west, and Pokhran/Jaisalmer Mari High in the south (Khalid et al., 2020; Aadil et al., 2011, Sohail & Aadil, 2014). The Thar platform lies further south of the Punjab Platform (Khan & Liu, 2019) and across the Jacocabad Mari High (JM High). Stratigraphically the Neoproterozoic to Cambrian sequences in the area are represented by the Salt range Formation and the overlying Jhelum Group comprise of Khewra, Kussak, Jutana and Baghanwala Formation (Table 2).

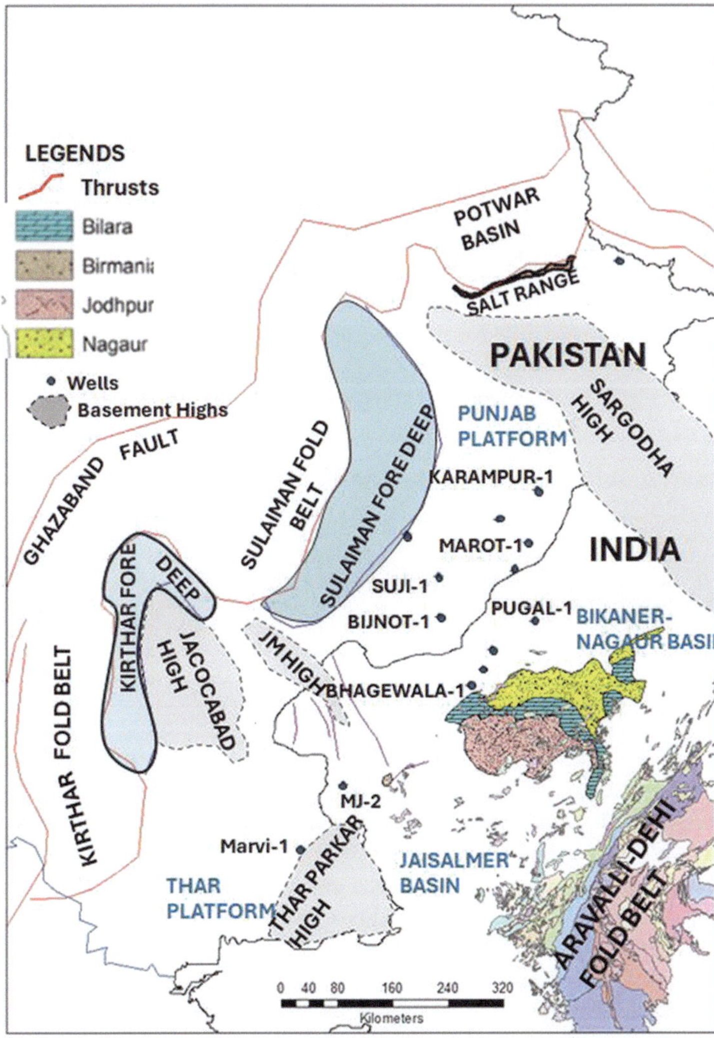

Fig. 4 Disposition of outcrops and subcrops of Neoproterozoic-Cambrian in Punjab platform and Bikaner-Nagaur basin (after Arif et al., 2009; Ghori et al., 2009; Yasin et al., 2022)

The Salt Range Formation is composed of red gypseous marl with rock salt, gypsum and dolomite in fluvial to restricted shelf conditions under (semi)arid and oxygenated conditions. The Early Cambrian Khewra Formation of the overlying Salt Range Formation consists predominantly of purple to brown fine-grained sandstone and flaggy shale associated with ripple marks and mud cracks. Studies indicate a depositional environment changing from near-shore to aeolian along with sediments derived from a Precambrian crystalline provenance (Saqab et al., 2009). Kussak Formation that conformably succeeds the Khewra Formation ranges in age from

Table 2 Stratigraphy, broad lithology and environment of deposition for Neoproterozoic-Cambrian in Punjab platform in Pakistan (after Cozzi et al., 2012; Choudhary et al., 2023)

Age		Name	Lithology	Environment
Cambrian	Jhelum Group	Baghanwala Formation	Variegated shales with interbedded thin flaggy sandstone	Shallow basin influenced by arid and strongly oxidizing conditions
		Jutana Formation	Massive to thick bedded sandy dolomite and dolomitic sandstone	Sub-tidal to Supra-tidal
		Kussak Formation	Sandstone and shales with thin dolomites	Shallow marine to Supra-tidal
		Khewra Formation	Fine grained sandstone and flaggy shale	Near-shore to aeolian
Neoproterozoic		Salt Range Formation	Red gypseous marl with rock salt, gypsum and dolomite	Fluvial to restricted shelf conditions under arid to semi-arid and oxygenated climate conditions

Late Early Cambrian to Early Middle Cambrian with a lithology of greenish gray to purple, glauconitic sandstone and shales with thin dolomites. Depositional environment varies from shallow marine to supra-tidal conditions with evidences of a meandering river to deltaic deposition. In subcrops, encountered in wells drilled in Punjab platform (drilled by Shell, AMOCO and OGDCL), a fining-up sequence is documented that indicates transgression (Abedin and Iqbal, 2014; Ghazi et al., 2020). The Jutana Formation of Early Middle Cambrian overlying the Kussak Formation is constituted of massive to thick bedded sandy dolomite and dolomitic sandstone with interlayered sandstone and shales. The overall sequence is deposited in subtidal to supratidal conditions (Ahmad et al., 2013). The Baghanwala Formation crops out in the eastern part of Salt Range and is also encountered as subcrops is a number of wells in Potwar basin and Punjab Plains (at depths of 2500 m). From its conformable contact with the underlying Jutana Dolomite, it is assigned a Late Middle Cambrian age. Lithologically, Baghanwala Formation consists of variegated shales with interbedded thin flaggy sandstones that deposited in a shallow basin influenced by arid and strongly oxidizing conditions. Presence of cubic pseudomorphs of halite in Baghanwala Formation is observed implying a high salinity of the basin during deposition (Khan and Ali et al., 2022; Khan, 1979). The stratigraphic equivalent of the Baghanwala Formation in the Khishor Ranges in Trans-Indus area is represented by the Khisor Formation. The Khisor Formation constitutes of gypsum, dolomites and shales with some silts and sandstones that deposited presumably in a local depression, possibly created by coastal ponds, under an evaporitic condition.

In the Thar Platform of Lower Indus basin, in well Marvi-1, unmetamorphosed Cambrian-Late Neoproterozoic (~550 Ma) sedimentary sequence consisting of >1000 m of Cambrian and 360 m of Infracambrian sediments believed to be equivalent to the Jhelum Group (Siddiqui, 2012). Further east of the well Marvi-1, in the continuation of the Thar Platform in the Indian part, in the well MJ-B, based on palynofloral assemblage, a Cambrian age has been assigned to the quartz wacke encountered in the well.

2.3 Bikaner-Nagaur Basin (Rajasthan, India)

In the western Rajasthan area (India), which probably forms the eastern fringe of the Punjab platform, an Ediacaran to Early Cambrian sedimentary sequence of ~1500 m thickness is observed in both outcrops and subcrops (Fig. 4). This sequence forming the Marwar Supergroup overlies the Malani Volcanics or Erinpura Granite suite of rocks in a broadly NNW–SSE elongated basin spanning >100000 km^2. Especially in wells Karampur-1, Marot-1, Fort Abbas-1, Suji-1, Bijnot-1 and Baghewala-1 in the Punjab Platform, the basement rocks encountered below Infracambrian are rhyolites, with interbedded flows of basalts and trachytes (Yasin et al., 2022). Stratigraphically, the Marwar sequence is constituted of Jodhpur Group, Bilara Group and the Nagaur Group (Table 3). The basal Jodhpur Group comprises of cross-bedded reddish sandstones and maroon shales deposited in a fluvio-marine environment. The Bilara Group is constituted of limestones and dolomites with sporadic clay beds overlying the Jodhpur Group. This dominantly carbonate sequence grades laterally to Hanseran Evaporite Group as observed in several wells in the central and northern parts of the Bikaner-Nagaur basin. The Hanseran Evaporite sequence is constituted of a maximum of seven evaporitic cycles composed of dolomites, magnesites, anhydrites, halites, polyhalites and clay bands in ascending order (with Sylvite being reported from some areas) that are considered to be deposited in a marine basin, restrictively connected with open sea. The Bilara and the equivalent Hanseran Evaporite Group are overlain by reddish to brown, medium- to coarse-grained, cross-bedded sandstone belonging to the Nagaur Group deposited in a fluvial environment (Cozzi et al., 2012).

Similar to the Ghaba and South Oman salt basins, the Bikaner-Nagaur basin was created by rifting along the Najd Fault system where the Jodhpur Group formed a syn-rift fluvio-deltaic deposit. The faults affecting the basin floor presumably resulted from the initial phase of rifting during the Late Proterozoic. These faults are oriented in an *en-echelon* pattern and forms a series of horst-graben structures with ~50 m throw that increases to a maximum of ~200–400 m at the rift boundary with anticlinal structures associated with NNE-SSW faults. These anticlines are the areas of hydrocarbon accumulation in the basin with dimensions ranging in between 10–50 km^2 (Mandal et al., 2022). The overall deposition of the Jodhpur and the Bilara Formation and their stratigraphic equivalents in the Punjab platform is dominantly controlled by the basement topography with the Infracambrian sediments that thicken

Table 3 Stratigraphy, broad lithology and environment of deposition for Neoproterozoic-Cambrian in Marwar Supergroup of Bikaner-Nagaur basin (after Choudhary et al., 2023)

Age		Formation	Lithology	Environment
Cambrian	Marwar Super Group	Nagaur Group	Reddish to brown, medium- to coarse-grained, cross-bedded sandstone	Fluvial
Neoproterozoic		Bilara Group laterally grading to Hanseran evaporite	Limestone and dolomite with sporadic clay beds seven evaporitic cycles and clay bands	Marine basin, restrictively connected with open sea
		Jodhpur Group	Cross-bedded, reddish sandstone and maroon shale	Fluvio-marine

on either side of the basement ridges extending from Pakistan to India (Sheikh et al., 2003). Seismic data indicates that the Infracambrian sequences in the 2500–3200 m deep Punjab platform are less deformed and appear to be relatively undeformed and unfolded. On the other hand, the equivalent rocks in the Bikaner-Nagaur basin (Jodhpur, Bilara and Nagaur Formation) are both folded and faulted (Sohail & Aadil, 2014) suggesting more transpression than what happened in the Panjab platform.

2.4 Birmania Basin

The Neoproterozoic-Cambrian rocks are represented in between Jaisalmer and Barmer basins, further south of Bikaner-Nagaur, as outcrops of Birmania and Randha Formation constituting the Birmania Basin (Rajasthan, India, Fig. 3a). This basin appears as an elliptical isolated remnant of Marwar basin in western Rajasthan (Maheswari et al., 2002). Though based on stratigraphic position, the Birmania basin is assigned an Edicaran to Lower/Middle Cambrian age, studies based on assemblage of multicellular algal fossils provide a tighter constrain of the age of the entire basin stratigraphy to be Edicaran (Shukla & Tewari, 2022).

Stratigraphy of the Birmania basin (Table 4) is broadly constituted of Lower Randha Formation dominantly comprising of coarse-grained calcareous quartzose sandstones with shale intercalations deposited in continental to near-shore environment. The overlying Birmania Formation is represented by a repetitive sequence of siliciclastics, cherty dolomitic limestone with phosphorite facies connoting a deposition characterized by a marine reducing environment with a restricted circulation.

Table 4 Stratigraphy, broad lithology and environment of deposition for Neoproterozoic-Cambrian of Birmania basin, Rajasthan (after Singh, 2006)

Age	Formation	Lithology	Environment
Mid-lower Cambrian	Birmania Formation	Siliciclastics, cherty dolomitic limestone with phosporite facies	Marine reducing environment with a restricted circulation
Neoproterozoic	Randha Formation	Coarse grained calcareous quartzose sandstone with shale intercalations	Continental to near shore

2.5 *Ganga Basin (GB)*

Most of the above basins with Neo-Proterozoic-Cambrian sequences as part of their stratigraphic column in both the Arabian as well as Indian plate (Salt basins of Oman, Infra-Cambrian basins of Punjab platform and Bikaner-Nagaur basin) have been thoroughly explored and proven to be hydrocarbon bearing from Infra-Cambrian, at least in parts. The Ganga Basin (GB) in the northern part of the Indian plate and located further east of the Bikaner Nagaur basin, being separated from it by the Aravalli-Delhi orogenic belt, is also characterized by a Neoproterozoic-Cambrian sequence as a part of its sedimentary column. Tectonically, the GB is delimited in the west by the NE-SW aligned Aravalli Delhi Fold Belt (ADFB) and the Mahendragarh-Dehradun Fault (Patel et al., 2018) whereas toward the south, it is bounded by exposures of the Vindhyan Supergroup and Bundelkhand massif (Dutta et al., 2023). The northern boundary of the GB is demarcated by the Main Boundary Thrust and the Siwalik supergroup exposed along the NW–SE trending Sub-Himalayan ranges. The subsurface NNE-SSW Munger-Saharsa basement Ridge that extends as a promontory of the peninsular shield exposed to its south demarcates the eastern extent of the GB (Fig. 2 of Biswas et al., 2024). The Bundelkhand massif exposed at the south is believed to extend underneath the sediments of GB as Faizabad Ridge (Suresh et al., 2023) dividing it into a western Ganga basin with the Sarda Depression and an Eastern Ganga basin with Gandak and Madhubani depressions (Fig. 1).

Morphologically, the Ganga Basin forms a large-scale regional depression of a very vast expanse of about 3,00,000 sq. km with sediments ranging in age from Meso-Proterozoic to Recent. Since the basin is formed in the foreland of the Himalayan orogenic belt and its bounding thrusts, it is categorized as a foredeep basin generated as a flexure on the underthrusting Indian plate. This large scale depression acts as sediment sink and extends all the way up to the cratonic peripheral forebulge (Singh, 1996) that is spatially represented as the Vindhyan plateau (Kasana et al., 2021). The sediments filling up the foredeep depression is believed to initiate around Middle Miocene with Siwalik sediments. These were eventually overlain by a thick layer of alluvium deposited by a well-developed fluvial network constituted of the River Ganga and its tributaries. However, recently acquired seismic data covering time up to 8 s (approx. 8000 m) has imaged a substantial thickness of further deeper sequences

Table 5 Stratigraphy, broad lithology and environment of deposition for Neo-Proterozoic of Ganga basin (after Srinivas et al., 2014)

Age	Formation	Lithology	Environment
Neoproterozoic	Karanpur Formation	Fine to medium grained sandstone and non-calcareous fissile shale	Shallow marine
	Tilhar Formation	Grayish limestone interbedded with calcareous shale	Wide-spread semi-arid middle to outer shelf
	Ujhani Formation	Medium to fine grained sub-greywacke to arenite and shale with bands of dolomitic limestone	Littoral to shallow marine with basaltic flows

of Neo-Proterozoic below Tertiary suggesting initiation of basin formation at a much earlier stage. Around 21 wells drilled in the GB by ONGC, Cairn India and OIL, with a very low well density (one well per 11,000 km^2) indicates Cryogenian to Edicaran sediments constituting the Neoproterozoic sequence deposited in the Sarda and the Gandak depressions. These Neo-Proterozoic sediments are lithologically classified into Ujhani, Tilhar and Karanpur Formation (Table 5) that unconformably rests above a Meso-Proterozoic Baharaich Formation constituted of quartzitic sandstone, limestone, phyllite and basic rocks. Of these Neo Proterozoic sequences, Ujhani Formation is composed of medium to fine grained arenite to sub-greywacke along with shale and bands of dolomitic limestone suggesting littoral to shallow marine deposition with basaltic flows in some parts of the Gandak depression. This sequence shows a shift from the intertidal to the inner shelf environment, indicating gradual transgression. With a 27 Ma biostratigraphic break, the Tilhar Formation sits concordantly on top of the Ujhani Formation and forms a broad development of a semi-arid middle to outer shelf environment, consisting of grayish limestone interbedded with calcareous shale. The topmost unit, Karnapur Formation is primarily arenaceous and is composed of fine to medium-grained sandstone and non-calcareous fissile shale and was deposited in a shallow marine environment (Shukla et al., 1994) (Fig. 4b). Though oil indications were reported during drilling through these formations in these wells in GB, no commercial quantities of hydrocarbon have been encountered in any of the drilled wells. This implies the existence of a valid separate Neo-Proterozoic petroleum system limited to the Sarda and Gandak depressions.

3 Methodology

In this study we have attempted a structural analysis of the Ganga basin, mapping fault trends from an integrated study of seismic, gravity and morphostructural interpretations. The results of the structural analysis are then correlated with the tectonic trends and structural patterns of the Bikaner-Nagaur and Oman Salt basins in a plate tectonic reconstruction for the Cambrian-Proterozoic span.

Since the GB forms a foreland basin of an active thrust belt, the area is considered to be one of the most neotectonically active terrains (Table 6). Pre-existing structures were reactivated (Misra and Mukherjee 2015) during neotectonic reactivations. These are reflected in the geomorphic elements on topographic surfaces especially in terms of drainage system. These reactivated signatures mostly constitute localized modifications or anomalies in channel pattern and gradient changes in channels and imply presence of active subsurface structural elements.

Local drainage anomalies are categorized based on their geometries such as *(i)* channels joining approximately at right angles (rectangular drainages), *(ii)* drainage following a linear course (rectilinear drainages), *(iii)* asymmetrically distorted/ compressed meanders, and *(iv)* channels taking abrupt high-angle turns from their original course (drainage offsets). These anomalies are subsequently interpreted in terms of their causative faults (Fig. 5a) based on principles detailed in Mazumder et al., (2011, 2013a, b, 2016). These anomalies are verified by field checks where anomalous changes in drainage are substantiated by neotectonics-based field evidences, e.g., abnormal incisions, development of cliff in river bank and unpaired terracing (Fig. 5b).

To add a depth connotation to these surface morphostructural faults, fault interpretations were carried out from the available seismic data sets (up to 8000 m depth/

Table 6 Few neotectonic based studies carried out in Ganga basin

Authors	Specific locations	Key approaches	Conclusions
Sahu et al., (2010)	Patna and surrounding areas	Fluvial anomalies	Presence of active faults
Shukla et al., (2012)	Areas near Varanasi	Optically stimulated luminescence (OSL) dating of sediments	Channel migration due to active tectonics
Singh, (2021)	Area overlying Faizabad ridge	Morphotectonic parameters and landform studies	Area is neotectonically active influencing badland formation
Kumar et al., (2018)	Kalyani river basin located in the interfluve region of Gomati-Ghaghara river	Morphometric and landform analysis	Channel course is tectonically influenced
Kashyap et al., (2022)	Area overlying Moradabad fault	Geomorphological analysis, transverse-tectonic analysis, previous seismic records and Ground Penetrating Radar analysis	Area is neotectonically active, triggered by the Himalayan tectonics
Bajpai, (1989)	Kanpur-Unnao area	Surface and sub-surface evidences	Tectonic subsidence

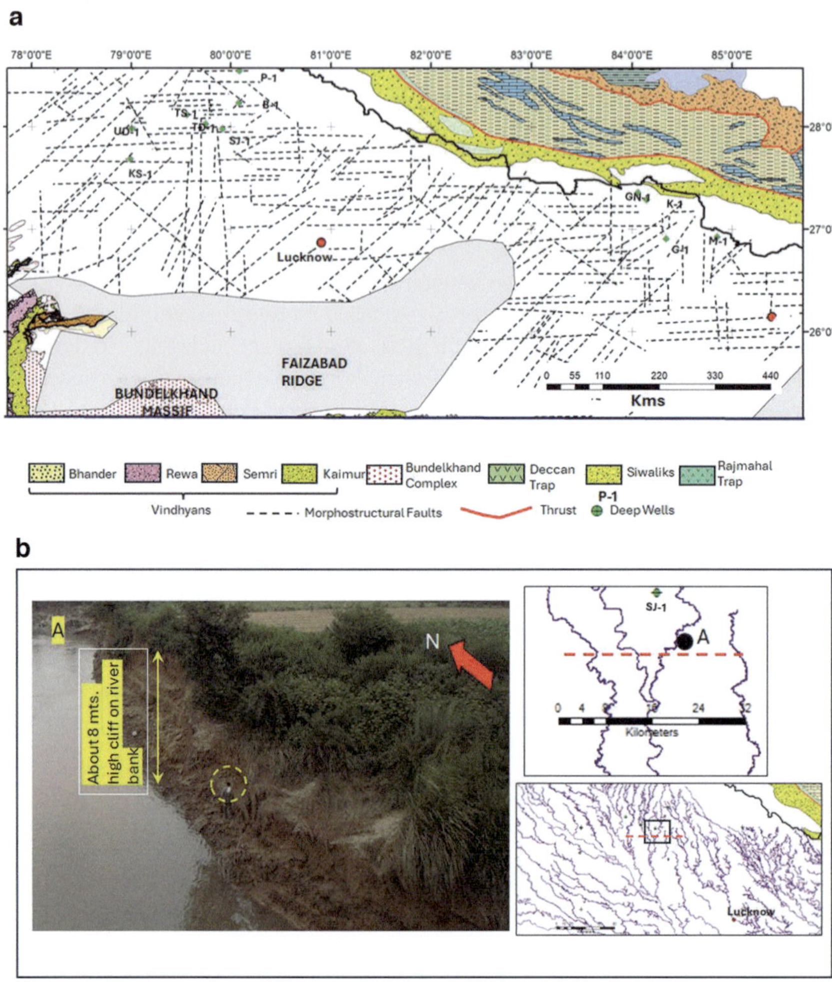

Fig. 5 **a** Morphostructural faults interpreted on basis of anomalous drainage geometries as a part of the present study (drainage map in inset of in this figure). **b** Field photo showing incision developed in bank of a channel in the Sarda depression, part of western Ganga Plain due to the effect of an E-W morphostructural fault (see inset maps)

granitic basement). The data-set indicates distinct ENE-WSW fault trends mapped atop the time-structure map of the Neoproterozoic Ujhani Formation interpreted based on displacement of delineated horizons. These faults are found to spatially correlate and broadly overlie the E-W surface morphostructural trends suggesting that these faults at least continue down to top of Ujhani Formation (Fig. 6a). Additionally, residual gravity data in the basin (KDMIPE, 2017) indicates regional ENE-WSW trends in the area correlating with the Neoproterozoic seismic faults and surface

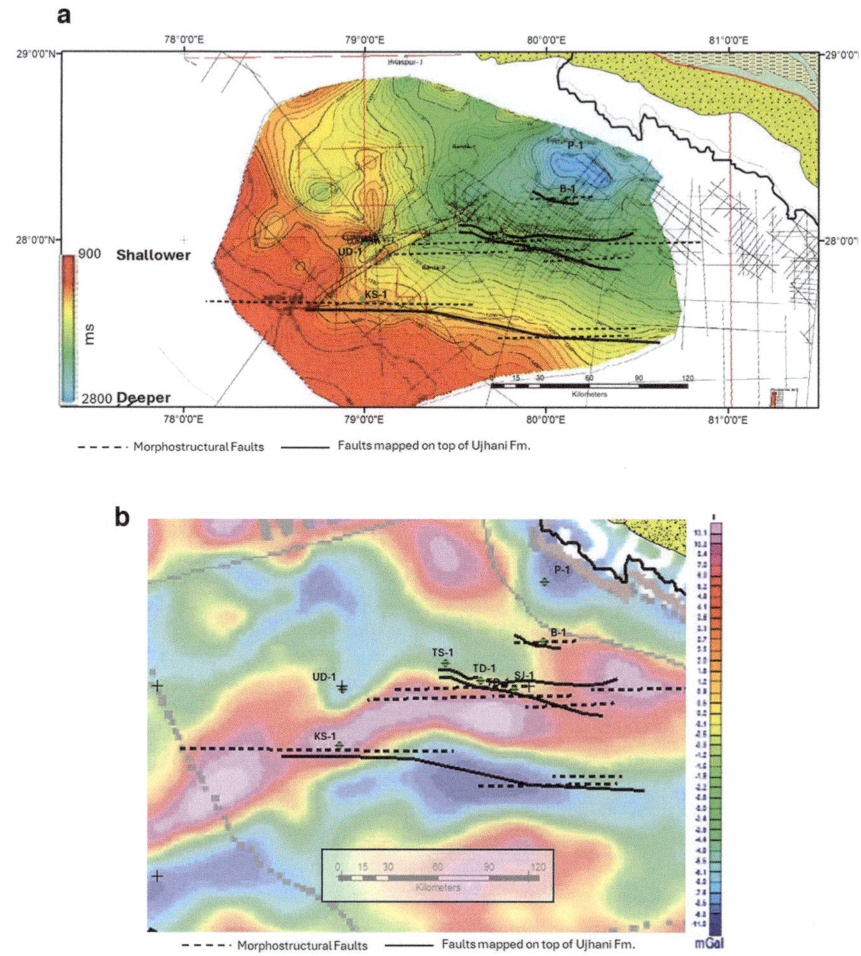

Fig. 6 **a** E-W trending morphostructural faults spatially correlating with faults mapped on the time-structure map on Ujhani Top. **b** E-W trending morphostructural faults and faults mapped on the Ujhani Top spatially correlating with residual gravity trends

morphostructural faults attributing them as their alias with a deep-seated crustal nature (Fig. 6b).

A qualitative plate tectonic reconstruction of the Arabian and the Indian plate was attempted in a GIS platform (ArcGIS 10.5, 2016) at the end of Neoproterozoic (based on model by Kroner & Stern, 2004). The Indian plate rotated ~60° clockwise with respect to its present position. The reconstruction depicts that the NF in the Arabian plate are tectonically aligned with the E-W faults mapped at the end of Neoproterozoic in Bikaner-Nagaur-Jaisalmer basins and in the GB. This implies a continuity of the NF in the GB (Fig. 7).

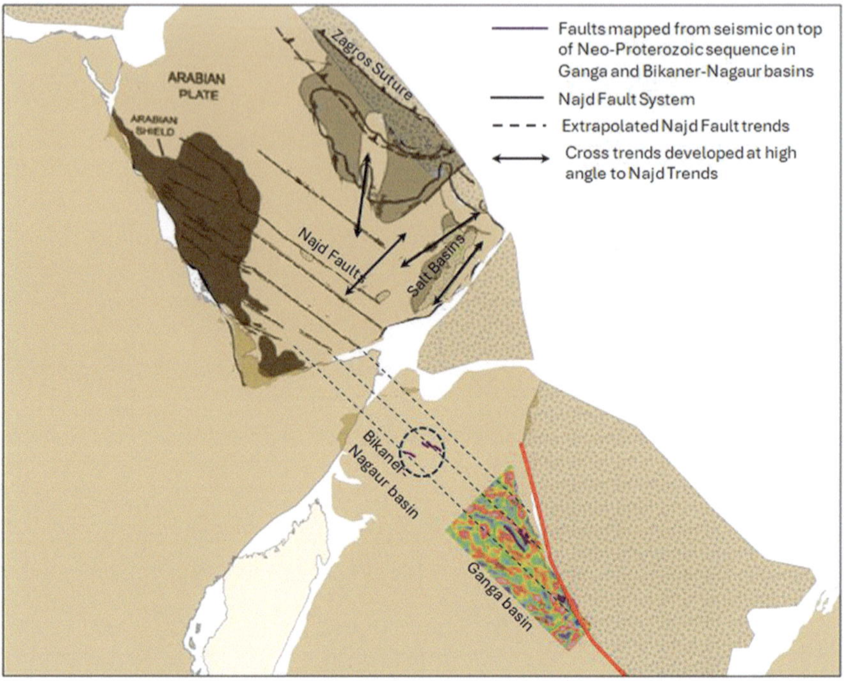

Fig. 7 Qualitative plate tectonic reconstruction at the end of Proterozoic (based on Kröner and Stern 2004) depicting similar tectonic alignment of the Najd Faults and E-W faults mapped at Neoproterozoic level in Ganga and Bikaner-Nagaur-Jaisalmer suggesting continuity of Najd faults on Indian plate. E-W trends in residual gravity data used as an alias to E-W trends (after Perotti et al., 2011)

4 Discussions

Sentinel 2 Level-2A composite images of bands 4, 3 and 2 of 12 m resolution that provides orthorectified Surface Reflectance (Bottom-Of-Atmosphere) corresponding to an area of ADFB exposures indicates E-W zones of faulting across the ADFB (Fig. 8a). These E-W strike-slip faults are spatially correlatable in terms of trend and continuity with the E-W fault trends mapped in the GB and suggest these fault trends are continuous across the ADFB from Bikaner-Nagaur basin on to the Ganga basin. Assuming the ADFB to be a boundary between the Marwar craton hosting the Bikaner-Nagaur basin and the Bundelkhand craton hosting the GB, this implies the possibility that the Najd Faults continue across the ADFB at least up to the Monghyr-Saharsa Ridge (Fig. 8b). The Najd Faults in the Arabian plate formed around Neoproterozoic (Nehlig et al., 2002) whereas the faults interpreted atop the Ujhani Formation in the GB having a similar trend are also attributed with a similar age of activation. This leads to a premise that the Najd Faults in the Arabian plate,

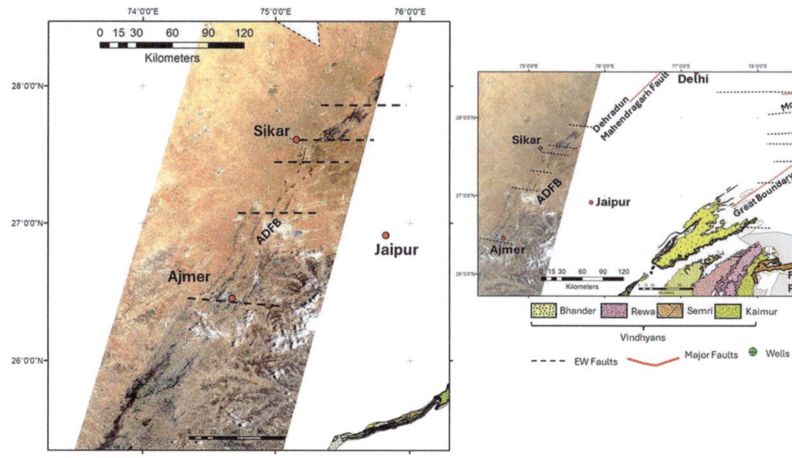

Fig. 8 E-W strike slip faults affecting Aravalli Delhi Fault Belt (ADFB) correlating with E-W faults mapped in Ganga Basin suggesting the continuity of Najd Faults across ADFB (after Mazumder & Misra, 2022)

faults mapped in Bikaner-Nagaur basin in western part of Indian plate and that in GB in eastern parts of Indian plate are parts of the same fault system.

Considering the Najd Faults to be continuous as E-W oriented faults in Ganga Basin, the faults in the Ganga basin can also be assumed to be associated with a similar strike-slip as in faults of the Arabian plate as well as in Bikaner Nagaur basin. The strike-slip movement associated with these E-W faults in turn leads to the development of NNE-SSW normal faults bounding rift systems in the above areas. Resultant grabens are believed to be ~100 km in southern Oman and ~200 km in Pakistan formed due to trans-tensional stresses resulting from Najd faulting.

In the GB, a density analysis of N-S morphostructural faults indicate two major clusters, one of which correlates with the Sarda Depression in the western GB whereas the other is a much wider density cluster in the Gandak Depression in the Eastern Ganga Basin (Fig. 9a). These N-S trends have been established by field checks along selected traverses (Fig. 9b). A NW–SE oriented 2D seismic section across part of the eastern cluster indicates signatures of two rift-like broadly N-S normal fault-bounded depressions/grabens (Rift 2a,b) filled with Neoproterozoic sediments (based on lithology and log correlations that are separated by a horst block (Fig. 9c). The eastern shoulder of the graben is formed by the Monghyr Saharsa Ridge, which demarcates the eastern limit of the GB. This implies that the N-S morphostructural trends probably are the reactivated manifestations of the Neoproterozoic rifts. Rift basins are commonly associated with volcanics. In the case of GB, basaltic layers in parts of the Gandak depression within the Ujhani Formation have been ascertained.

The high-angle, basement rooted N-S/NNW-SSE/NNE-SSW trending faults defining the rift zones are associated in some areas within the rifts with compressional structures such as antiforms. These fault-bound antiformal structures in Bikaner-Nagaur basin are limited to the Neoproterozoic-Cambrian strata only and

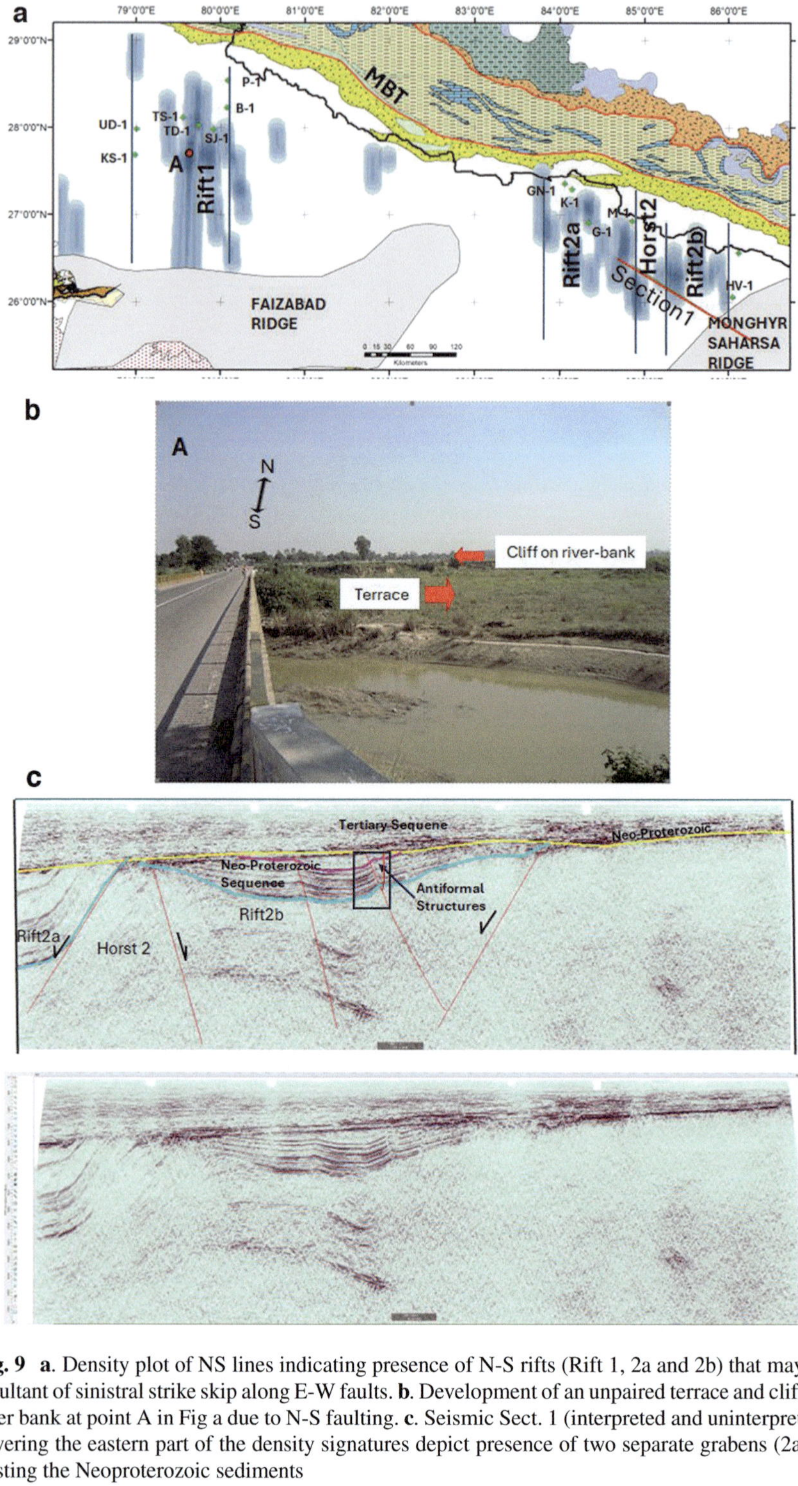

Fig. 9 **a**. Density plot of NS lines indicating presence of N-S rifts (Rift 1, 2a and 2b) that may be resultant of sinistral strike skip along E-W faults. **b**. Development of an unpaired terrace and cliff on river bank at point A in Fig a due to N-S faulting. **c**. Seismic Sect. 1 (interpreted and uninterpreted) covering the eastern part of the density signatures depict presence of two separate grabens (2a, b) hosting the Neoproterozoic sediments

are presumably related genetically with the Neoproterozoic-Cambrian compression/transpression (Mandal et al., 2022). Considering an *en-echelon* pattern of the E-W faults in the GB, compressional/transpressive antiformal structures can also be envisaged with N-S alignment or at a high-angle to the E-W strike-slip faults. These transpressive antiformal structures are believed to be a combination of thrusting and wrenching (Casas et al., 2001) and are predominantly associated with upward spreading splays of high-angle reverse faults that merge at depth with a sub-vertical basement rooted fault segment (Huang & Liu, 2017).

The transpressive antiformal structures with a broadly N-S axis in the GB were reactivated due to the Himalayan orogeny in the Tertiary causing them to be imprinted on the topographic surface as geomorphic highs. Such geomorphic highs are developed when the course of a drainage is obstructed by a concealed sub-surface structure and the drainage deviates from its normal pathway and move as a set of curved streams along the periphery of the concealed structure (Fig. 10a). Since the coverage of seismic data in the GB is limited to 2D lines and few 3D regions, the geomorphic highs can be used as an alias to premise these antiformal areas (Fig. 10b). An association of these geomorphic highs with N-S morphostructural trend establishes them to be a product of transpression/shortening due to E-W strike slip. Structural restoration attempted along another broadly NW–SE seismic section in the GB indicated that post-restoration of erosion along Neoproterozoic sequences, the reconstructed Neoproterozoic structures emerge as an asymmetric anticline generated by reactivation of basement associated sub-vertical faults, primarily in transpressive stress regime analogous to that in Bikaner Nagaur basin (Fig. 11a–c).

In order to validate these faults and further add value to the geomorphic-highs/antiformal units, these have been spatially correlated with results of adsorbed gas surveys carried out in the area. In case, the faults act as a conduit for hydrocarbon, the lighter factions/alkanes (especially C2-C4) seep out along the faults and get adsorbed on the soil near the fault showing a higher concentration than the surrounding (Srinivas et al., 2014), e.g., in areas near Muzaffarpur (Bihar). The geomorphic highs that serve as manifestation of the Neo proterozoic antiformal structures are also found to associated with the adsorbed gas maxima implying a thermogenically mature source rock of an active petroleum system that has the potential of generating and expelling hydrocarbons (Fig. 12a). An analogy can be observed in Bhagewala-Bijnot area of Bikaner-Nagaur-Punjab platform area where anomalous intensities of C2-C4 hydrocarbons are observed in surface adsorbed gas samples above the antiformal structures (Yasin et al., 2022).

The areas overlying such hydrocarbon pools also are places of micro-seepages resulting from vertical movement of lighter components of hydrocarbons from reservoir to surface through permeable routes in the overlying rock. These micro-seepages create a reducing environment inducing chemical and mineralogical changes in the surface cover sediments that overlies the reservoir area which in the present case is the area overlying the transpressive antiforms. This alteration due to the reducing environment causes an anomalous enrichment of ferrous iron minerals and higher concentrations of clay and carbonate minerals in the covering soils/rocks. These areas can be delineated by multispectral remote sensing where the mineralogically

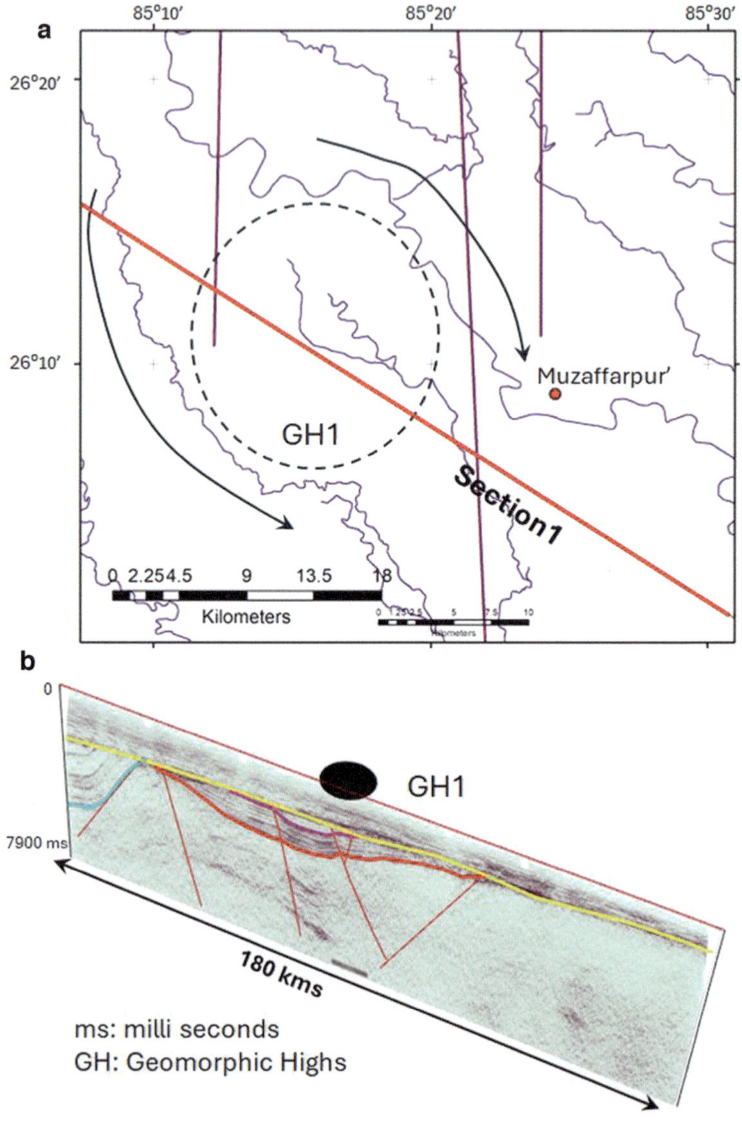

Fig. 10 Geomorphic High (GH1) associated with a N-S morphostructural trend correlating with a transpressive antiform delineated in Sect. 1b

anomalous zones appear as spectral anomalies (Garain et al., 2019) as in the Motihari and the Muzzafarpur areas. Landsat-8 OLI 6-band multispectral satellite image data of 06-May-2023 corresponding to the VNIR-SWIR channels forms the primary data-set for such analysis (Misra & Maurya, 2020). This data was digitally processed to highlight the zones of alteration using standard spectral mineral indices developed

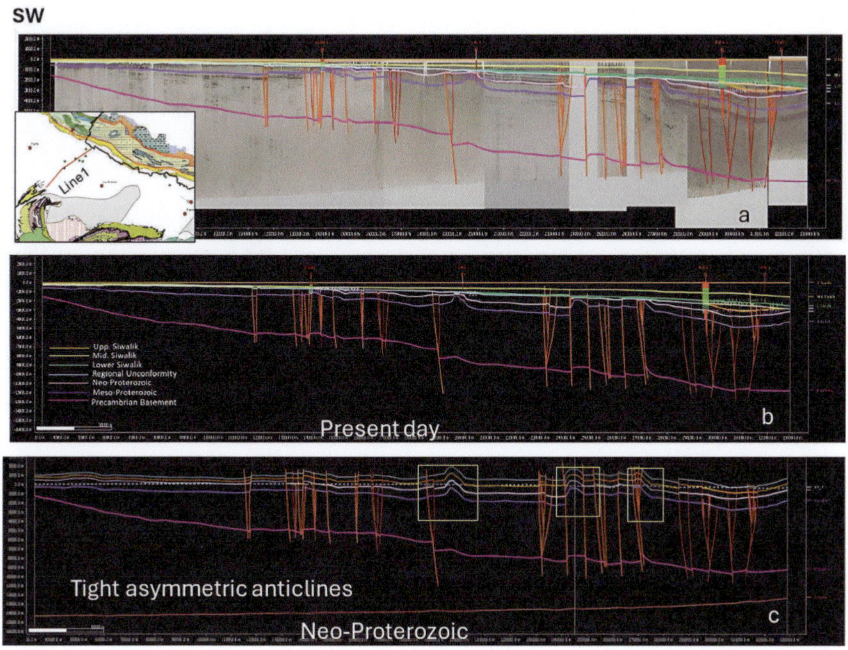

Fig. 11 **a**. Present day SW-NE seismic profile across the Sarda Depression of the GB. Inset: line layout. **b**. Present-day structural configuration across the seismic profile showing sub-vertical basement related faults. **c**. Erosion reconstructed above the Neoproterozoic. The reconstructed structures emerge as tight asymmetric anticlines formed by reactivation of basement associated sub-vertical faults similar to the Bikaner-Nagaur basin. Yellow rectangles: tight asymmetric anticlines

to highlight spectral signatures of key alteration mineral species. Three band false color composites (FCC) of different spectral indices were tested to map distribution of key alteration mineral species in the study area.

Of these, FCCs corresponding to Ferric Oxide Content- Ferrous (2+)-Clays as Red–Green–Blue composite (RGB) were found to effectively spectrally capture most of the alteration mineral distribution in the region. The mineralogically anomalous zones suggesting areas of micro-seepages are found to be demarcated in a light yellow tone in the FCC (Fig. 12b). A good degree of correlation is observed between the spectral anomalies and the adsorbed gas anomalies implying that this procedure can be used as an alternative in areas where adsorbed gas surveys does not exist. Geomorphic high GH1 is found to correlate with such an instance of spectral anomaly and adsorbed gas maxima indicating the area overlying a transpressive antiformal structure may host a hydrocarbon reservoir.

A 1D modeling of the GB indicates onset of oil generation for the Ujhani Formation similar to that of the Huqf Supergroup of Oman, Proterozoic-Cambrian of Indus Basin and Bikaner-Nagaur basin. Additionally, studies indicate heavy oil from Bilara and Jodhpur Formation in Baghewala-1 well of Bikaner-Nagaur basin (India) and Salt Range Formation in Karampur-1 well of the Punjab Platform (Pakistan) resemble

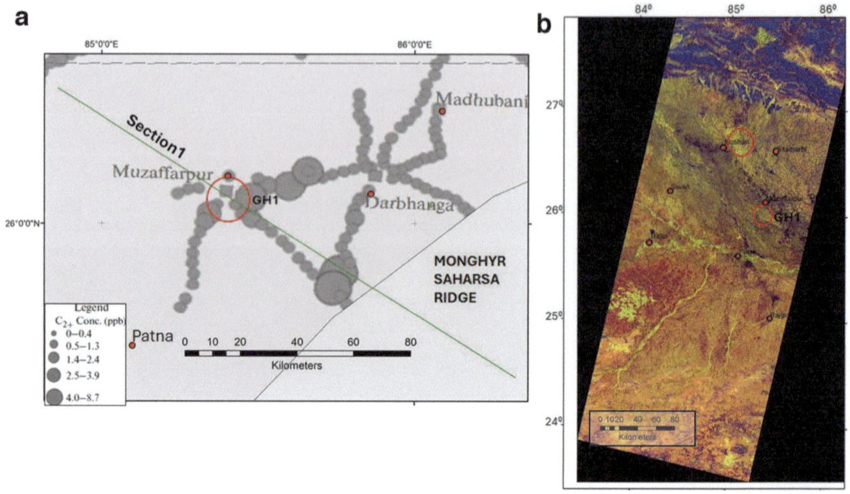

Fig. 12 **a** Adsorbed gas anomalies for C^{2+} concentrations in eastern part of Ganga basin (after Srinivas et al., 2014) with geomorphic high GH1 correlating with an area of anomalous high concentration near Muzaffarpur. **b** False color composite (FCC) corresponding to Ferric Oxide Content-Ferrous (2+)-Clays as RGB delineating mineralogically anomalous zones suggestive of micro-seepages demarcated by lighter yellow color. Geomorphic high GH1 near Muzaffarpur correlating with adsorbed gas maxima corresponds to one such anomalous zone. A similar zone is also identified northward near Motihari

geochemically with the oil from the Infracambrian Huqf Group (Yasin et al., 2022). Overall, it can be premised that the hydrocarbon generation in the GB probably was coeval and genetically similar to that of the Huqf Supergroup in Oman, the Salt Range Formation in the Indus basin and the Marwar Supergroup in the Bikaner-Nagaur basin. They were limited to the Ujhani-Tilhar-Karanpur plays in the Sarda and the Gandak depressions, with shales within Ujhani Unit acting as a source. A vertical migration along the N-S faults is assumed with a subsequent structural entrapment within the fault associated antiformal structures. This play model indicates that the areas of geomorphic highs associated with N-S faults and attributed with adsorbed gas and alteration related spectral anomalies to be exploration-worthy.

5 Conclusions

The Najd Fault system with a dominant strike slip component is established to extend from the Arabian plate on to the Punjab platform and the Bikaner-Nagaur basin part of the Indian plate. Analysis carried out as a part of this study premises that this fault system cuts across the Aravalli Delhi Fold Belt and further continues across the Ganga basin at least up to its eastern boundary bounded by the Monghyr Saharsa

Ridge. The constituent faults of the Najd Fault system in the Ganga basin is manifested in the gravity and the seismic data, and in the present-day geomorphology as present day *en-echelon* E-W trends. Strike slip activity along these trends developed two broadly N-S oriented rift systems correlating with the Sarda and the Gandak depressions. These depressions served as depocenters of Neoproterozoic Ujhani, Tilhar and Karanpur Formation that are stratigraphically coeval to their counterparts in the western Indian plate and the Arabian plate. Transpression/strike slip associated compression developed asymmetric fault-associated antiformal structures with a N-S fold axis in these rift systems. These antiformal units can be envisaged to be structural entrapments for a Neoproterozoic play. Geomorphic highs interpreted from morphostructural analysis serves as an effective instrument to delineate these antiformal areas with adsorbed gas anomalies and remote sensing-based mineralogical alteration related spectral anomalies serving as an essential tool to establish prospectivity. In the present study, the Muzaffarnagar-Darbhnga-Motihari area forms one of the prospective areas for hydrocarbon exploration in Infracambrian sediments.

Acknowledgements Soumyajit Mukherjee (IIT Bombay) invited, handled and reviewed this work in three rounds. The views expressed in this manuscript are solely that of the authors and may not be similar to the views of the organization they represent. Mukherjee (2026) summarized this article.

References

Aadil, N., & Sohail, G. M. (2011). Stratigraphic correlation and isopach maps of Punjab Platform in Middle Indus Basin, Pakistan. *Search and Discovery* 10364. Greater Noida, New Delhi, India: Adapted from extended abstract presented at GEO-India.

Abedin, Z., & Iqbal, W. (2014). The sedimentology and sequence stratigraphy of early Cambrian Kussak formation, Salt Range Pakistan. In *76th EAGE Conference & Exhibition 2014 Amsterdam RAI*. The Netherlands.

Ahmad, N., Ahsan, N., Sameeni, S. J., Muhammad, M. M. A. F., & Khan, B. (2013). Sedimentology of the early Middle Cambrian Jutana Formation of Khewra area, Eastern Salt Range, District Chakwal, Pakistan. *Science International (Lahore), 25*(3), 551–558.

Al-Husseini, M. I. (2000). Origin of the Arabian plate structures: Amar collision and Najd rift. *GeoArabia, 5*(4), 527–542.

Ali, R. I., Ghazi, S., Khan, M. J., Dar, Q. U. Z., Ahmed, S., Baig, K. S., Ali, S. H. (2023). The Sedimentological and provenance analysis of the Cambrian succession in the Saiduwali Section, Trans Indus ranges, Pakistan. *Pakistan Journal of Hydrocarbon Research, 26*, 23–48.

Ali, S. H., Shoukat, N., Bashir, Y., Qadri, S. M. T., Wahid, A., & Iqbal, M. A. (2022). Lithofacies and Sedimentology of Baghanwala formation (Early-Middle Cambrian), Eastern Salt Range, Pakistan. *Pakistan Journal of Science & Industrial Research, 65A*(2), 159–168.

Allen, P. A. (2007). The Huqf Supergroup of Oman: Basin development and context for Neoproterozoic glaciation. *Earth Science Reviews, 84*, 139–185.

Arif, M., Ali, B., Khan, M. A., & Ghafoor, A. (2009). Petrography of sandstone of the Lumshiwal formation from the Samana range, Hangu, Northwestern Pakistan: Implications for provenance, diagenesis and environments of deposition. *Pakistan Journal of Hydrocarbon Research, 19*, 01–09.

Azmi, R. J., Joshi, D., Tiwari, B. N., Joshi, M. N., & Srivastava, S. S. (2008). A synoptic view on the current discordant geo- and biochronological ages of theVindhyan Supergroup, central India. *Himalayan Geology, 29*(2), 177–191.

Bajpai, V. N. (1989). Surface and subsurface evidence of neotectonics and the aquifer disposition in central gangetic alluvial terrain of Kanpur-Unnao region in Uttar Pradesh, India. *Journal Pf Indian Social Remote Sensing, 17*, 47–53.

Biswas, T., Bose, N., Dutta, D., & Mukherjee, S. (1988). Arc-parallel shears in collisional orogens: Global review and paleostress analyses from the NW Lesser Himalayan Sequence (Garhwal region, Uttarakhand, India). *Marine and Petroleum Geology, 138*, 105530. https://doi.org/10.1016/j.marpetgeo.2022.105530

Biswas, M., Raha, A., Mukherjee, S., & Kotak, V. S. (2024) Geomorphic imprints of active tectonics of the Bikaner-Nagaur petroliferous rift basin and its surroundings (western Rajasthan, India). *Journal of the Geological Society of India, 100*, 377–390.

Casas, A. M., Gapais, D., Nalpas, T., Besnard, K., & RomaÂn-Berdiel, T. (2001). Analogue models of transpressive systems. *Journal of Structural Geology, 23*, 733–743.

Choudhary, L. R., Srivastava, S. K., & Jakhar, S. R. (2023). Geology and petroleum prospects of Bikaner-Nagaur basin. *Dew Journal,* 41–70

Cozzi, A., Rea, G., & Craig, J. (2012). From global geology to hydrocarbon exploration: Ediacaran–early Cambrian petroleum plays of India, Pakistan and Oman. *Geological Society, London, Special Publications, 366*. https://doi.org/10.1144/SP366.14.

Domeier, M. (2018). Early Paleozoic tectonics of Asia: Towards a full-plate model. *Geoscience Frontiers, 9*(3), 789–862.

Dutta, T., Prasad, R. R., Nonia, B. P., Shankar, R., Ojha, S., & Joshi, G. (2023). Geological model and prospectivity perception of Ganga-Punjab basin: Exploration status, global analogy and opportunities. In *14th Biennial International Conference and Exposition, SPG 2023*. Kochi.

Garain, S., Mitra, D., Das, P. (2019). Detection of hydrocarbon microseepage-induced anomalies by spectral enhancements of Landsat 7 ETM+ images in part of Assam–Arakan Fold Belt, India. *Journal of Petroleum Exploration and Production Technology*. https://doi.org/10.1007/s13202-019-00747-w.

Ghazi, S., Ahmad, S., Riaz, M., & Zafar, T. (2020). Sedimentology and palaeo-environmental reconstruction of the early Cambrian Kussak Formation, Salt Range, Pakistan. *Current Science, 119*(10, 25), 1671–1684.

Ghori, K. A. R., Craig, J., Thusu, B., Lüning, S., & Geiger, M. (2009). Global Infracambrian petroleum systems: A review. *Geological Society, London, Special Publications, 326*, 109–136. https://doi.org/10.1144/SP326.6

Gorain, S., & Kumar, A. (2021). Seismic sequence stratigraphy and Paleo-structure analysis of Proterozoic sediment within Ganga basin. In *EAGE Annual 82nd Conference and Exhibition, Amsterdam, Netherlands.*

Gorain, S., Kumar, A., & Reddy, C. L. (2023). Application of paleo-structural and seismic sequence stratigraphy analysis to determine the potential plays within the Proterozoic sequence of the Ganga Basin, India. *Journal of Asian Earth Sciences, X9.*

Goswami, P. K., Pant, C. C., & Pandey, S. (2009). Tectonic controls on the geomorphic evolutionof alluvial fans in the Piedmont Zone of Ganga Plain, Uttarakhand, India. *Journal of Earth System Science, 118*(3), 245–259.

Hallett, D., & Clark-Lowes, D. (2016). Chapter 2—Plate Tectonic History. In *Petroleum Geology of Libya* (Second ed.). Elsevier. https://doi.org/10.1016/B978-0-444-63517-4.00002-8.

Hefferan, K., Soulaimani, A., Samson, S. D., Admou, H., Inglis, J., Saquaque, A., Latifa, C., & Heywood, N. (2014). A reconsideration of pan African orogenic cycle in the Anti-Atlas mountains, Morocco. *Journal of African Earth Sciences, 98*, 34–46.

Heidorn, R., Salem, H., Shuaili, S., Khattak, A., & Pentland, C. (2018). Recent advances in the understanding of the salt tectonic evolution of the eastern Flank at regional and field scale and its relationship to the South Oman Salt Basin of the Sultanate of Oman. In *Paper presented at the*

Abu Dhabi International Petroleum Exhibition & Conference, Abu Dhabi, UAE. Paper Number: SPE-192611-MS. https://doi.org/10.2118/192611-MS.

Huang, L., & Liu, C. (2017). Three types of flower structures in a divergent-Wrench fault zone. *Journal of Geophysical Research: Solid Earth, 122*(10), 478–497. https://doi.org/10.1002/2017JB014675

Husseini, M. I. (1988). The Arabian lnfracambrian extensional system. *Tectonophysics, 148*(1988), 93–103.

Husseini, M. I., & Husseini, S. I. (2014). Origin of the Infracambrian salt basins of the Middle East. *Geological Society, London, Special Publications, 50*, 279–292. https://doi.org/10.1144/GSL.SP.1990.050.01.14.

Kasana, P., Singh, V., & Devrani, R. (2021). Investigation of flexural bulge of the Indian plate for controls on landscape evolution. In *Conference: American Geophysical Union*. https://doi.org/10.1002/essoar.10509428.1.

Kashyap, R., Pati, P., Jaiswal, M. K. (2022). Tectonic geomorphology along the transverse basement fault (Moradabad fault) in the Ganga basin: A nexus between the Himalayan tectonics and the pre-Himalayan transverse basement structure Catena. /https://doi.org/10.1016/j.catena.2021.105981.

KDMIPE (2017). Residual gravity anomaly atlas of Indian Sedimentary Basins. KDMIPE, ONGC.

Khalid, P., Akhtar, S., & Khurram, S. (2020). Reservoir Characterization and multiscale heterogeneity analysis of Cretaceous Reservoir in Punjab platform of Middle Indus Basin. *Pakistan, 45*, 4871–4890.

Khan, M., & Liu, Y. (2019). Geodynamic evolution of the offshore Indus Basin Pakistan: The western Indian plate passive continental margin. *Geophysical Journal International, 217*(2), 1366–1386.

Khan, M. A., & Khan, M. J. (1979). Stratigraphy of the Baghanwala formation, Khewra Gorge, Khewra, Jheum District; Punjab: Pakistan, National Centre of Excellence in Geology, New Campus, Peshawar. *Geological Bulletin, 12*, 21–31. http://nceg.uop.edu.pk.

Kroner, A., & Stern, R. J. (2004). *Encyclopedia of Geology, 1*, 1–12. https://doi.org/10.1016/B0-12-369396-9/00431-7

Kumar, D., Singh, D. S., Prajapati, S. K., et al. (2018). Morphometric parameters and neotectonics of Kalyani River Basin, Ganga Plain: A remote sensing and GIS approach. *Journal of Geological Society of India, 91*, 679–686. https://doi.org/10.1007/s12594-018-0923-9

Kusky, T. M., & Matsah, M. (2003). Neoproterozoic dextral faulting on the Najd fault system, Saudi Arabia, Preceded Sinistral Faulting and Escape Tectonics Related to Closure of the Mozambique Ocean. *Geological Society London Special Publications, 206*(1), 327–361.

Loosveld, J. H. R., Andy, B., & Terken, J. J. M. (1996). The tectonic evolution of interior Oman. *GeoArabia, 1*(1), 28–51.

Maheshwari, A., Sial, A. N., & Mathur, C. (2002). Carbon isotope fluctuations through the Neoproterozoic-lower Cambrian Birmania basin, Rajasthan, India. *Carbonates and Evaporites, 17*(1), 53–59.

Mandal, A., Saha, D., & Kumar, A. (2022). Structural analysis and seismic stratigraphy for delineation of Neoproterozoic-Cambrian petroleum system in central and eastern part of Bikaner-Nagaur basin, India. *Journal of Petroleum Exploration and Production Technology*. https://doi.org/10.1007/s13202-021-01432-7

Mazumder, S., & Misra, A. (2022). GIS based integration of multi-spectral remote sensing and morphostructural analysis for delineating potential areas of hydrocarbon exploration in Western Vindhyan Basin, India. *Energy Geoscience, 3*(2), 191–201.

Mazumder, S., Adhikari, K., Mitra, D. S., Mahapatra, S., & Pangtey, K. K. S. (2016). A Neotectonic based geomorphic analysis using remote sensing data to delineate potential areas of hydrocarbon exploration: Cachar area, Assam. *Journal of Geological Society of India, 88*, 87–97.

Mazumder, S., Dave, H. D., Samal, J. K., & Mitra, D. S. (2011). Delineation of shallow structures as exploratory input in Ganga Basin based on Morphotectonic studies. *ONGC Bulletin, 46*, 190–198.

Mazumder, S., Pangtey, K. K. S., & Mitra, D. S. (2013a). Delineation of a possible subsurface ridge in Onshore Palar Basin based on morphotectonic studies and its implications. In *Proceedings of 10th Biennial International Conference & Exposition*. Kochi: SPG.

Mazumder, S., Tep, B., Pangtey, K. K. S., & Mitra, D. S. (2013b). A Morphotectonic based approach to derive tectonic framework and possible areas of hydrocarbon accumulation in onshore Palar Basin. *ONGC Bulletin, 48*, 113–123.

Meyer, S. E., Passchier, C., Abu-Alam, T., & Stuwe, K. (2014). A strike-slip core complex from the Najd fault system, Arabian shield. *Terra Nova, 26*(5), 387–394

Misra, A., & Maurya, P. C. (2020). Principal component analysis of multispectral remote sensing data for spectral characterization of hydrocarbon seepage induced alteration of subaerial regolith in a part of Son Valley Vindhyan Basin, Kochi: SPG.

Misra, A. A., & Mukherjee, S. (2015). Tectonic inheritance in continental rifts and passive margins. In *Springerbriefs in earth sciences*. ISBN 978-3-319-20576-2.

Mohamed, A., & Wahed, A. E. (2010). The role of the Najd fault system in the tectonic evolution of the Hammamat molasse sediments, Eastern Desert, Egypt. *Arabian Journal of Geoscience, 3*, 1–26.

Montemagni, C., Fulignati, P., Iaccarino, S., Marianelli, P., Montomoli, C., & Sbrana. A. (2016). Deformation and fluid flow in the Munsiari Thrust (NW India): A preliminary fluid inclusion study.

Mukherjee, S. (2026). Introduction to "Structural Geology & Tectonics". In S. Mukherjee (Ed.), *Structural Geology & Tectonics*. Springer. ISBN: 978-981-95-4743-2.

Mukherjee, S., Talbot, C. J., & Koyi, H. A. (2010). Viscosity estimates of salt in the Hormuz and Namakdan salt diapirs, Persian Gulf. *Geological Magazine, 147*(4), 497–507.

Nehlig, P., Genna, A., Asfirane, F., Dubreuil, N., Guerrot, C., Eberlé, J. M., Kluyver, H. M., Lasserre, J. L., Goff, E., Le, N. N., Salpeteur, N., Shanti, M., Thiéblemont, D., & Truffert, C. (2002). A review of the pan-African evolution of the Arabian Shield. *GeoArabia, 7*(1), 103–124.

Nicholas, C. J., & Gold, S. E. P. (2012). Ediacaran-Cambrian Sirab formation of the Al Huqf region, Sultanate of Oman. *GeoArabia, 17*(1), 49–98.

Patel, N. K., Pati, P., & Verma, A. (2018). Seismotectonic study of the basement fault in the western Ganga plain, India: A neotectonic approach. In *Conference: American Geophysical Union At: Walter E.* Washington, DC, USA: Washington Convention Center.

Patel, N. K., Pati, P., Verma, A. K., Dash, C., Gupta, A., & Sharma, V. (2020). Seismicity around the Mahendragarh-Dehradun basement fault in the western Ganga plain, India: A neotectonic perspective. *International Journal of Earth Sciences, 109*, 689–706. https://doi.org/10.1007/s00531-020-01826-8

Perinçek, D., Saner, S., & Al-Hinai, K. G. (2006). Rejuvenation of the pre-Cambrian Najd Fault system and its importance in the oil province of Saudi Arabia. *Mineral Research and Exploration of Bulletin, 132*, 73–82.

Perotti, C. R., Carruba, S., Rinaldi, M., Bertozzi, G., Feltre, L., & Rahimi, M. (2011). The Qatar–South Fars arch development (Arabian Platform, Persian Gulf): Insights from seismic interpretation and analogue modelling. In U Schattner (Ed.), *New Frontiers in Tectonic Research—At the Midst of Plate Convergence*. https://doi.org/10.5772/20299.

Pollastro, R. M. (1999). Ghaba salt basin province and Fahud salt basin Province, Oman. In *Geological Overview and Total Petroleum Systems*. Report, U.S. Department of the Interior U.S. Geological Survey, Open-File Report 99-50-D.

Raju, S. V., Mathur, N., & Sarmah, M. K. (2014). Geochemical characterization of Neoproterozoic heavy oil from Rajasthan, India: Implications for future exploration of hydrocarbons. *Current Science, 107*(8), 1298–1305.

Sahu, S., Raju, N. J., & Saha, D. (2010). Active tectonics and geomorphology in the Sone-Ganga alluvial tract in mid-Ganga Basin, India. *Quaternary International, 227*(2), 116–126.

Saqab, M. M., Murtaza, G., Khan, M. A., Ahmad, T., & Rahim, H. (2009). Sedimentology and reservoir potential of the early Cambrian Khewra Sandstone, Khewra Gorge, Eastern Salt Range, Pakistan. In *Conference: PAPG ATC 2009*.

Schroder, S., Schreiber, B. C., Amthor, J. E., & Matter, A. (2004). Stratigraphy and environmental conditions of the terminal Neoproterozoic-Cambrian Period in Oman: Evidence from sulphur isotopes. *Journal of the Geological Society, 161*(3), 489–499.

Sheikh, R. A., Jamil, M. A., McCann, J., & Saqi, M.I. (2003). Distribution of infracambrian reservoirs on Punjab platform in Central Indus Basin of Pakistan. In *AAPG Search and Discovery Article #90150©2012 PAPG/SPE Annual Technical Conference*. Islamabad, Pakistan.

Shukla, R., & Tewari, M. (2022). Proterozoic microfossil assemblage from the cherts of Birmania formation, Western Rajasthan, India. *Journal of Earth System Sciences,* 131–141. https://doi.org/10.1007/s12040-022-01890-5.

Shukla, S. N., Mudiar, B., & Misra, M. M. (1994). Lithostratigraphy of Ganga Basin, India. *Indian Journal of Petroleum Geology, 3*, 69–90.

Shukla, U. K., Srivastava, P., & Singh, I. B. (2012). Migration of the Ganga River and development of cliffs in the Varanasi region. *India during the Late Quaternary: Role of Active Tectonics, 171–172*, 101–113.

Siddiqui, N. K. (2012). A prospective Neoproterozoic-Cambrian hydrocarbon/exploration play in the Kirthar Fold Belt, Pakistan. *Geological Society London Special Publications, 66*(1), 123–130. https://doi.org/10.1144/SP366.13

Siddiqui, N. K. (2016). A prospective Neoproterozoic–Cambrian hydrocarbon/exploration play in the Kirthar Fold Belt, Pakistan. In G. M. Bhat, J. Craig, J. W. Thurow, B. Thusu & A. Cozzi (Eds.), *Geology and Hydrocarbon Potential of Neoproterozoic–Cambrian Basins in Asia* (Vol. 366, pp. 123–130). Geological Society, London: Special Publications. First published online March 14, 2012. https://doi.org/10.1144/SP366.13.

Singh, I. B. (1996). Geological evolution of the Ganga Plain, an overview. *Journal of the Paleontological Society of India, 41*, 99–137.

Singh, N. (2021). Morphotectonic signatures of the neotectonic control on the badland formation, Chapter 6. idr-lib.iitbhu.ac.in/xmlui/bitstream/handle/123456789/1932/10.

Singh, N. P. (2006). Mesozoic lithostratigraphy of the Jaisalmer basin, Rajasthan. *Journal of the Palaeontologic Society of India, Volume., 51*(2), 1–25.

Sohail, G. M., & Aadil, N. (2014). Infracambrian play in Punjab Platform, Pakistan, with reference to Bikaner-Nagaur Basin, India. *Arabian Journal of Geosciences, 8*(8). https://doi.org/10.1007/s12517-014-1621-6.

Srinivasa, C., Madhavi, T., Lakshmi, M., Mani, D., Kalpana, M. S., Patil, D. J., Dayal, A. M., & Raju, S. V. (2014). Geochemical characterization of adsorbed light gaseous alkanes in near surface soils of the Eastern Ganga Basin for hydrocarbon prospecting. *Geochemistry International, 52*(1), 68–76.

Stern, B. (1985). The Najd fault system, Saudi Arabia and Egypt: A late Precambrian rift-related transform system? *Tectonics, 4*(5), 497–511.

Suresh, M., Mamglik, A., & Thiagrajan, S. (2023). Basement and lithospheric structure of the central Ganga Basin between the Bundelkhand craton and the Sharda deep by magnetotellurics. *Tectonophysics, 863*.

Yasin, Q., Baklouti, S., Sohail, G. M., Asif, M., & Xufei, G. (2022). Evaluation of Neoproterozoic source rock potential in SE Pakistan and adjacent Bikaner-Nagaur Basin, India. *Scientific Reports*. https://doi.org/10.1038/s41598-022-14831-5

Yue, L., & Vandeginste, V. (2020). Neoproterozoic Nafun group sediments from Oman affected by an active continental margin. *Minerals, 10*(6), 511. https://doi.org/10.3390/min10060511

Preliminary Study of Artificial Slope Facets from Coal Dump Sites

Adrija Raha, Suvashree Das, Mery Biswas, and Soumyajit Mukherjee

Abstract Coal dump sites offer useful perspectives to understand stability of artificial slopes. Such studies require geotechnical research, which can be aided by numerical simulation model-based stability evaluations. Anthropogenic geomorphology can be well studied at the coal dump sites. Four coal dump sites close to Durgapur and Dhanbad (eastern India) have been examined in this study in order to better understand the different aspects of slopes and their instability risk. The studied anthropogenic landforms have clearly demonstrated the different slope segments e.g. free face, convex and concave. The majority of slumping occurs with >50° slopes. Natural vegetation partly stabilised the studied slopes. Improved management of the dump slope will contribute to the accomplishment of the Sustainable Development Goals (SDG), which include safeguarding land life and providing clean water and sanitation.

Keywords Dump-site geomorphology · Slope stability · Geomorphology · Neogeomorphology · Anthropogenic geomorphology

Abbreviations

DEM	Digital Elevation Model
NCL	Northern Coalfields Limited
OCP	Open Cast Project
PAHs	Polycyclic Aromatic Hydrocarbons
TS07	Total Station 07

A. Raha · S. Das · M. Biswas
Department of Geography, Presidency University, Kolkata, West Bengal, India

S. Mukherjee (✉)
Department of Earth Sciences, Indian Institute of Technology Bombay, Mumbai, Maharashtra, India
e-mail: smukherjee@iitb.ac.in

S. Mukherjee (ed.), *Structural Geology and Tectonics*,
https://doi.org/10.1007/978-981-95-4743-2_19

Highlights

I. Coal dump sites provide ideal place to study anthropogenic landforms
II. The slope of the dump sites are distinctly varied in terms of concavity or convexity
III. Vegetation plays an important role in the slope stability.

1 Introduction

In recent years, there has been a noticeable increase in environmental awareness amongst the public, driven by growing recognition of human's substantial impact on natural ecosystems (Dávid, 2008). Anthropogenic geomorphology/ neogeomorphology is a relatively new field within the Quaternary science that has gained much focus in the last two decades or so (Lóczy & Süto, 2011). The subject examines surface alterations resulting from human activities, predicts the repercussions of disrupting natural equilibria, and proposes strategies to mitigate adverse impacts.

Mine dump is one of the human-induced landforms (review in Mohanty et al., 2024). The morphologic components of these accumulated macroforms typically consist of individual mesoforms e.g. plateaus and slopes. These formations arise from the accumulation of economically (so called) valueless materials, often produced during mining operations (Lóránt, 2012). Various shapes of dumps can develop-curved, fan and round-shaped, created during the bankfill operations in quarrying. The shape of a positive form is influenced by factors e.g. the original ground surfaces, methods of accumulation and the physical properties of the dumped materials. Cone-shaped, truncated cone-shaped and terraced dumps are amongst the most common configurations encountered (Lóránt, 2012). Table 1 provides several examples of studies on coal dump from a geoscientific perspective.

For more than 200 years, there has been widespread coal mining in the Paschim Barddhaman district of West Bengal (India), especifically in the Asansol-Raniganj tract. In 1774, commercial coal mining began in the Raniganj block of present-day Asansol, India. Besides few privately run coal mines, West Bengal has got 107 coal mines run by the Eastern Coalfield Limited (a division of Coal India Limited). Open cast mining eventually resulted in a significant dumping ground in nearby areas of mines. With time they formed a typical anthropogenic landform that got reframed by subsequent human activities in conjunction with natural erosion induced by rainfall and slumping. The primary objective of this work is to understand the geomorphologic attribute of anthropogenic landforms in the dump sites.

Organic matter from coal waste is the main source of organic compounds found in coal-waste dump water in thermally active locations. Phenolic chemicals e.g. ketones, organic acids, and Polycyclic Aromatic Hydrocarbons (PAHs)—most concentrated in coal-waste dump water—are the soluble byproducts (Fabianska et al., 2020). Al, B, Ba, Ca, Cd, Cr, Cu, Fe, K, Li, Mg, Mn, Na, P, Pb, S, Si, Sr and other elements have been found in coal-waste waters in many cases around the world. These concentrations are

Table 1 Examples of slope stability studies on coal-waste dump sites

Serial no.	Authors	Study area	Key approach	Key conclusion
1	Chaluya et al. (1999)	Dhanbad, Jharkhand (India)	Numerical modelling to assess the increase in stability and comprehend how plants stabilises the dump slope	Grass roots considerably raise the dump slope's safety factor from 1.2 to 1.4, contribute favourably to long-term stability
2	Chauliya et al. (2000)	Katras area of Bharat Coking Coal Limited, Jharkhand (India)	Measurement of dump geometry was conducted using an electronic distance metre (EDM). Standard procedures were used in both field and lab investigations to ascertain the physico-chemical characteristics of the dump material. Jack shear testing was used to determine the *in-situ* shear strength characteristics of the dump material both before and after revegetation	The stability of coal mine overburden dump slopes greatly depends on the presence of grass and tree roots. Roots provide mechanical support to the waste materials. The root matrix also strengthens the dump material's shear strength, which raises the dump slopes' long-term stability
3	Kasmer and Ulusay (2006)	Central Anatolia, Turkey	Limit equilibrium analysis is performed to analyse slope stability	This circumstance suggests that when dumping is done in line with current practice and when the creation of excessive water pressure in the piles is not permitted, spoil pile instability would not be expected. Critical floor inclination in the south pit is ~5° based on the stability of the spoil pile
4	Verma et al. (2011)	Wardha Valley Coalfield, Nagpur, Maharashtra (India)	Rock mass classification schemes, slope mass rating, and kinematic analysis of slopes using stereonet	The rock mass is partially stable to unstable according to the SMR investigation at various sites, with a failure probability of 0.4 to 0.6 (SMR class III to IV), which is further supported by kinematic analysis

(continued)

Table 1 (continued)

Serial no.	Authors	Study area	Key approach	Key conclusion
5	Kainthola et al. (2011)	Nagpur, Maharashtra (India)	The physico-mechanical parameters of the dump material were ascertained by testing representative loose samples that were taken from the site. For the study of complicated geometries, stress modelling and material behaviour, the finite element code (FEM) was utilised	When the dump angle was reduced while maintaining a constant dump height of 75 m, the factor of safety increased logarithmically. Due to the dump's poor geo-mechanical strength, it was advised to maintain the flatter slope of 25° with a height of 75 m. The dump slope produced a *factor of safety* (FOS) of 1.3 when at 25° and at 85 m height
6	Muthreja et al. (2012)	(Numerical modelling)	Application of geo-grid reinforcement techniques, waste dump numerical and physical modelling for stability analysis	Higher dumps that can hold significant overburden can be created with the use of geo grids
7	Tripathi et al. (2012)	Dhanbad, Jharkhand (India)	Biological and chemical analysis methods, bio-stability analysis method, finite difference method (FDM) are applied for stability analysis	Long-term revegetation was shown to have an indirect effect on soil fertility status in mine spoil and a direct influence on dump stability. Major contributions to the ecological regeneration of the mine spoil are made by microbial and plant biomass
8	Rai et al. (2012)	Modelling, of no specific study areas	Sensitivity analysis has been done for the dump slope's geometrical and geotechnical factors	Based on the findings, it can be inferred that friction angle is a more sensitive characteristic than dump material cohesiveness. The slope angle and height of dragline dump are extremely sensitive Height of main dump and thickness of coal rib are the least sensitive factors, while gradient of the seam and thickness of coal rib are moderately sensitive

(continued)

Table 1 (continued)

Serial no.	Authors	Study area	Key approach	Key conclusion
9	Verma et al. (2013a, b)	Wardha Valley Coal Field, Maharashtra (India)	Finite Element Method (FEM)	The Factor of Safety (FOS) decreases significantly as the dump slope's height rises. FOS depends on the kind of material and its geomechanical characteristics; however, in this case, the dump material mostly consists of shale, carbonaceous shale, and sandstone
10	Verma et al. (2013a, b)	Wardha Valley Coal Field, Maharashtra (India)	On a 32 m high cut slope, a 2D numerical simulation is utilised. Hoek–Brown strength parameters were developed and included within the model	The factor of safety was determined by the 2D analysis to be within a narrow range (0.931–1.067). The analysis has made the cut slope unstable to extremely critical, as evidenced by the rock fragments from the cut slope constantly spalling
11	Cho and Song (2014)	Heungjeon-ri, Dogye-eup (South Korea)	To examine the behaviours of both the waste dump slope and the naturally occurring slope underneath, wire sensors, a rain gauge and inclinometers were positioned at the atop the waste dump slope and in the naturally occurring slope beneath the trash dump	The monitored data showed that due to rainwater seepage into the ground during, the deformation at the peak of the waste dump's slope gradually grew until eventually converging. Furthermore, accumulation of rainwater had an impact on the natural slope underneath the waste dump's horizontal distortion
12	Rahul et al. (2015)	No specific study area	Artificial neural network (ANN) used to determine coal mine's FOS for the dump slope	ANN may predict the dump slope's safety factor more accurately than numerical modelling

(continued)

Table 1 (continued)

Serial no.	Authors	Study area	Key approach	Key conclusion
13	Liu et al. (2016)	Shanxi province (China)	The Normalised Difference Vegetation Index (NDVI), mixed pixel decomposition, variation trend computation and slope stability understanding transition matrix were used	The post-reclamation vegetation covering in the dumps is rather stable, as per the stability study of plant coverage succession and its recovery following a fire
14	Koner and Chakravarty (2016)	Wardha valley coal field (India)	The gathered samples were studied for geotechnical factors-grain size distribution, soil permeability, compaction parameters, shear and normal strength characters, etc.	The examinations of the three distinct types of overburden geo-materials showed that the shear strength characteristics for the running dump materials were the lowest of all the 11 sites studied
15	Behera et al. (2016)	Bhubaneswari mines, Odisha (India)	In order to comprehend slope instability, the Limit Equilibrium Method (LEM) was utilised	Rainfall infiltration-related increases in pore water pressure account for the majority of dump failures in the study region
16	Bednarczyk (2017)	Zloczew (Poland)	For the purpose of designing open pits and spoil dumps, slope stability analysis was conducted using the Shear strength reduction method and Flac/slope v.7 codes	Low strength clayey soils, paleo-landslide zones and structural surfaces may all be potential locations for ground movements The results portrayed that the factor of safety (Fs) ranges 0.75–1.65 for the pit and 1.12–1.60 for the dump

(continued)

Table 1 (continued)

Serial no.	Authors	Study area	Key approach	Key conclusion
17	Martín-Moreno et al. (2018)	Poveda de la Sierra–Peñalén Mining District (east-central Spain)	Slope stability analysis was conducted through topography of waste dumps and characterisation of substrata, quantification of silt captured by check dams, quantification of erosion from waste sites, and examination of precipitation and its connection to the process of filling check dams	Long, steep slopes with highly erodable materials were validated by topographic and substrate property analysis. It became out that the check dams were ineffective in reducing sediment loads
18	Rajak et al. (2021)	Mand Raigarh coalfield, Chhattisgarh (India)	Different percentages of fly ash were combined with overburden material to assess different geotechnical properties. In order to determine the ideal dump height and evaluate the stability of fly ash mixed overburden dump, numerical modelling utilising the finite difference approach was conducted	When 20% fly ash is added to the overburden (OB) maximum unconfined compressive strength and cohesion values are shown. The stability study indicates that, with 10% fly ash added to the OB material, the ideal dump height is 55 and 50 m for 35° and 37° slopes, respectively
19	Igwe and Chukwu (2019)	Enyigba (Southeastern Nigeria)	Using GeoStudio® 2012, a program created by Geoslope International Limited, stability analysis was performed and the geotechnical characteristics of the slopes were examined	One of the elements impacting all types of mass wasting is clay plasticity on moisture absorption, which is shown by the presence of swelling mineral (illite) in soil material, as proven by X-Ray Diffraction (XRD)

(continued)

Table 1 (continued)

Serial no.	Authors	Study area	Key approach	Key conclusion
20	Li et al. (2020)	The Wulanhada (WLHD) coal mine, the Liulingou (LLG) coal mine, and the Jinzhengtai (JZT) coal mine, in Jungar Banner (China)	Wiener index (H′), Simpson index (D), Pielou index (J) and Margalef index (Mg) were applied for the ecological stability analysis	Plant community and the recovery of soil properties could benefit from ecological restoration in mining areas
21	Gupta et al. (2021)	Review paper		Multiple failure modes can coexist within the same dump due to the heterogeneous nature of the material Numerical modelling enables accurate simulation of input parameters e.g. rainfall, groundwater, earthquakes, blasting, material variability, complex and uneven geometry, interfaces, discontinuities, external loading
22	Raha et al. (this work)	Dhanbad and Durgapur, Jharkhand and West Bengal (India)	Total station survey is conducted on the dumpsite for terrain analysis and field photographs are interpreted	The dumpsites portray various segments of slopes- free face, convex and concave with the events of slope failure. Vegetation is a slope-stabilizing factor

typically much higher than the legally permissible concentration levels in wastewater discharged into surface waters and the ground (Djinovic & Popovic, 2007; Lewińska-Preis et al., 2021).

Adibee et al. (2013) report that parts of the downstream farmlands around Karmozd coal mine, Iran, have been devastated by recent flooding caused by dumping waste. The loss of agriculture and the ensuing risks to public health/environment are evident. As per Bian et al. (2009), following a period of continuous rain, inner vapour pressure built up in the mining waste dump of Pingdingshan coal mining company's No. 4 coal mine (China), which detonated on 15-May-2005. Eight locals lost their lives in the explosion, 123 others suffered injuries and 18 residents' homes were destroyed.

A common blueprint for peace and prosperity for people and the planet, both now and in the future, is provided by the 2030 Agenda for Sustainable Development, which was accepted by all United Nations Member States in 2015 (https://www.coursehero.com/file/226174032/ASSIGNMENT-HRMdocx/). The 17 Sustainable Development Goals (SDGs), which represent an urgent call to action for all nations—developed and developing—in a global partnership, are at the centre of it.

Understanding the harmful environmental issues in dump-site locations is crucial for achieving the SDGs of clean water and sanitation, and protecting life on land.

2 Study Area and Dataset

Four dump sites near Damagoria Colliery Bharat Coking Coal Limited (BCCL) and one near Mugma were studied (Fig. 1). Field surveys were conducted to gather primary data. Total station surveys were employed to study slopes in the dump sites.

3 Methodology

The dump site at the studied three locations (Fig. 2) was surveyed with total station (Leica TS07) in three segments. Latitude, longitude and elevation data were collected. The data has been processes on ArcGIS 10.4 (2016) platform and Surfer 11.0 (2013) to prepare digital elevation models (DEMs) and draw cross sections.

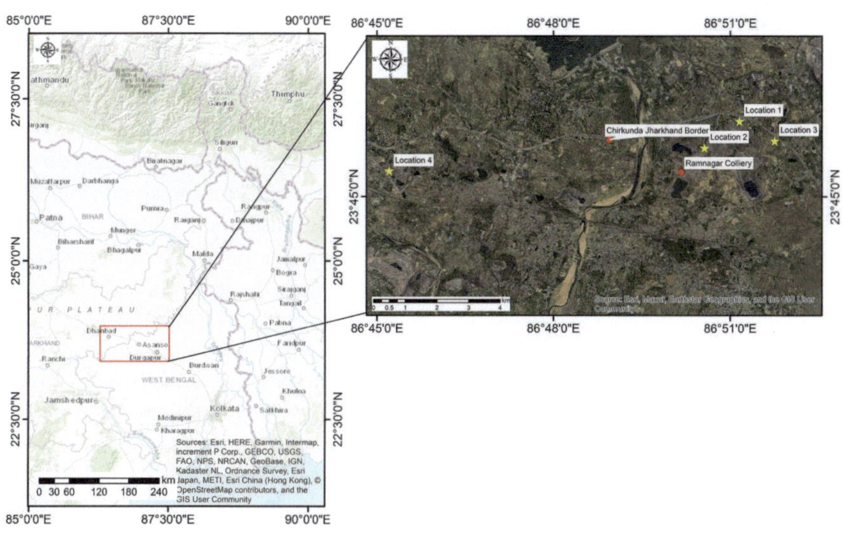

Fig. 1 Location map of the study area; locations 1–4 have been plotted

Fig. 2 Dump site at location 1. **a**. Slumping along the slope. Slumped material deposited at the base. **b**. Slab failure at the toe of the slope. Minor gullies developed along the slope. **c**. Overall geometry of the slope is convex. Two mining sites noted. **d**. Predominance of angular- boulders at the dump site

4 Results

The dump site at location 1 (Fig. 2) typically exhibits a convex slope with a steep gradient of 70° (Fig. 2a). The loosely packed materials on this slope are prone to slumping, particularly at the base of the slope. The material that slumps accumulates further downhill in a heap. Rain has created several gullies on the slope (Fig. 2b). Steepening of the slope exacerbates slope failure at its base. Interestingly, the dump site is still being utilised for mining activities. Such unauthorised and unplanned quarrying is making slopes unstable, leading to more frequent slab failures and slumping. Angular boulders are predominantly observed throughout the area.

The dump site at location 2 (Fig. 3) also exhibits flow paths/gullies on its slope. Notably, coarser materials dominate the lower parts of the slope, while finer materials prevail in the upper segment where shrubs/grasses exist. No slumping/slab failure are observed. In contrast, location 3 (Fig. 4) displays a different topographic expression. This dump site is partially covered with thick grasses and shrubs, indicating a stabilised slope and is devoid of slab failure/slumping. Several flow paths are documented across the slope.

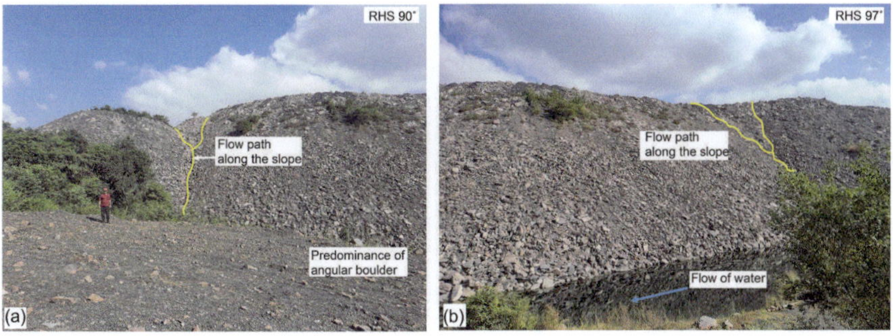

Fig. 3 Dump site of location 2. **a**. Mostly composed of angular bouders. Convex slope is obsereved with the development of flow path along the slope. **b**. Vegetation grown over the dump site along with the development of flow path on the slope

Fig. 4 Dump site at location 3. **a**. Thick vegetation on the slope with multiple flow paths. No evidence of slumping/slab failure. **b**. Convex slope with ~50° gradient

The dump site at location 4 (Fig. 5), is situated close to a mining site. The dump site is relatively new and demonstrates a less altitude. The topography is a convex slope, which got stabilised due to the growth of thick vegetation. The dumped materials exhibit poor sorting, with angular boulders being predominant. Segment 1 at location 1's DEM illustrates various slope segments (Fig. 6a, b). Cross sections (Fig. 6c, d) depict steep slopes on the flanks. Slopes range between 70–80°.

At the second segment of location 2, there is a contrast in slope profile characteristics. The DEM (Fig. 1, Repository) reveals a dome-shaped terrain with a slightly concave slope on the flank. The AB section (Fig. 1, Repository) delineates three distinct slope segments from the top to the toe. Starting with a convex slope at the top, it transitions into a free face at an altitude of 141.5 m, followed by a gentler convex slope, and finally a concave slope at the bottom. The CD profile (Fig. 1, Repository) demonstrates a steeper slope (60°). Segment 3 at location 2 (Fig. 2, Repository) indicates a higher altitudinal profile with steeper gradient. The AB profile forms near a horizontal plane at the top, followed by steep (70°) free face slope segment of a height of ~50 m. The base of the slope shows a convex-upward geometry, which can

Fig. 5 **a**. Dump site of location 4. **b**. Materials are dumped in the adjoining areas and got covered by grasses. **c**. Convex slope of ~50°, covered with thick vegetation. **d**. Poorly sorted materials defining the convex slope with gentle gradient (30°)

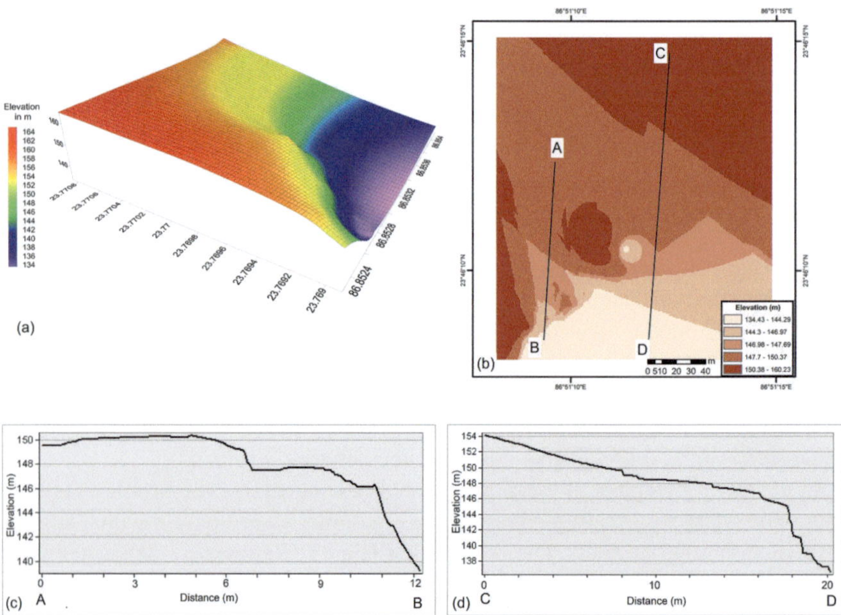

Figs. 6 Topographic expression of segment 1 of the dump site at Location 1 (23.768911°N/ 86.852588°E). **a**. 3D view of the slope segment and surface expression prepared with Surfer 11.0. **b**. Digital elevation model. **c**. Cross-section along AB. **d**. Cross section along CD

be attributed to the deposition of slumped materials from the free face segment. The CD section (Fig. 2, Repository) exhibits similar pattern of slope segments. The flat top is followed by steep (80°) free face slope, which is susceptible to slumping/slab failure.

5 Discussions

This geomorphic study helps to understand the nature of an anthropogenic landform. The primary natural microforms found on dumps are rainwater channels etched into slopes, typically arranged in radial patterns on cone-shaped or truncated cone-shaped dumps. The gradient of slope additionally provides insight into the mechanism of slumping and slab failure. Vegetation developed at places has stabilised slopes.

During the monsoon season, the channels that have grown across the waste slopes transport water. It is important to note that the water carries the finer debris down to the slope. There is strong possibility that this water will combine with the neighbouring clean source of water (as pointed out in another area by Wohl (2006). Therefore, it is critical to take the required steps to safeguard and conserve the clean water sources in order to achieve the SDG of clean water and sanitation.

Locals in the dump sites under study continue to mine in order to survive. To prevent any tragedies, it is crucial to understand the slope failure at dump sites. The casualties (Indian review from Mohanty et al., 2022) and land degradation are to be checked in order to fulfil the SDG of life on land.

At Nigahi Open Cast Project (OCP), Northern Coalfields Limited (NCL), grass bedding and seed ball stabilisation of overburden dumps have been actively pursued for the past few years by the Coal India Limited, Govt. of India.

In order to stabilise the slopes of the dump sites, we propose the following landform guidelines- (i) reduce the gradient of the linear slopes of the four main waste dumps by removing materials from their upper reaches; (ii) avoid mining at the base of the slope to prevent slumping (community awareness is required in this regard); (iii) adopt natural slope stabilising methods, such as planting native vegetation on the slope.

This preliminary investigation on the dump slope expands the possibilities for a more precise analysis of the slope failure by utilising model simulation (Bednarczyk, 2017; Mutherja et al., 2012; Verma et al., 2013a, b), comprehending the properties of the sheer stress of the materials dumped (Chaluya et al., 2000), identifying the type of grass (Li et al., 2020) and measuring the lengths of grass/vegetation roots required to stabilise the dump slope (Chaluya et al. 2020).

Acknowledgements IIT Bombay provided research grant to SM. Mukherjee (2026) summarised this work.

References

Adibee, N., Osanloo, M., & Rahmanpour, M. (2013). Adverse effects of coal mine waste dumps on the environment and their management. *Environment and Earth Science, 70*, 1581–1592.

Bednarczyk, Z. (2017). Slope stability analysis for the design of a new lignite open-pit mine. *Procedia Engineering, 191*, 51–58.

Behera, P. K., Sarkar, K., Singh, A. K., Verma, A. K., & Singh, T. N. (2016). Dump slope stability analysis–A case study. *Journal of the Geological Society of India, 88*, 725–735.

Bian, Z., Dong, J., Lei, S., et al. (2009). The impact of disposal and treatment of coal mining wastes on environment and farmland. *Environmental Geology, 58*, 625–634.

Chaulya, S. K., Singh, R. S., Chakraborty, M. K., & Dhar, B. B. (1999). Numerical modelling of biostabilisation for a coal mine overburden dump slope. *Ecological Modelling, 114*, 275–286.

Chaulya, S. K., Singh, R. S., Chakraborty, M. K., & Srivastava, B. K. (2000). Quantification of stability improvement of a dump through biological reclamation. *Geotechnical & Geological Engineering, 18*, 193–207.

Cho, Y. C., & Song, Y. S. (2014). Deformation measurements and a stability analysis of the slope at a coal mine waste dump. *Ecological Engineering, 68*, 189–199.

Dávid, L. (2008). Quarrying: An anthropogenic geomorphological approach. *Acta Montan. Slovaca, 13*, 66–74.

Djinovic, J. M., & Popovic, A. R. (2007). In situ influence of coal ash dump on the quality of neighboring surface and ground waters by applying correlation statistic analysis. *Fuel, 86*, 218–226.

Fabiańska, M. J., Nádudvari, Á., Ciesielczuk, J., Szram, E., Misz-Kennan, M., & Więcław, D. (2020). Organic contaminants of coal-waste dump water in the lower- and upper Silesian coal basins (Poland). *Applied Geochemistry, 122*, Article 104690.

Gupta, G., Sharma, S. K., Singh, G. S. P., & Kishore, N. (2021). Numerical modelling-based stability analysis of waste dump slope structures in open-pit mines-A review. *Journal of The Institution of Engineers (India): Series D 102*, 589–601.

Igwe, O., & Chukwu, C. (2019). Slope stability analysis of mine waste dumps at a mine site in Southeastern Nigeria. *Bulletin of Engineering Geology and the Environment, 78*, 2503–2517.

Internet ref-1. Retrieved January 20, 2025, from https://www.coursehero.com/file/226174032/ASSIGNMENT-HRMdocx/.

Kainthola, A., Verma, D., Gupte, S. S., & Singh, T. N. (2011). A coal mine dump stability analysis—A case study. *Geomaterials, 1*, 1.

Kasmer, O., & Ulusay, R. (2006). Stability of spoil piles at two coal mines in Turkey: Geotechnical characterization and design considerations. *Environmental & Engineering Geoscience, 12*, 337–352.

Khandelwal, R., Rai, R., & Shrivastva, B. K. (2015). Evaluation of dump slope stability of a coal mine using artificial neural network. *Geomechanics and Geophysics for Geo-Energy and Geo-Resources, 1*, 69–77.

Koner, R., & Chakravarty, D. (2016). Characterisation of overburden dump materials: A case study from the Wardha valley coal field. *Bulletin of Engineering Geology and the Environment, 75*, 1311–1323.

Lewińska-Preis, L., Szram, E., Fabiańska, M. J., et al. (2021). Selected ions and major and trace elements as contaminants in coal-waste dump water from the lower and upper Silesian Coal Basins (Poland). *Int J Coal Sci Technol, 8*, 790–814.

Li, X., Lei, S., Liu, F., & Wang, W. (2020). Analysis of plant and soil restoration process and degree of refuse dumps in open-pit coal mining areas. *International Journal of Environmental Research and Public Health, 17*, 1975.

Liu, X., Zhou, W., & Bai, Z. (2016). Vegetation coverage change and stability in large open-pit coal mine dumps in China during 1990–2015. *Ecological Engineering, 95*, 447–451.

Lóczy, D., & Süto, L. (2011). Human activity and geomorphology. In K. J. Gregory & A. S. Goudie (Eds.), *The Sage Handbook of Geomorphology* (pp. 260–278). Sage Publications.

Lóránt, D. (2012) Introduction to anthropogenic geomorphology. In Studies on environmental and applied geomorphology. IntechOpen. https://doi.org/10.5772/30583.

Martín-Moreno, C., Martín Duque, J. F., Nicolau Ibarra, J. M., Muñoz-Martín, A., & Zapico, I. (2018). Waste dump erosional landform stability—A critical issue for mountain mining. *Earth Surface Processes and Landforms, 43*, 1431–1450.

Mohanty, M., Sarkar, R., & Das, S. K. (2022). Probabilistic assessment of effects of heterogeneity on the stability of coal mine overburden dump slopes through discrete element framework. *Bulletin of Engineering Geology and the Environment, 81*, 1–28.

Mohanty, M., Sarkar, R., & Das, S. K. (2024). A critical review on static and dynamic performance of coal mine overburden dump slopes: Present status and way forward. *Journal of the Geological Society of India, 100*, 1271–1286.

Mukherjee, S. (2026). Introduction to "Structural Geology & Tectonics". In S. Mukherjee (Ed.), *Structural Geology & Tectonics*. Springer. ISBN: 978-981-95-4743-2.

Muthreja, I. L., Yerpude, R. R., & Jethwa, J. L. (2012). Application of geo-grid reinforcement techniques for improving waste dump stability in surface coal mines: Numerical modeling and physical modeling. *International Journal of Engineering Inventions, 1*, 16–23.

Rai, R., Kalita, S., Gupta, T., & K Shrivastva, B. (2012). Sensitivity analysis of internal dragline dump stability: Finite element analysis. *Geotechnical and Geological Engineering 30*, 1397–1404.

Rajak, T. K., Yadu, L., Chouksey, S. K., & Dewangan, P. K. (2021). Stability analysis of mine overburden dump stabilised with fly ash. *International Journal of Geotechnical Engineering, 15*, 587–597.

Tripathi, N., Singh, R. S., & Chaulya, S. K. (2012). Dump stability and soil fertility of a coal mine spoil in Indian dry tropical environment: A long-term study. *Environmental Management, 50*, 695–706.

Verma, D., Kainthola, A., Gupte, S. S., & Singh, T. N. (2013a). A finite element approach of stability analysis of internal dump slope in Wardha valley coal field, India, Maharashtra. *American Journal of Mining and Metallurgy, 1*, 1–6.

Verma, D., Kainthola, A., Thareja, R., & Singh, T. N. (2013b). Stability analysis of an open cut slope in Wardha valley coal field. *Journal of the Geological Society of India, 81*, 804–812.

Verma, D., Thareja, R., Kainthola, A., & Singh, T. N. (2011). Evaluation of open pit mine slope stability analysis. *International Journal of Earth Sciences and Engineering, 4*, 590–600.

Wohl, E. (2006). Human impacts to mountain streams. *Geomorphology, 79*, 217–248.

Potential Location for Relief Center for Earthquake Hazards Using Bipolar Fuzzy Graphs Example from Kutch Area, Gujarat, India

Megh Kanvinde, O. T. Manjusha, and Soumyajit Mukherjee

Abstract Seismic hazards are severe natural disasters for which we must stay prepared, especially at locations the hazard has its highest chance. The Kutch area in western India is one such terrain. In this work, we apply the bipolar fuzzy graph theory to locate the best possible relief centers in Kutch in the context of its future major seismicity event. Considering several geographic issues, Bhuj and Mandvi are found to be the two locations where such relief centers can be established. Bhuj being a city much better established than Mandvi, we apply "*human judgement*" and choose the former as the best possible location for the relief center.

Keywords Numerical modeling · Tectonics · Seismicity · Deformation of rocks · Rift basin · Rock mechanics

1 Introduction

Tectonic and structural geological modeling analyses spatial events and structures numerically. In a few simpler cases dealing with connection issues for several nodes, one can apply the graph theory (e.g., Mukherjee, (2019), Sanderson et al., (2019)). The theory works when there is a complete surety of interconnection of nodes/vertices through edges. Zadeh (1965) introduced the concept of fuzzy sets, which was applied to solve several scientific and engineering research problems where uncertainty was involved. In the geoscientific context, Justman et al. (2020) presented the broad applications of fuzzy set theory to deal with tectonic problems. Anand et al. (2021) presented a fuzzy set theory to explain the kinematics of the ductile shear zones. Manjusha et al. (2023) utilized fuzzy graph theory to demonstrate how to deal with

M. Kanvinde · S. Mukherjee (✉)
Department of Earth Sciences, Indian Institute of Technology Bombay, Mumbai, Maharashtra, India
e-mail: smukherjee@iitb.ac.in

O. T. Manjusha
Government Arts and Science College, Kondotty, Kerala, India

S. Mukherjee (ed.), *Structural Geology and Tectonics*,
https://doi.org/10.1007/978-981-95-4743-2_20

uncertainties in fluid flow through fractured media. Klügel (2008) reviewed several numerical techniques used in seismic data analysis where (fuzzy) graph theory has been referred.

This article applied the concept of bipolar fuzzy graphs (BFG) to find the best possible locations for relief centers for future earthquakes that may hit the Kutch area in western India.

2 Bipolar Fuzzy Graphs (BFG)

Zhang (1994) presented the concept of bipolar fuzzy sets (BFS) by generalizing the concept of fuzzy sets. BFS has its membership degree ranging -1 to $+1$. A membership degree $= 0$ in a BFS connotes that the element is irrelevant to the corresponding property. Membership degree of an element to be [0, 1] indicates that it somewhat satisfies the property. Membership degree equal to $[-1, 0]$ indicates that the element approximately satisfies the implicit counter-property. Positive information implies that what is granted is plausible. On other hand, negative information denotes impossibility. The control set of vertices in the BFG is the dominant set. It refers to a critical position in analyzing fuzzy graphs. Appendix 1 provides the preliminaries of BFG. Geoscientific applications of BFGs have recently been reviewed by Gong et al. (2021).

3 Geology and Tectonics

The Kutch district, whose boundaries completely enclose the Kutch basin (Shaikh et al., 2020), is highly prone to earthquakes. As per India's seismic zonation, the district falls within the most destructive Zone V (e.g., Walling & Mohanty (2009), Rastogi et al., (2012)). We choose this region in this work keeping in mind its extreme seismicity.

The Kutch basin is an asymmetric graben tilted toward the south, flanked by the North Kathiavar Fault (NKF) at the southern margin and the Nagar Parkar Fault (NPF) at the north (Biswas, 2005). Further, the rift basin contains E-W trending uplifts bound by six sub-parallel intra-rift faults—Kutch Mainland Fault (KMF), Katrol Hill Fault (KHF), South Wagad Fault (SWF), Gedi Fault (GF), Gora Dongar Fault (GDF) and Island Belt Fault (IBF) (Biswas, 2005; Thakkar, 2017). The KMF remains the prime zone of activation, with most strain occurring at its intersection with the SWF. The 7.1 magnitude 2001 Bhuj Earthquake and its aftershocks concentrated around this fault junction (Biswas, 2005; Raval, 2001).

Kutch has also witnessed several major earthquakes, with significant earthquake events since the 1800s (Table 1). The Government of Gujarat State set up the Gujarat State Disaster Management Authority to prepare programs for disaster preparedness, especially for future earthquakes. Since this agency is nodal in nature, selecting

Table 1 Known earthquake events in the Gujarat region with magnitudes >5.5 on the Richter scale (reproduced from Rastogi et al., 2012)

Year	Location	Magnitude
1668	Samaji, Indus	7.8
1819	Kachchh	7.8
1845	Lakhpat	6.3
1845	Lakhpat	5.7
1848	Mount Abu	5.7
1856	Surat	5.7
1864	Ahmedabad	5.7
1903	Kachchh	5.6
1919	Ghogha (Bhavnagar)	5.7
1921	Indus, Kachchh	5.5
1935	Surat	5.7
1938	Paliyad	5.7
1956	Kachchh	6
1969	Mount Abu	5.5
2001	Kachchh	7.7
2006	Gedi	5.7
2006	Lakadia	5.6

appropriate towns in the Kutch district is essential for earthquake preparedness, be it for emergency response teams or emergency equipment warehouses.

4 The Present Work

In this work, we use the theory of dominating sets in BFGs to select the most appropriate towns in Kutch district for such emergency measures. Figure 1 presents the Kutch district with the taluka boundaries. Each taluka's main town or city was chosen as a node/vertex in the graph, with the assigned vertex labels as in Table 2. The selected towns are considered representatives of the respective talukas. The goal is to choose a small number of towns to build emergency/relief centers by using a BFG. Roads that connect these cities (nodes/vertices) are considered edges of the graph.

Each node/town was assigned a bipolar score keeping in mind various uncertainties and qualities of the towns. The main criteria for these assignments are:

(i) The seismic hazard associated with the town. The scores for this were borrowed from the preliminary seismic hazard map of the Kutch district from Pancholi et al. (2022). Pancholi et al. (2022) used a GIS-based multi-criteria approach, considering various aspects such as peak ground acceleration, amplification, geology/geomorphology, liquefaction potential of the soil, tsunami hazard,

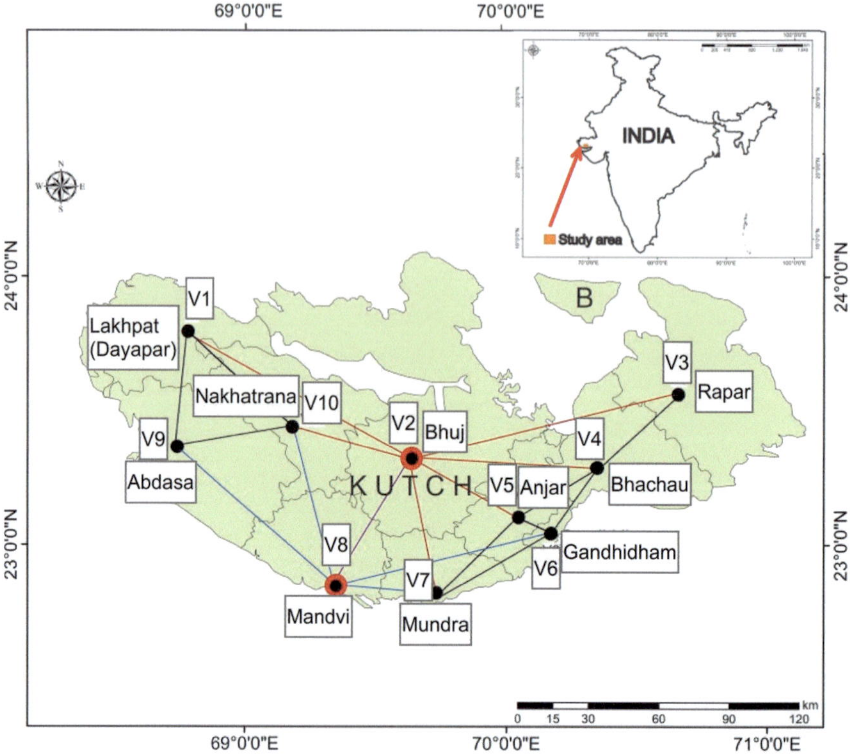

Fig. 1 Taluka map of the Kachchh district, Gujarat, India. The towns considered in this study are marked. Bhuj- district capital. The graph with the various towns connected by direct roads (edges). Vertices marked in bold: major towns (Bhuj and Mandvi). Edges marked in red and blue: roads connecting Bhuj and Mandvi with the other towns, respectively. Purple edge: road connecting Bhuj and Mandvi. Black edges: roads connecting towns with a common taluka boundary

Table 2 Town and vertex correlation table

Town	Vertex
Lakhpat	V_1
Bhuj	V_2
Rapar	V_3
Bhachau	V_4
Anjar	V_5
Gandhidham	V_6
Mundra	V_7
Mandvi	V_8
Abdasa	V_9
Nakhatrana	V_{10}

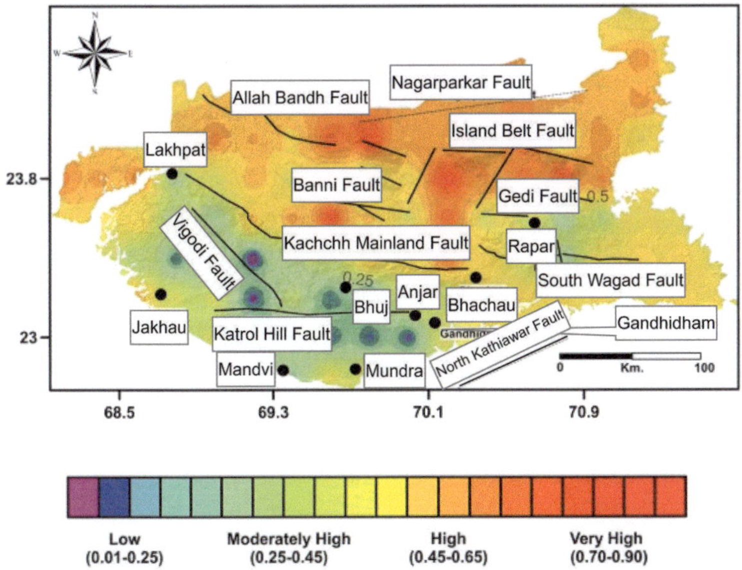

Fig. 2 A macro-scale seismic hazard zonation map of Kachchh district. Reproduced from Pancholi et al. (2022)

groundwater depth, amplification, and shear wave velocity to construct a macro-scale seismic zonation map for the Kutch district (Fig. 2).

(ii) The scores pertaining to the infrastructure of the towns were assigned based on the number of households in the town. Higher the number of households, the better the score. The number of households for each town was taken from the Indian census data.

(iii) The number of public healthcare centers (PHCs) plus community health centers (CHCs) in each taluka was considered to gauge the health facilities, and scores were assigned based on them. Data was taken from the Gujarat Government health website.

Based on these criteria, each selected town was given a score in the intervals of [0, 1] and [−1, 0]. Negative values are given to properties that are unfavorable or have a negative impact, while positive values are given in the opposite situation. While the seismic hazard for the town were represented by the final negative score, the number of healthcare centers and households were denoted by positive scores.

Table 3 presents the combined scores for each town. Based on these scores, Bhuj and Mandwa (plotted in Fig. 1) were considered the dominating towns/nodes in the graph. It was also checked whether each town or node in the graph is connected with at least one of these two locations by a direct road. See Appendix 2 for the associated work.

In due course, a graph was constructed with each town/node connected to at least one of Bhuj or Mandwa (Fig. 1). Edges representing the road connections were also

Table 3 The weights assigned to the towns based on the three criteria. The earthquake hazard score contributes to the negative polarity assigned, and the number of households and healthcare scores contribute to the positive polarity assigned equally. The red boxes mark the highest earthquake hazard score, number of households and number of PHCs + CHCs, respectively (PHC: Public Healthcare Centers, CHC: Community Healthcare Centers). Final weights are marked in orange

Town	Seismic hazard score [16]	Seismic hazard polarity	Number of Households (2011)	Households polarity	Number of CHC+PHC	Healthcare polarity	Negative values	Positive values
Lakhpat	0.275	-0.46	12155	0.17	6	0.4	-0.5	0.3
Abdasa	0.325	-0.54	24070	0.34	5	0.33	-0.5	0.3
Mandvi	0.275	-0.46	31508	0.45	11	0.73	-0.4	0.6
Mundra	0.425	-0.71	35192	0.50	10	0.66	-0.7	0.6
Bhuj	0.325	-0.54	64496	0.92	13	0.86	-0.5	0.9
Anjar	0.375	-0.62	33032	0.47	8	0.53	-0.6	0.5
Gandhidham	0.45	-0.75	16882	0.24	3	0.2	-0.7	0.2
Bhachau	0.575	-0.95	32746	0.46	9	0.6	-0.9	0.5
Rapar	0.45	-0.75	36630	0.52	11	0.73	-0.7	0.6
Nakhatrana	0.375	-0.62	28608	0.40	7	0.46	-0.6	0.4

assigned bipolar scores based on the following criteria. Based on these, the scores were assigned to the roads or the graph's edges.

(i) Time taken to travel or the length of each road was considered. The time to traverse each road or edge considered in the graph was taken from Google Maps, considering minimal traffic conditions. The longer the time taken, the lower the score assigned to this criterion. The time taken was converted to minutes to compare all the edges and give a relative score for each edge.

(ii) The number of lanes each road has was considered as this decides the traffic movement rate, especially during an emergency. If a road or edge has a variable number of lanes along its length, the effective or weighted average number of lanes was calculated for each road using the length of each 2-, 4- or 6-laned segment, which constitutes the road.

(iii) Since bridges are major weak points during earthquakes, the number of bridges on each road was also considered a criterion for assigning scores. The number of bridges on each road was counted using satellite data from Google Maps and compared to give a score. Bridges going over the roads were also counted.

Higher the number of bridges encountered on the road, the higher the chance of the road closing after an earthquake.

The edges were given scores in a similar way to nodes, i.e. within the interval of [−1, 0] and [0, 1]. While the number of bridges and time taken were accounted by the negative weight in equal measures, the positive weight was assigned based on the number of lanes solely. Table 4 presents the initial weights assigned to the edges.

In the definition of bipolar fuzzy graphs, we use two values (positive values and negative values) for representing the weights of vertices and edges. In case of vertices, we can give any value between zero and one as positive node weight and any value

Table 4 Initial weights assigned to each road connection/edge. Edges marked in orange and blue: roads connecting Bhuj and Mandvi with the other towns in the graph, respectively. Green edge: road connecting Bhuj and Mandvi. Other edges are for roads connecting towns with a common taluka boundary. Final scores are marked in yellow. Red boxes mark the highest values for each criterion

Edge	Road connecting towns	No. of bridges	Bridge polarity	Equivalent no. of lanes	Lane polarity	Time without traffic (mins)	Time Polarity	Negative Values	Positive Values
V_1V_9	Lakhpat - Abdasa	6	-0.3	2.00	0.33	124	-0.62	-0.5	0.3
V_1V_{10}	Lakhpat - Nakhatrana	7	-0.35	2.00	0.33	97	-0.49	-0.4	0.3
V_1V_2	Lakhpat - Bhuj	10	-0.5	2.00	0.33	186	-0.93	-0.7	0.3
V_9V_{10}	Abdasa - Nakharatna	11	-0.55	2.00	0.33	62	-0.31	-0.4	0.3
V_8V_9	Abdasa - Mandvi	12	-0.6	2.00	0.33	76	-0.38	-0.5	0.3
V_8V_{10}	Mandvi - Nakhatrana	6	-0.3	2.00	0.33	83	-0.42	-0.4	0.3
V_2V_{10}	Bhuj - Nakhatrana	3	-0.15	2.00	0.33	66	-0.33	-0.2	0.3
V_2V_8	Bhuj - Mandvi	2	-0.1	2.00	0.33	73	-0.37	-0.2	0.3
V_2V_7	Bhuj - Mundra	10	-0.5	2.00	0.33	78	-0.39	-0.4	0.3
V_2V_5	Bhuj - Anjar	3	-0.15	2.00	0.33	62	-0.31	-0.2	0.3
V_7V_8	Mundra - Mandvi	10	-0.5	3.63	0.61	63	-0.32	-0.4	0.6
V_6V_7	Gandhidham - Mundra	7	-0.35	3.39	0.56	66	-0.33	-0.3	0.6
V_5V_7	Anjar - Mundra	3	-0.15	3.13	0.52	62	-0.31	-0.2	0.5
V_5V_6	Anjar - Gandhidham	2	-0.1	3.33	0.56	29	-0.15	-0.1	0.6
V_4V_5	Bhachau - Anjar	5	-0.25	3.50	0.58	52	-0.26	-0.3	0.6
V_4V_6	Bhachau - Gandhidham	10	-0.5	5.56	0.93	46	-0.23	-0.4	0.9
V_3V_4	Rapar - Bhachau	6	-0.3	3.71	0.62	69	-0.35	-0.3	0.6
V_2V_4	Bhuj - Bhachau	10	-0.5	2.74	0.46	105	-0.53	-0.5	0.5
V_2V_3	Bhuj - Rapar	18	-0.9	2.47	0.41	189	-0.95	-0.9	0.4
V_8V_6	Mandvi - Gandhidham	9	-0.45	3.36	0.56	125	-0.63	-0.5	0.6

between negative one (−1) and zero as negative node weight. But the edge weights are taken based on the positive value and negative value of its end vertices. The initial edge weights given in Table 4 were changed according to the following rule.

Consider two vertices, V_1 and V_2. Suppose they are connected by an edge called 'e', and we give the positive weight of V_1 as 0.6 and the negative weight as −0.2. Similarly, we give the positive weight of V_2 as 0.5 and the negative weight as −0.4.

Hence, we can give the positive weight of edge 'e' as ≤ 0.5 (as we have to take the weight less than or equal to the minimum among the vertex weights) and the negative weight of edge 'e' as ≥ -0.2. This is because we need to assign the weight greater than or equal to the maximum weight among the vertices. Subsequently, the weights given to the edges were modified (Table 5).

5 Discussions

This work has followed in this article Gong et al.'s (2021) and Mandal et al.'s (2021) approach to locating the relief center in the study area. Gong et al. (2021) developed the model from the Chinese Yunnan Province. They considered major cities as the vertices and ignored smaller cities. Likewise, Mandal et al. (2021) applied the BFG theory for seismicity in western India and proposed that the locations Kandle and/or Mount Abu can act as relief centers. We note that Mount Abu is a hilly area with narrow meandering roads. Therefore, even though Mandal et al.'s (2021) numerical work reveals Mount Abu as a possible relief center, choosing Kandle as the best location will be better.

In a similar line, as per the present work in the Kutch area, the cities of Bhuj and Mandvi can be considered for the location of relief centers. Beyond mathematical considerations, it can be stated that Bhuj is better developed as a city than Mandvi. Therefore, we can inform the policymakers to make a relief center in Bhuj.

We note two different cities have come up as per Mandal et al. (2021) and this work to establish the relief station while considering parts of western Indian terrain. The reason is that Mandal et al. (2021) considered a different portion of western India than what has been considered in this work. Therefore, BFG theory must be applied cautiously for regional and local cases to develop the best possible relief center. Human judgment will further be needed to negate a few possible centers where cities are not well established (e.g., Mandvi in this work) or where the location is at an adverse geographic location (e.g., Mount Abu in Mandal et al., (2021)).

6 Conclusions

In this study, we used the mathematical concept of dominating set in BFG theory in assigning scores. We selected the most appropriate towns in the Kutch district to set up emergency response measures in the case of a major earthquake. Based on

Table 5 The final weights assigned to each edge/road connecting the nodes in the graph

Road connecting towns	**Edge**	**Negative values**	**Positive values**
Lakhpat - Abdasa	V_1V_9	-0.5	0.3
Lakhpat - Nakhatrana	V_1V_{10}	-0.4	0.3
Lakhpat - Bhuj	V_1V_2	-0.5	0.3
Abdasa - Nakharatna	V_9V_{10}	-0.4	0.3
Abdasa - Mandvi	V_8V_9	-0.4	0.3
Mandvi - Nakhatrana	V_8V_{10}	-0.4	0.3
Bhuj - Nakhatrana	V_2V_{10}	-0.2	0.3
Bhuj - Mandvi	V_2V_8	-0.2	0.3
Bhuj - Mundra	V_2V_7	-0.4	0.3
Bhuj - Anjar	V_2V_5	-0.2	0.3
Mundra - Mandvi	V_7V_8	-0.4	0.6
Gandhidham - Mundra	V_6V_7	-0.3	0.2
Anjar - Mundra	V_5V_7	-0.2	0.5
Anjar - Gandhidham	V_5V_6	-0.1	0.2

(continued)

Table 5 (continued)

Bhachau - Anjar	V_4V_5	-0.3	0.5
Bhachau - Gandhidham	V_4V_6	-0.4	0.2
Rapar - Bhachau	V_3V_4	-0.3	0.5
Bhuj - Bhachau	V_2V_4	-0.5	0.5
Bhuj - Rapar	V_2V_3	-0.5	0.4
Mandvi - Gandhidham	V_8V_6	-0.4	0.2

the scores assigned to the nodes and the road connectivity, Bhuj and Mandvi are suitable locations for establishing relief centers. Further, based on human judgment, we can negate Mandvi since Bhuj is a better-developed city. Note that this work is a preliminary study, and more parameters can be included to make such a selection.

Acknowledgements We thank the Chief Editor, the Associate Editor, reviewers, publisher and the proofreaders. Mukherjee (2026) summarized this work

Appendix 1

Definition 1 Fuzzy set: Let V be a non-empty set. A **fuzzy subset** of V is a mapping σ: $V \rightarrow [0, 1]$ which assigns to each element $x \in V$, a degree of membership, $\sigma(x)$, such that $0 \leq \sigma(x) \leq 1$.

Definition 2 Let V and U be two sets, and let σ and τ be fuzzy subsets of V and U, respectively. A **fuzzy relation** μ from σ into τ is a fuzzy subset of $V \times U$ such that $\mu(x, y) \leq \sigma(x) \wedge \tau(y)$, $\forall x \in V$ and $y \in U$.

Definition 3 G denotes a fuzzy graph: (V, σ, μ) where V is a set of objects (nodes), σ is a fuzzy subset of V, and μ is a fuzzy relation on σ. i.e. $\mu(x, y) \leq \sigma(x) \wedge \sigma(y)$, $\forall x, y \in V$. We assume that V is finite and nonempty, μ is reflexive (i.e. $\mu(x, x) = \sigma(x)$, $\forall x$) and symmetric (i.e. $\mu(x, y) = \mu(y, x)$, $\forall (x, y)$).

Also, we denote the underlying crisp graph by G^*: (σ^*, μ^*), where $\sigma^* = \{u \in V: \sigma(u) > 0\}$ and $\mu^* = \{(u, v) \in V \times V: \mu(u, v) > 0\}$. We assume $\sigma^* = V$ and σ^* and μ^* are **'the node set'** and **'the arc set'** of the fuzzy graph G, respectively. Also, $\sigma(x)$ and $\mu(x, y)$ represent, respectively, the **degree of membership** (**membership values**) of nodes and arcs in G. If $\mu(u, v) > 0$, then u and v are called **neighbors** (**adjacent nodes**). The set of all neighbors of u is denoted by $N(u)$.

Definition 4 Let X be a non-empty set. A bipolar fuzzy set B in X is an object having the form $B = \{(x, \mu_B^P(x), \mu_B^N(x))/x \in X\}$, where $\mu_B^P : X \to [0, 1]$, $\mu_B^N : X \to [-1, 0]$ are mappings. We use the positive membership degree $\mu_B^P(x)$ to denote the satisfaction degree of an element x to the property corresponding to a bipolar fuzzy set B and the negative membership degree $\mu_B^N(x)$ to denote the satisfaction degree of an element x to some implicit counter-property corresponding to a bipolar fuzzy set B.

Definition 5 Let X be a non-empty set. Then we call a mapping $A = (\mu_A^P, \mu_A^N) : X \times X \to (-1, 1) \times (-1, 1)$ a bipolar fuzzy relation on X such that $\mu_A^P(x, y) \in [0, 1]$ and $\mu_A^N(x, y) \in [-1, 0]$.

Definition 6 (*Bipolar fuzzy graph*) By a bipolar fuzzy graph, we mean a pair $G = (A, B)$ where $A = (\mu_A^P, \mu_A^N)$ is a bipolar fuzzy set in V and $B = (\mu_B^P, \mu_B^N)$ is a bipolar fuzzy relation on V such that $\mu_B^P(x, y) \le \min\{\mu_A^P(x), \mu_A^P(y)\}$ and $\mu_B^N(x, y) \ge \max\{\mu_A^N(x), \mu_A^N(y)\}$.

Definition 7 An edge (u, v) is said to be a *strong edge* in BFG, $G = (A, B)$ if $\mu_B^P(x, y) = \min\{\mu_A^P(x), \mu_A^P(y)\}$ and $\mu_B^N(x, y) = \max\{\mu_A^N(x), \mu_A^N(y)\}$.

Definition 8 Let $G = (A, B)$ be a BFG on A. Let $u, v \in A$. We say that u dominates v in G if there exists a strong edge between them.

Definition 9 A subset S of A is called a *dominating set* in Gif for every $v \in A \backslash S$ there exists $u \in S$ such that u dominates v.

Definition 10 A dominating set of a BFG is said to be a *minimal dominating set* if no proper subset of is a dominating set.

We use the concept of dominating sets in bipolar fuzzy graphs to set an experiment to select the most appropriate towns in Kachchh district for setting up emergency response centers.

(1) Time taken to travel or the length of each road was considered. The time to traverse each road or edge considered in the graph was taken from Google Maps, considering minimal traffic conditions. The longer the time taken, the lower the score assigned to this criterion. The time taken was converted to minutes to compare all the edges and give a relative score for each edge.
(2) The number of lanes each road has was considered as this decides the traffic movement rate, especially during an emergency. If a road or edge has a variable number of lanes along its length, the effective or weighted average number of lanes was calculated for each road using the length of each 2-, 4- or 6-laned segment which makes up the road.

(3) Since bridges are major weak points during earthquakes, the number of bridges on each road was also considered a criterion for assigning scores. The number of bridges on each road was counted using satellite data from Google Maps and compared to give a score. Bridges going over the roads were also counted. Higher the number of bridges on the road, the higher the chance of the road closing after an earthquake.

Appendix 2

Calculations to assign scores for each criterion accounting for the final node scores are made as follows (Table 3):

(i) The highest seismic hazard score assigned by Pancholi et al. (2022) is 0.575 to Bhachau. It was considered that a score of 0.6 would be equivalent to a seismic hazard polarity of −1. Bhachau was hence given a seismic hazard polarity of −(0.575/0.6) = −0.95. The same method was used for the other towns.
(ii) The highest number of households is in Bhuj (= 64,496). It was considered that Bhuj has a household polarity of 64,496/70000 = 0.92. The same method was used for other towns.
(iii) The highest CHCs + PHCs are in Bhuj = 13. Bhuj has a Healthcare polarity of 13/15 = 0.86. The same method was used for the other towns.

Calculations to assign scores for each criterion accounting for the final edge scores are made as follows:

(i) The highest number of bridges is on the Bhuj-Rapar road = 18. Thus, it was considered that it has a bridge polarity of −(18/20) = −0.9. The same method was used for the other edges.
(ii) The highest number of lanes is on Bhachau-Gandhidham road, = 5.56. Thus, it was considered that it has a lane polarity of (5.56/6) = 0.93.
(iii) The highest amount of time taken for traveling is on the Bhuj-Rapar = 189 min. Hence, it was considered it has a time polarity of −(189/200) = −0.95.

References

Anand, P., Chakraverty, S., & Mukherjee, S. (2021). Fuzzy set concept in structural geology: Example of ductile simple shear. *Journal of Earth System Science, 130*, 1–9.
Biswas, S. K. (2005). A review of structure and tectonics of Kutch basin, western India, with special reference to earthquakes. *Current Science, 88*, 1592–1600.
District and Taluka-wise List of 1474 PHCs. 2018. Health and Family Welfare Department, Government of Gujarat. gujhealth.gujarat.gov.in/images/pdf/1474-list-18.pdf.
Gong, S., Hua, G., & Gao, W. (2021). Domination of bipolar fuzzy graphs in various settings. *International Journal of Computational Intelligence Systems, 14*, 1–14.

Justman, D., Creason, C. G., Rose, K., & Bauer, J. (2020). A knowledge-data framework and geospatial fuzzy logic-based approach to model and predict structural complexity. *Journal of Structural Geology, 141*, Article 104153.

Klügel, J. U. (2008). Seismic hazard analysis—Quo vadis? *Earth Science Reviews, 88*, 1–32.

Mandal, S., Patra, N., & Pal, M. (2021). Covering problem on fuzzy graphs and its application in disaster management system. *Soft Computing, 25*, 2545–2557.

Manjusha, O. T., & Mukherjee, S. (2023). Application of Fuzzy graph theory in brittle plane network analysis—A potential method for carbon sequestration models. *The Journal of Indian Geophysical Union, 27*, 109–117.

Mukherjee, S. (2019). Using graph theory to represent brittle plane networks. In A. Billi, A. Fegereng (Eds.), *Developments in Structural Geology and Tectonics* (Vol. 5, pp. 259–271). Elsevier.

Mukherjee, S. (2026). Introduction to "Structural Geology & Tectonics". In S. Mukherjee (Ed.), *Structural Geology & Tectonics*. Springer. ISBN: 978-981-95-4743-2.

Pancholi, V., Bhatt, N., Singh, P., & Chopra, S. (2022). Multi-criteria approach using GIS for macro-level seismic hazard assessment of Kachchh Rift Basin, Gujarat, western India—First step towards earthquake disaster mitigation. *Journal of Earth System Science, 131*, 3.

Rastogi, B. K., Kumar, S., & Aggarwal, S. K. (2012). Seismicity of Gujarat. *Natural Hazards, 65*, 1027–1044.

Raval, U. (2001). Earthquakes over Kutch: A region of'trident'space–time geodynamics. *Current Science India, 81*, 809–815.

Sanderson, D. J., Peacock, D. C., Nixon, C. W., & Rotevatn, A. (2019). Graph theory and the analysis of fracture networks. *Journal of Structural Geology, 125*, 155–165.

Shaikh, M. A., Maurya, D. M., Mukherjee, S., Vanik, N. P., Padmalal, A., & Chamyal, L. S. (2020). Tectonic evolution of the intra-uplift Vigodi-Gugriana-Khirasra-Netra Fault System in the seismically active Kachchh rift basin, India: Implications for the western continental margin of the Indian plate. *Journal of Structural Geology, 140*, Article 104124.

Thakkar, M. G. (2017). Geomorphological Field Guide Book on Kachchh Peninsula. In A. Kar (Ed.), *Indian Institute of Geomorphologists* (Vol. 5). Allahabad.

Walling, M. Y., & Mohanty, W. K. (2009). An overview on the seismic zonation and microzonation studies in India. *Earth Science Reviews, 96*, 67–91.

Zadeh, L. A. (1965). Fuzzy sets. *Information Control, 8*, 338–353.

Zhang, W. R. (1994). Bipolar fuzzy sets and relations: a computational framework for cognitive modeling and multiagent decision analysis. In *NAFIPS/IFIS/NASA'94. Proceedings of the First International Joint Conference of The North American Fuzzy Information Processing Society Biannual Conference* (pp. 305–309). IEEE: The Industrial Fuzzy Control and Intelligence.

MIX
Papier aus verantwortungsvollen Quellen
Paper from responsible sources
FSC® C105338

If you have any concerns about our products,
you can contact us on
ProductSafety@springernature.com

In case Publisher is established outside the EU,
the EU authorized representative is:
Springer Nature Customer Service Center GmbH
Europaplatz 3, 69115 Heidelberg, Germany

Printed by Libri Plureos GmbH
in Hamburg, Germany